Proceedings of the Fifth International Congress on Mathematical Education

Edited by
Marjorie Carss

Springer Science+Business Media, LLC

Editor
Marjorie Carss
Education Department
University of Queensland
St Lucia, Brisbane, Australia 4067

This book was typeset at the Prentice Computer Centre,
University of Queensland, Australia.
Birkhäuser Boston, Inc., would like to thank Marjorie
Carss who supervised the production of the manuscript.

Library of Congress Cataloging-in-Publication Data

International Congress on Mathematical Education (5th :
    1984 : Adelaide, S. Aust.)
    Proceedings of the Fifth International Congress on
Mathematical Education.
    Held in Adelaide, S. Aust., Aug. 24-30, 1984.
    Includes bibliographies and index.
    1. Mathematics — Study and teaching — Congresses.
I. Carss, Marjorie. II. Title. III. Title:
Proceedings of the 5th International Congress on
Mathematical Education.
QA11.A1I46 1986   510'.7'1   85-30700
ISBN 0-8176-3330-8

CIP-Kurztitelaufnahme der Deutschen Bibliothek

ICME:
Proceedings of the ... International Congress
on Mathematical Education.
    ISBN 978-0-8176-3330-1   ISBN 978-1-4757-4238-1 (eBook)
    DOI 10.1007/978-1-4757-4238-1

ISBN 978-0-8176-3330-1

CONTENTS

FOREWORD

International Congresses on Mathematical Education (ICMEs), under the auspices of the International Commission on Mathematical Instruction, are held every four years. Previous Congresses have been held in France (Lyons), England (Exeter), the Federal Republic of Germany (Karlsruhe), and the United States of America (Berkeley).

The Fifth International Congress on Mathematical Education (ICME 5) was held in Adelaide, Australia, from August 24-30, 1984. More than 1800 participants from over 70 countries participated in the Congress, while some additional 200 people attended social functions and excursions.

The program for ICME 5 was planned and structured by an International Program Committee, and implemented by the National Program Committee in Australia. For the main body of the program, Chief Organisers, assisted by Australian Coordinators, were invited to plan and prepare the individual components of the program which addressed a wide range of topics and interest areas. Each of these teams involved many individuals from around the world in the detailed planning and preparation of the working sessions for their area of program responsibility. For the actual working sessions at the Congress, the smallest group had some 60 members, while the largest had well over 300. In addition to the working sessions, there were three major plenary addresses, several specially invited presentations, and over 420 individual papers in the form of short communications, either as posters or brief talks. As well, there was a variety of exhibits, film and video presentations, and workshops. In all, more than half of those attending the Congress made a direct contribution to the scientific program.

The National Program Committee for ICME 5 considered it essential that an international meeting seeking to address the major issues in mathematics education should provide as many opportunities as possible for discussion and exchange of ideas. The program was comprehensive, challenging and exciting, enabling participants to canvass the widest range of individual interests. By encouraging Organisers to provide and plan for opportunities for participants to discuss their own work and that of others, share experiences, reflect on similarities and differences in their respective countries and locales, we hope that further work on a collaborative, international basis will evolve from this and future ICMEs.

The opportunities for discussion were only possible because the Organisers accepted the challenge and devoted two years to the planning and preparation of sessions that would encourage discussion and small group interaction. This was facilitated by the preparation and delivery of a microfiche of preliminary papers to each registrant some six months prior to the actual meeting. Again, it was the Organisers who prepared over 200 typed pages of text (often with further documentation provided for their special groups on arrival in Adelaide), which enabled all members to participate fully in the discussions of the Congress. The microfiche was translated into Japanese by the Japan

Society of Mathematical Education and into French by the French Embassy in Canberra, Australia. To both those groups the Organisers would like to express their deep appreciation.

This publication is a record of the proceedings of the Congress. As the major activity for the Congress sessions was discussion rather than the presentation of papers, the bulk of this document is a report of the sessions rather than a set of prepared papers. As such, it reflects closely the spirit of those sessions, and the structure and major components of each topic in mathematics education as perceived and planned by the Organisers. Thus much of this publication is a report of actual work done.

Most of the presentations and papers prepared for ICME 5 are not included here. The copies we have for all presentations sum to over 3000 typed pages, many of them only in 'note form'. Several groups have arranged for the separate publication of collections of papers prepared for their group's sessions.

The listing of sessions not formally reported here gives some idea of the extent and variety of other opportunities available to Congress participants, both during formal sessions and in the breaks between. During each timeslot, simultaneous interpretation was available among English, French and Japanese for one session.

Several conferences of special interest to groups of participants were held in Australia both prior to and immediately following the Congress. There were many meetings, both formal and informal, held during ICME, and the hospitality shared each evening provided an opportunity to extend Congress activities in a relaxed social setting.

As with all such large international gatherings, the goodwill of many organisations and the sterling efforts of many individuals were necessary to make it all happen. It is impossible to thank all those organisations and individuals here, but it would be remiss of us indeed not to single out the outstanding cooperation and support given by the University of Adelaide, the South Australian Institute of Technology, and the South Australian College of Advanced Education. We also wish to thank the Australian Academy of Science and the International Commission for Mathematical Instruction, co-hosts for the Congress with the University, and the Government of South Australia, without whose financial support the Congress would not have been possible. The assistance of the Australian Federal Government and the Commonwealth Foundation is also gratefully acknowledged.

A special note of thanks must go to the Australian Organisers for Theme and Action Groups and their employing institutions for support in the preparation of documents and communications leading up to the Congress, and in hosting the Chief Organisers for a week after the Congress to prepare the reports published here.

The preparation, planning and execution of the Congress was largely dependent on the personal efforts and commitment of members of the Australian Association of Mathematics Teachers. Special thanks must go to the members of national and local committees who undertook specific responsibilities and tasks on behalf of the Congress.

For the care and attention to the needs of participants both before and during the Congress, indeed from the time of indicating interest in attending, we must thank Elliservice Convention Management for their cheerful and helpful support.

Lastly, we would like to express our thanks to the University of Queensland, particularly to the Department of Education and the Prentice Computer Centre, both of whom have made their facilities available to support the work of the Congress. Two individuals must be singled out for special mention: Dianne Reid, who typed all correspondence related to the Program, prepared the microfiche document and completed all data entry for this publication in her usual, careful, intelligent and cheerful manner, and Barry Maher, who prepared the final copy for publication and gave invaluable and patient advice in the overall preparation.

Relevant documents have been checked by each of the Australian Organisers for editorial changes. We believe that the editing of these Proceedings reflects the essence of the Congress sessions.

On behalf of the organising committees for ICME 5,

Marjorie C. Carss,
Editor,
University of Queensland,
St. Lucia, Q. 4067
Australia.
July, 1985

KOALA BEAR

ICME 5

SOCIO–CULTURAL BASES
FOR MATHEMATICAL EDUCATION

Ubiratan D'Ambrosio
State University of Campinas
Brazil

Mathematics Education is going through one of the most critical periods in its long history, a history recorded since western classical antiquity. The prominent role of mathematics in Greek civilisation placed mathematics in a unique position in the roots of modern science and technology, and in the development of the universal model of an industrial society. Even the most critical periods in the history of mathematics education (which were the invention of writing and the adoption of the Hindu-Arabic number system in Europe), had a less dramatic and global effect on society than the period we are living through. This current era is witnessing both the emergence of what might be called the electronic era and profound changes in the social, political and economic texture of the world. Through the universal concept of mass education in a fast changing world, 'mathematics for all' reaches an unprecedented dimension as a social endeavour, and it makes it urgent to question, in a much deeper and broader way than ever before, its socio-cultural roots, and, indeed, the place of mathematics education in societies as a whole. Being in such a privileged position in western thought, mathematics may be, at the same time, an essential instrument in building up modern societies and a strong disrupting factor in cultural dynamics. As well, mathematics exerts a strong influence on the unbalancing factor which threatens the needed equilibrium between those who have and those who have not. An equilibrium has to be achieved if we want to look at our species as behaving in a more dignified way than it has in its long history. If we hope for a better world, in the utopian view, without human beings massively exploiting and killing each other, we have to examine the role of mathematics education in bringing a new human dimension into relations between individuals, societies and cultures.

This presentation is an attempt to bring socio-cultural dimensions into mathematics education and, to the vast literature on human behaviour, it adds the work I have carried on for a couple of decades in diverse cultural environments, with special reference to mathematical perception, abilities and uses. I ask for your tolerance since I will mention mathematics only late in this presentation.

When I say 'perception, abilities and uses', I am placing myself in the position of looking at 'reality' as it is perceived by individuals who use their abilities in the form of strategies, to perform actions which invariably have their uses in modifying reality. I am talking of human behaviour as a cyclic model connecting *reality — individual — action — reality* as characteristic of human beings. We speak also of an hierarchical order of human behaviour which goes from the individual to collective (or social) to cultural behaviour, and finally to cultural dymanics which is the result of transcultural behaviour. Each of these hierarchical steps is characterised by an instrument of interaction between several individuals which can be easily explained in the context of reification and of language uses, of education, and of communication and information, as the decisive steps in this hierarchical view of behaviour.

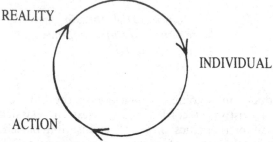

Children, as well as mankind, exhibit evolutionary behaviour in their learning which goes from individual to social to cultural behaviour, and, we add, at an increasingly rapid pace, to transcultural behaviour. Quite often a child, raised in a rural area, moves to urban areas, an event which occurs frequently in developing countries. As well as the building of new factories, new farms, new social benefits bring to different nations new patterns of behaviour which, again, are increasingly seen in developing countries. But this happens equally in the most developed countries, giving to this transcultural concept an important dimension in understanding the cycle, 'reality — individual — action — reality'.

This is basic in the conception of education as action, a view which fits particularly in this cyclic model.

We also look at knowledge as action in the framework of the cyclic model which we used to characterise human behaviour in its several hierarchical levels: individual, collective, cultural, and transcultural. Indeed we have to understand the role knowledge plays in allowing human behaviour to be a thought as well as an action, which is impacted upon by reality and which brings about an action which modifies reality. We insist that action is inherent in human beings, in particular in children. There are no still or inactive moments, if we understand action in its most general sense, be it a material action or a purely reflective, intellectual, cognitive action. As far as there is life, there is action.

We look to action as a modifier of reality, which is continually changing. Again, we talk of reality in a most broad and general way, both material reality or purely cognitive reality, that is, an intellectual, psychic and emotional reality. We look to action as a modifier of reality in very broad and general terms.

Now let us look at human behaviour and knowledge, also in the context of this cyclic model, allowing for an action which will have an impact upon reality. Since we know more, we can have more influence as modifier of a reality. Children feel this as they grow.

In what sense am I using the concept of knowledge? What does it mean 'to know' in this broad context of cultural dynamics?

To know has a dual sense if we look into the concept in several cultural contexts. To know has always been understood as to clarify the cosmic and psychic order, that is to know in the most popular acceptance of the term, which is the root of the idea of science. But at the same time, to know is to create, to do something, which is the root of the idea of art. This duality is well illustrated in the first four chapters of the Book of Genesis, which is an important tool in understanding the evolution of Western thought to what may be its most strikingly characteristic endeavour, western science and western art or technic, and its brainchild which is technology.

To understand cosmic and psychic order and to create does not mean a dichotomy. They lead to science, which is a pure act of knowing, and to art and to technic, which are acts of doing. Science does not materialise in the same way that art never becomes art if it is not conveyed. This complementarity of science and art, which has its greatest impact in technology as far as the modern world is concerned, is indeed the complementarity of knowing and doing, or to know and to do. If one knows, one does, and to do you must know. This is a high level of consciousness of the individual, as homo sapiens. Regrettably, much of the attempts to make children behave in a certain way have, in recent decades, disregarded this. This has had a damaging effect, particularly in mathematics education, and, unfortunately, it is still going on. Although we may be very worried about the course which society is taking, our main concern as educators is the individual, child or adult. The individual is a complex of reactions, both sensual, rational and emotional or psychic. Even children are this complex, which is sometimes forgotten by mathematics educators. Children are immersed in a reality. But which reality?

We consider reality as both environmental, which comprises the natural and the artificial, and as a pure intellectual, emotional, psychic, cognitive reality, which is the very intimate abstract reality of ideas. Thoughts are part of a reality which affect all individuals in a very intimate way, as well as emotions. The individual is not alone, but is part of a society. Reality is also social. The interplay of the environmental, of the abstract, and of the social is a key issue in mathematics education, again unfortunately often disregarded. The interplay of the natural and the artificial in building up environmental reality is probably one of the most critical areas in which mathematics education has a major role to play. The equilibrium between natural and artificial has much to do with the future of mankind, hence with education, and with mathematics education. Environmental equilibrium deserves the special concern of mathematics educators, and fits perfectly into the cycle, 'reality — individual — action — reality'.

Let us return to the concept of knowledge as action which involves the perception of reality through the senses and through memory. This involves performing action through strategies and models, and causes a modification of reality, through the introduction into reality of objects, of things, of ideas. These are results of the actions of individuals, which have an impact upon reality. They are incorporated into the reality in which every individual is immersed, and which, through the mechanisms of the senses plus the emotions (that is the sensual) and through memory, leads the individual to design strategies and models for action.

This comprises, in a global way, what has become known as art, technic and science as modifiers of reality, and the mechanisms of information and codification. Although art, technic and science have been the traditional domain of education (that is, the typical contents), we want to concentrate a little more on information and codification, which indeed converge to give to knowledge the possibility of action.

Information in this sense, which is bringing to the individual, through the senses, and through the information mechanisms which are the essence of what we call memory, both genetic and acquired memory, is to me the crux of what is going on in education. Let us relate information and education, a relationship which seems to be particularly appropriate in the era in which we are living, where information (through the concept of informatics) has become a key issue. Let us discuss education, both formal and non-formal, that is education taking place in school environments and in out-of-school environments.

Recall that, in the course of history, information has gone through an evolution from spoken language to written language, and, in the more technological models for disseminating information, through printed material to electronics, which indeed is a joint process of information and processing the information. Formal education is still dominated by written material and printed material, while non-formal education has a domineering role in helping individuals to speak and in the modern world, mainly through the media, in generating skills in absorbing processed information. This is particularly important in mathematics education and it seems to be urgent that formal education comes to recognise the increasing pressure of our society for information processing devices and technology. This is probably the greatest challenge mathematics educators still face in both developed and developing countries.

A full understanding of the evolution of the mechanisms of information systems, as in the history of mankind and in the evolution of children, seems to be decisive if we want to look at the relationship between education and curriculum.

Education has its key strategy in the curriculum. We adopt a concept of curriculum which brings into consideration the classical objectives, contents and methods, but in an integrated way. It is impossible to consider each one separately and probably the main reason for the many failures identified in the so-called 'modern math' movement has its roots in the breaking up of the curriculum components into independent domains of research.

Curriculum, as has been generally agreed, should reflect what is going on in society, that is, curricular dynamics always ask 'where' and 'when' does a certain curriculum take place. The key problem in curricular dynamics is to relate the societal moment, time and locality, to the curriculum, in the form of objectives, contents and methods in an integrated view.

The societal moment is more than simply time and locality, or when and where. I bring to the picture an extra dimension of a much more complex nature, that is, cultural diversity. The same place, same instant, but different cultural background makes the situation entirely different. If you have in a classroom at a certain moment, a child from a family of working parents, or a child from a family of a professional father and a non-working mother, things are completely different. Not to mention when you have different ethnic backgrounds, which happens so often in both developed and developing countries. The big challenge I see in education, in rapidly changing societies, is how to bring this cultural diversity into curriculum design. This is particularly true when we look into mathematics as a subject for all in rapidly changing societies. In other words, the key issue in curriculum design for the years to come seems to me to be meeting this challenge.

Cultural diversity is so complex, it is a mesh of attitudes and behaviours which have not been sufficiently understood in education, and especially in mathematics education. I would dare to say they have practically never been recognised as important factors in mathematics education. Attitudes such as modes of thought, jargon, codes, interests, motivation, myths, build up to generate very definitive cultural roots, modes of production, modes of poverty, class conflicts, sense of social security, human rights and so on. These are the factors which comprise society, but are usually ignored in mathematics education.

We are now faced with a concept of society which grows from individual behaviour, and which is the key issue in our recent concerns about mathematics education, that is, the relationship between mathematics and society.

But I will go further. We have built a concept of society out of cultural attitudes and cultural diversity, that is, different groups of individuals behave in a similar way, because of their modes of thought, jargon, codes, interests, motivation, myths. They constitute what we call societal groups, with clearly defined cultural roots, modes of production and property, class structure and conflicts, and senses of security and of individual rights. All these constitute societal background for children. Several studies we have conducted of social behaviour of children identify what we might call 'children's societal arrangements'. We are also concerned, in mathematics education, with this kind of society, which is the ground on which we work, as well as with societies in the general sense. Both have, as a result of the interaction of their individuals, developed practices, knowledge, and, in particular, jargons (the way they speak) and codes, which clearly encompass the way they mathematise, that is, the way they count, measure, relate and classify, and the way they infer. This is different from the way all these things are done by other cultural groups. Hence, we have the question, in dealing with the relationship of mathematics and society: which mathematics? Are we interested in the relationship between learned mathematics and society, or between ethno-mathematics and society, where 'ethnos' comes into the picture as the modern and very global concept of ethno both as race and/or culture, which implies language, codes, symbols, values attitudes and so on, and which naturally implies science and mathematical practices.

We have to look more carefully at this concept of ethno-mathematics in this context.

Boat construction in an Amazon Indian community is done totally without formal schooling. The presence of children during the process of building assumes the transmission of the practice from one generation to another. In house-building among Yawanwa Indians in the Brazilian Amazon, from the early stages of preparing the wood, the presence of children is spontaneous. There is no doubt of the strong presence of mathematics in both examples. These are the practices which are the essence of building up a body of knowledge associated with action, associated with doing. In yet another example of a set of sieves, which are also from the Amazon region, the step towards symbolic meaning of a clearly geometrical code which generates the pattern in the weaving is very significant. It is indeed a language, in the sense that each pattern recalls divinities and intentions.

These are mathematics practices identified with cultural groups, and which are taught, perfected, reflected upon, in a non-formal education system. We could easily multiply this set of examples with the situations drawn from developed societies, from industrial and commercial environments.

We can look at the ways learned mathematics feeds itself with new knowledge, mainly in the course of mathematics curricula in schools. It is indeed a closed body of knowledge, feeding itself with ideas taken from this same body of knowledge, while society has little or no influence in the evolution or building up of mathematical knowledge. In other words, innovation, which is a key element in education, in particular in mathematics education, practically ignores the results of the evaluation of mathematics practice vis-a-vis societal impact. In talking of learned mathematics, evaluation of the impact of what is learned upon societal activities has had practically no effect on innovation, or if there is an effect, there is an enormous time-lag in this interaction. Of course, keeping alive the interest of children in new ideas, new concepts, innovation in general, is a very difficult step, and the results are far from satisfactory. The enormous time-lag works against motivation. On the other hand, ethno-mathematics has an almost nonexistent barrier with respect to society. This is like a porous system with permanent interaction. Evaluation of what is the result of an ethno-mathematics practice results from immediately changing it into societal practice, which in turn feeds the body of knowledge, in this case ethno-mathematics, with innovation. In other words, the relationship between ethno-mathematics and society is characterised by a fast reaction, through a self-regulating system. This self-regulating system manifests itself in the building-up of motivation, an essential component in education.

Indeed, this self-regulating system activates the basic cycle, 'reality — individual — action — reality', upon which we have based our remarks on mathematics education, and its more dynamic relation with rapidly changing societies. It seems to me that to generate this dynamic is the key issue in mathematics education in the years to come.

REFLECTION AND RECURSION

Jeremy Kilpatrick
University of Georgia
U.S.A.

Where is the wisdom we have lost in knowledge?
Where is the knowledge we have lost in information?

T.S. Eliot, *The Rock*

Each age defines education in terms of the meanings it gives to teaching and learning, and those meanings arise in part from the metaphors used to characterise teachers and learners. In the ancient world, one of the defining technologies (Bolter, 1984) was the potter's wheel. The student's mind became clay in the hands of the teacher. In the time of Descartes and Leibniz, the defining technology was the mechanical clock. The human being became a sort of clockwork mechanism whose mind either was an immaterial substance separate from the body (Descartes) or was itself a preprogrammed mechanism (Leibniz). The mind has also, at various times, been modelled as a wax tablet, a steam engine and a telephone switchboard.

We live today in the age of the electronic computer, and that technology has served to shape much of our thinking about how education can and should proceed. We speak of thinking as 'information processing' and of teaching and learning as 'programming', 'assembly', or 'debugging'. A whole field of cognitive science has developed that attempts to capitalise on the power of the computer metaphor for understanding cognition. In recent years, mathematics educators have given much attention to how the computer as technology might be used in instruction. Attention is also needed, however, as to how the computer as metaphor affects our understanding of the processes of learning and teaching. In this paper, I explore some consequences of the metaphor.

A theme that echoes through recent discussions of problems of mathematics education is that of *self-awareness*. Some theorists are suggesting that for the learning and teaching of mathematics to become more effective, students and teachers alike will need to become more conscious of what they are doing when they learn or teach. Much of students' apparent inability to apply mathematical procedures that they have been taught may be attributable, at least in part, to their failure to recognise when a procedure is appropriate or to keep track of what they are doing as they carry it out. Similarly, many teachers' apparent reluctance to depart from their tried-and-true teaching practices may stem, at least in part, from a lack of commitment to looking back at those practices and making a careful assessment of them.

In this paper, I have borrowed two terms from mathematics — reflection and recursion — to illustrate the emerging view that more attention should be given to turning cognitive processes back on themselves. These terms can be used metaphorically as strands on which to thread some ideas about self-awareness as it relates to mathematics education. Much of the interest in self-awareness comes from recent work in cognitive science that uses an information-processing conception of cognition. I hope to suggest how that conception illuminates, within limits, various facets of thinking and learning about mathematics.

Reflection and Recursion in Mathematics

Reflection and *recursion* have reasonably clear meanings as mathematical terms. Some attention to these meanings may be helpful before the terms are applied metaphorically to issues of mathematics education.

Reflection
The term *reflection* comes from geometry but originated in physics. Its original reference was to the change in direction that light, heat or sound undergoes when it strikes a surface and turns back, while remaining in the same medium. In geometry, it has both static and dynamic connotations. Statically, it refers to a correspondence between points symmetric in a point, line, plane or other geometric figure; it is often used simply to describe the set of image points. One thinks of two sets of points: the original set and its reflected image — as in a mirror. Dynamically, reflection refers to a transformation in which points are mapped into their symmetric images. One can think of a single set of points moving to a new position — as when a paper is folded. These two connotations may come to much the same thing mathematically, but they are not the same pedagogically. Children respond differently depending on whether reflections are presented with a mirror or with paper folding (Genkins, 1975). Freudenthal (1983) has stressed the importance of using reflections to help children learn that in today's mathematics, mappings do not take place in time; they occur 'at one blow'. He advocates the extensive use of a mirror to present the relation between point and image, noting, however, that 'the actual use of the mirror, if maintained too long and too rigidly, can block the development of the mental operation and the mental object *reflection*' (p. 345).

When applied to human thought, the concept of reflection may be somewhat limited. As Hofstadter and Dennett (1981, p. 193) note, reflection is not a rich enough metaphor to capture the active, organising, evolving, self-updating qualities of a human representational system — the mind is not a mirror. Nonetheless, the image of reflecting on an idea, turning it over in one's mind, is a powerful device for thinking about thinking, and for thinking about one's own thought. John Locke (1706/1965) defined reflection as 'that notice which the mind takes of its own operations, and the manner of them' (p. 78). For Wilhelm von Humboldt, the essence of thinking consisted in 'reflecting, that is, in distinguishing the thinking from that which is thought about' (Rotenstreich, 1974, p. 211). And Cardinal Newman (1870/1903) pictured multiple reflections: 'The mind is like a double mirror, in which reflexions of self within self multiply themselves till they are undistinguishable, and the first reflexion contains all the rest' (p. 195).

Recursion
Recursion comes from number theory and mathematical logic. It, too, has its roots in physics: one of the early users of the term in English was Boyle, who employed it to characterise the action of a pendulum. It has some of the same connotations of turning back, of movement backward, that reflection has, but it has taken on other associations as it has been used increasingly in computer science. Peter (1967, p. 241) has pointed out that as a mathematical procedure, recursion goes back to Archimedes, who calculated π as the limit of a recurrent sequence.

Recursion has no generally accepted meaning; it is applied to several related concepts: recursion relations, recursive functions, recursive procedures, and recursive conditional expressions (McCarthy, 1983). One can think of recursion as a method of defining a function 'by specifying each of its values in terms of previously defined values, and possibly using other already defined functions' (Cutland, 1980, p. 32). Or one can think somewhat more generally of a recursive function or procedure as one that calls itself (Cooper and Clancy, 1982, p. 236).

This self-referential aspect of recursion is its most intriguing feature. The *Dictionary of Scientific and Technical Terms* (1978) defines recursion as 'a technique in which an apparently circular process is used to perform an iterative process'. More broadly still, Hofstadter (1979) characterises recursion as 'nesting, and variations on nesting' (p. 127). For Hofstadter, recursion is a metaphor for organising the world.

As an aside, it should be noted that proposals are being made that ideas related to recursion be given greater emphasis in the school curriculum in line with its enhanced role in mathematics. Hilton (1984) has observed that the computer is changing mathematics in several ways, including giving a new prominence to iteration theory. Maurer (1983) argues that teachers should present 'the modern precise idea of an algorithm, and some of its particular techniques such as recursion, as among the great ideas in human intellectual history' (p. 161).

Reflection and Recursion in Thinking and Learning

Reflection
When a triangle is reflected in a line, its image has a reverse orientation. A scalene triangle cannot be made to coincide with its image by moving it within the plane; instead, it must be picked up and turned over. Similarly, in three dimensions an object like your right hand cannot be made to concide with its mirror image; the mirror image has a different orientation. By analogy, you would need four dimensions in order to superimpose your right hand and its image. Möbius (1885/1967, pp. 171–172) used exactly this argument to conclude that geometric objects that are mirror images in three dimensions can be superimposed in four dimensions. (He went on, by the way, to conclude that since a four-dimensional space cannot be thought about, the superimposition would be impossible.)

Möbius's argument suggests metaphorically that we somehow move into another dimension when we reflect on what we have done. In reflecting on our experience, we move out of the plane of our everyday existence. We give meaning to experience by getting outside the system.

Dewey (1933) defined reflective thinking as 'the kind of thinking that consists in turning a subject over in the mind and giving it serious and consecutive consideration' (p. 3). For Dewey, learning was learning to think:

> Upon its intellectual side education consists in the formation of wide-awake, careful, thorough habits of thinking.
> Of course intellectual learning includes the amassing and retention of information. But information is an undigested burden unless it is understood. It is *knowledge* only as its material is *comprehended*. And understanding, comprehension, means that the various parts of the information acquired are grasped in their relations to one another — a result that is attained only when acquisition is accompanied by constant reflection upon the meaning of what is studied. (pp. 78-79)

Constant reflection? Think of the last school mathematics classroom you observed. Presumably learning was supposed to be going on. But was there constant reflection? Was any opportunity provided for reflection?

Dewey was not the only thinker to emphasise the need for reflection in learning. Piaget (1974/1976, p. 321) distinguished between the unconscious process of 'reflexive abstraction' and its culmination in a conscious and conceptualised result, which he termed a 'reflected abstraction' — a product of the 'reflecting' process. He pointed out that the operational structures of intelligence, although logico-mathematical in nature, are not present in children's minds as conscious structures. They direct the child's reasoning but are not an object of reflection for the child. 'The teaching of mathematics, on the other hand, specifically requires the student to reflect consciously on these structures' (Piaget, 1969/1971a, p. 44).

Skemp (1979, p. 174) has distinguished between two levels of intellectual functioning — intuitive and reflective. Reflective intelligence is the 'ability to make one's own mental processes the object of conscious observation, and to change these intentionally from a present state to a goal state. ... By the use of reflective intelligence, trials and the correction of errors become contributors to a goal-directed progress towards optimal performance' (pp. 175-176). Reflective intelligence, therefore, operates at a higher level of consciousness than intuitive intelligence. Skemp sees his formulation as differing somewhat from that of Piaget. For Skemp, conceptual structures, or schemas, exist in their own right, independent of action (p. 219), whereas Piaget saw conceptual structures as closely linked to a class of action sequences. In Skemp's theory, the schema is an active agent for acquiring and integrating knowledge, with reflective intelligence providing a means for organising one's schemas.

Recursion
Recursion, too, requires that the user step outside the system. In my experience, students who are learning a programming language that allows recursion find the programming of recursive procedures exceedingly difficult to follow, even when they have some idea of what a recursive function is. They are perplexed by a procedure that calls itself, feeling that the procedure is somehow operating on two levels at once — which it is. To see how recursion works, one needs to get outside the process itself and, so to speak, look down on it from above.

Recursion has been used less frequently as a metaphor to describe learning than reflection has, but learning does seem to have a recursive quality. If mental growth and development occur in stages, as thinkers such as Whitehead (1929) and Piaget (1956, 1971b) and Pinard and Laurendeau (1969) have claimed, then each stage must be built on the foundation of the preceding one. Later stages reproduce earlier stages, but with a difference. To capture this quality, Whitehead referred to 'the rhythm of education'. Mathematics itself may have an 'iterative or monotonous character' (Hawkins, 1973, p. 128), but the learning of mathematics proceeds in a rhythm in which repetition is combined with variation. The repetition can take on the quality of a recursion when old knowledge is used as a substrate for the construction of new knowledge. Freudenthal (1978) contends that there are levels in the process of learning mathematics, and it often happens 'that mathematics *exercised* on a lower level becomes mathematics *observed* on the higher level' (p. 61).

An example of a theory in which mathematics learning can be seen as a recursion is the theory of van Hiele and van Hiele–Geldof (1958) concerning the levels of thought in geometry. As students learn geometry, they move from level to level, returning to the same geometric concepts with a different language that gives new meaning to the concepts and makes explicit what was implicit before (Fuys, Geddes and Tischler, 1984, p. 246).

As Vergnaud (1983) has noted, the hierarchy of competence in mathematics cannot be totally ordered, as stage theory suggests; it has at best a partial order:

> Situations and problems that students master progressively, procedures and symbolic representations they use, from the age of 2 or 3 up to adulthood and professional training, are better described by a partial-order scheme in which one finds competences that do not rely upon each other, although they may all require a set of more primitive competences, and all be required for a set of more complex ones. (p. 4)

The partial order does not negate the claim that learning is recursive; it implies only that the recursion may occur across sets of competences rather than at a more molecular level.

Consciousness, Control and Metacognition

Both reflection and recursion, when applied to cognition, are ways of becoming conscious of, and getting control over, one's concepts and procedures. To turn a concept over in the mind and to operate on a procedure with itself can enable the thinker to think how to think, and may help the learner learn how to learn.

Over the last decade, as psychology has turned away from behaviourism toward a more cognitive orientation, and as concepts from computer science have filtered into the psychologist's vocabulary and world view, two related movements have occurred — a resurrection of the concept of consciousness and a recognition of the importance of executive procedures to guide thinking.

The Return to Consciousness
Explicit attention to the phenomenon of consciousness has been coming back into favour in psychology since about 1960, and more recently has come a renewed interest in the consciousness of consciousness (Jaynes, 1976).

Consciousness is at once 'both the most obvious and the most mysterious feature of our minds' (Hofstadter and Dennett, 1981, pp. 7–8). When we examine our consciousness we become aware of our awareness, or, as Bartlett (1967) put it, 'the organism discovers how to turn round upon its own 'schemata'' (p. 208). Hofstadter (1979) has characterised consciousness in terms of a 'Strange Loop, an interaction between levels in which the top level reaches back down towards the bottom level and influences it, while at the same time being itself determined by the bottom level. ... The self comes into being at the moment it has the power to reflect itself' (p. 709).

For Locke, and others who favoured introspection, self-consciousness permitted the inspection of the content of one's mind. After the behaviourists outlawed introspection, however, consciousness went underground in psychology — while remaining above ground in phenomenology (Levinas, 1963/1973; Merleau–Ponty, 1964/1973).

The construct that has resurfaced in psychology is somewhat more sophisticated than the older conceptions (Mandler, 1975, Natsoulas, 1983). Although one is tempted to think of consciousness as self-evident, as pervading our mental life, and as located somewhere in our heads, Jaynes (1976) claims that it can be more truly described as 'a metaphor–generated model of the world [invented] on the basis of language [that parallels] the behavioral world even as the world of mathematics parallels the world of quantities of things' (p. 66). Although Bartlett saw consciousness as a unitary phenomenon, it is now viewed as having a variety of forms (Rapaport, 1957, p. 169). Consciousness is not necessary in order for us to learn, and some would claim that instances of 'incubation', where, for example, the key idea of a proof might come in a flash, demonstrate that our reasoning processes are not always conscious (Jaynes, 1976, pp. 36–44).

> Consciousness is a much smaller part of our mental life than we are conscious of, because we cannot be conscious of what we are not conscious of. How simple that is to say; how difficult to appreciate! It is like asking a flashlight in a dark room to search around for something that does not have any light shining upon it. The flashlight, since there is light in whatever direction it turns, would have to conclude that there is light everywhere. And so consciousness can seem to pervade all mentality when actually it does not (Jaynes, 1976, p. 23)

Our conscious memory, for example, is a construction, not a storing up of sensory images. To illustrate how experience is reconstructed, introspect on the moment when you entered the room you are in and picked up this paper. Your image is likely to be more like a bird's eye view of yourself from the outside looking in than a series of sensory images of what you actually saw, heard, and felt. With some effort, of course, you can recreate how things looked, sounded, and felt from your perspective, but your imagery is primarily a construction of images that you never actually had of yourself except in your mind's eye. Nijinsky claimed that while he danced, he was not conscious of what movements he was making but only how he looked to others. It was as though he were in the audience rather than on the stage (Jaynes, 1976, p. 26). All of us have had this feeling at times. In a sense, each of us constructs a movie about ourselves where we are both the star and the director.

Our consciousness is a presentation to ourselves, though we may seldom conceive of it in those terms. And as educators, we seem to have given little attention to helping students become more aware of how their consciousness functions and how they might use it to monitor and control their mental activity.

The Rise of the Executive

When a researcher attempts to program a computer to simulate the performance of some cognitive task such as writing an equation for a 'word problem' or proving a theorem, he or she quickly gets caught up in devising executive procedures to coordinate various parts of the task. The information-processing metaphor has drawn psychologists' attention to the need people have to keep track of and direct their thinking. It has led to the incorporation of executive processes into models of cognition (Campione and Brown, 1978; Carroll, 1976; Case, 1974). This notion of monitoring and controlling has been taken up by mathematics educators interested in research on problem solving (Schoenfeld, 1983; Silver, 1982a, 1982b) partly because it appears that students often possess all the concepts and skills needed to solve a mathematical problem without being able to marshal them to come up with a solution.

Cognition About Cognition

A phenomenon related to both consciousness and executive control has been the growing attention to metacognition — knowledge about and control of one's cognitive processes (A.L. Brown, 1978, in press; Flavell, 1979). Although the term *metacognition* has been given a variety of meanings by different authors (Thomas, 1984), most usages encompass at least the three aspects of knowledge about how one thinks, knowledge of how one is thinking at the moment (monitoring), and control over one's thinking. For example, Flavell (1976) defined metacognition as 'one's knowledge concerning one's own cognitive processes and products or anything related to them', and then proceeded to characterise cognitive monitoring and control as aspects of metacognition — 'the active monitoring and consequent regulation and orchestration of [one's cognitive] processes in relation to the cognitive objects or data on which they bear' (p. 232). Flavell (1979) argues that metacognitive *experiences* (e.g., feeling that you will fail in this task, recalling that you have successfully solved a similar problem before) are necessarily conscious, but one's metacognitive *knowledge* about a situatiuon may or may not become conscious. Ann Brown (1978), defending the use of the term *metacognition*, appears to see it as concerned primarily with consciousness, contending that 'in the domain of deliberate learning and problem solving situations, conscious executive control of the routines available to the system is the essence of intelligent activity' (p. 79).

Most of the research on metacognition has been conducted by developmental psychologists who have been especially interested in documenting differences between younger and older children in how they view and direct their thinking. Young children, for example, often fail to realise that they don't understand a teacher's explanation, whereas older children are more likely to be aware of their failure to comprehend and to attempt to get clarification. Developmental psychologists concerned with metacognition have not always been careful to distinguish between knowledge about cognition and the regulation of cognition (A.L. Brown and Palincsar, 1982). Self-regulation, as is clear from Piaget's theory, is characteristic of any active learning attempt

— even very young children can detect and correct their errors in learning — whereas the ability to reflect on one's own thought seems to develop at adolescence (Piaget, 1974/1978, p. 217).

The topic of metacognition, of course, is not new. Much of what is studied today as 'strategies for self-interrogation and self-regulation' would be familiar to an older generation of teachers as 'study skills' (A.L. Brown, 1978, p. 80). But there is a new emphasis on analysing the components of metacognition so that instruction can be directed at those components. Further, some researchers have observed that metacognitive processes, rather than being imposed on top of acquired knowledge, interact with knowledge as it is being acquired (Gitomer and Glaser, in press).

One's metacognitions include one's beliefs about oneself as a doer of mathematics, as a mathematician, but they do not — strictly speaking — include other beliefs that may have an important bearing on how one does mathematics. These other beliefs concern such matters as the effectiveness of one's teacher (Garofalo and Lester, 1984), whether a problem might have more than one solution (Silver, 1982a), or whether the task itself is worth doing (Schoenfeld, in press-a).

Putting all of these observations together, one gets a picture of human cognition with respect to a restricted domain such as mathematics as a network of stored information being operated on by various processes under the control of some sort of executive system. Much of the knowledge is tacit, and most of the processing is unconscious, but when critical decisions must be made, control is given over to consciousness, and metacognitive knowledge may be invoked. Beliefs about oneself, others, and the tasks one is confronted with serve to shape the responses one makes.

This oversimplified and admittedly inadequate characterisation of human cognition rests, one should recall, on the metaphorical use of the computer as a model of the human mind. Before considering some implications that metacognitive knowledge and other manifestations of self-awareness might have for the learning and teaching of mathematics, let us first examine some forms the model has taken.

Models of Mind

The preceding observations concerning reflection and recursion in learning have tacitly assumed the dominant viewpoint of cognitive science; namely, that cognition is carried out by a central processing mechanism controlled by some sort of executive system that helps cognition get outside itself so the mind can become conscious of what it is doing. Recent models of mind have been like most recent models of computers — a general-purpose computer with a single-stream processor (R.M. Davis, 1977) capable of storing and executing programs expressed in a high-level language.

Cognitive science has devoted considerable attention over the past several decades to attempts to develop general models of intelligent performance — in particular, models of human problem solving (Newell and Simon, 1972). In such models, the mind is seen as essentially unitary, and mental structures tend to be viewed as primarily 'horizontal' in nature, cutting across a variety of contents of thought. For example, our naive, unanalysed conceptions of our memory, our judgement, our perception, etc., are likely to be horizontal; we think of a single mechanism being brought to bear on a wide range of problems. Theorists such as Piaget and Skemp have adopted a similar view, claiming that the mental operations of thought are essentially invariant across content. As Simon (1977) has phrased it, 'the elementary processes underlying human thinking are essentially the same as the computer's elementary information processes, although modern fast computers can execute these processes more rapidly than can the human brain' (p. 1187).

The Return of Faculty Psychology

More and more, however, there are signs that a counter movement is developing. There is a growing recognition that many processes of the nervous system are modelled better as occurring simultaneously rather than sequentially. Some theorists seem to be following Chomsky's (1980) lead and viewing mind as composed of relatively independent subsystems, such as a language system, that are 'vertical' in the sense of being domain specific. These theorists challenge the claim that the mind is composed of all-purpose mechanisms. Instead, they are turning to the idea that there is a 'society of mind' (Minsky, 1980). In such a view, the mind is a collective of partially autonomous smaller minds, each specialised to its own purpose, that operate in parallel rather than sequential fashion. These smaller minds, or 'mental organs' (Chomsky, 1980, p. 39), are often seen as essentially preprogrammed, or 'hardwired'. They may develop in specific ways according to their own genetic programs, much as our physical organs develop. They may be sensitive to only selected types of input, and their operation may not be open to conscious examination or control. We are witnessing a revival of faculty psychology (Fodor, 1983).

One sign of this new movement is that researchers in cognitive science seem to have abandoned the search for general models of intelligence and are giving their attention to the simulation of relatively specialised processes of imagery, language production, etc. What might be called the 'Wagnerian phase' (Fodor, 1983, p. 126) of research in artificial intelligence has ended.

An example of the new type of model is a cognitive model of the planning process devised by Hayes–Roth and Hayes–Roth (1979) in which various cognitive 'specialists' collaborate opportunistically rather than systematically to come up with a course of action to solve a problem. The researchers tested the model by asking people to think aloud as they planned a day's errands and compared their plans with those from a computer simulation based on the model. The specialist model seemed to account fairly well for how people plan when confronted with certain kinds of problem situations.

More sweepingly, Fodor (1983) has asserted 'the modularity of mind'. In Fodor's view, mental processes are organised into *cognitive modules* that are 'domain specific, innately specified, hardwired, autonomous, and not assembled' (p. 37). A central system draws information from these modules to make decisions, solve problems, and, in general, construct beliefs about the world. Because the central processor has no distinctive neuroarchitecture and is so global in character, it is likely, according to Fodor, to resist investigation; cognitive science will be limited to studying the modules of mind.

Gardner (1983) goes even further. He has developed a theory of multiple intelligences in which, for example, linguistic intelligence is separate from logical–mathematical intelligence which, in turn, is separate from spatial intelligence, and so on. Presenting some persuasive evidence that these different intelligences develop and break down in separate ways, Gardner eschews an executive — a central homunculus in the brain — that decides what to do. He sees no need to postulate a separate module that controls the other modules.

The models of mind as composed of separate entities operating in parallel, like a network of specialised computers, have some suggestive implications for the learning and teaching of mathematics. Let us take a closer look at two of the newer approaches.

Domains of Experience

Lawler (1981) sees the mind as composed not of modules or faculties but of *microworlds*. Microworlds are cognitive structures that reflect, in microcosm, 'the things and processes of that greater universe we all inhabit' (p. 4); they are related to the ideas of frame, schema and script (see R.B. Davis, 1983), terms that are used to describe structures in one's memory that organise and represent one's knowledge. To illustrate what is meant by microworld, Lawler reports the following three solutions given by his six year old daughter, Miriam, to the 'same' problem:

> I asked Miriam, 'How much is seventy-five plus twenty-six?' She answered, 'Seventy, ninety, ninety-six, ninety-seven, ninety-eight, ninety-nine, one hundred, one-oh-one' (counting up the last five numbers on her fingers). I continued *immediately*, 'How much is seventy-five cents and twenty-six?' She replied, 'That's three quarters, four and a penny, a dollar one'. Presented later with the same problem in the vertical form of the hindu–arabic notation (a paper sum), she would have added right to left with carries. Three different structures could operate on the same problem. (p. 4)

According to Lawler's analysis, Miriam was operating in a different microworld when she solved each form of the problem. She had a count microworld that analysed the problem in terms of counting numbers, including multiples of ten; a money microworld that dealt with the denominations of coins and the equivalence of various combinations of coins; and a paper sums microworld that governed her use of the standard addition algorithm. As another example, Miriam was able to add 90 and 90 — a result she knew from working with angle sums in a 'turtle geometry' microworld — well before she could calculate 9 plus 9.

Lawler is concerned with how the different microworlds of the child interact and become integrated. Like Gardner, he rejects the notion of a problem solving homunculus in the mind which decides what knowledge is appropriate to the solution of a problem. Abandoning the idea of an executive control structure, Lawler views the microworlds of mind as actively competing with each other, working in parallel to find problems to work on. As L. Hatfield (personal communication, 11 June 1984) has suggested, a child's microworlds may be less like a parliament where some executive officer attempts to keep order, and more like the floor of a stock exchange where the loudest voice gets heard. Which microworld provides an answer to a problem depends on how the problem is posed and what relevant knowledge the microworld has. The control structure ultimately grows out of the interaction of microworlds. Lawler sees his theory as an equilibration theory that will explain how the mind gets organised. More than other theorists who postulate a parallel-processing mind, Lawler has attempted to develop a computational theory of learning.

Bauersfeld (1983) has put forward the notion of *domains of subjective experiences*, by which he means something more than microworlds, which are confined to cognitive knowledge only. Domains of subjective experiences refer to the totality of experiences an individual has had. In Bauersfeld's view, the brain never subtracts. Experiences are built up layer by layer, so to speak. When the child is faced with a baffling problem, there is a tendency to regress to an earlier, more elaborated domain of experiences. Such a regression might help explain something of students' failure to apply techniques they have just been taught to the solution of novel problems.

The Development of Self-Awareness
The problem of how the mind becomes aware of itself has been addressed by Johnson–Laird (1983a, 1983b) who, in addition to assuming that higher mental processes operate in parallel, sees the mind as necessarily hierarchical. He contends that if a model of the human mind is to account for consciousness, it must possess:
(a) an operating system that controls a hierarchically organised system of parallel processors;
(b) the ability to embed models within models recursively; and
(c) a model of its own operating system.

The paradox of self-awareness is that you can think and be aware of yourself thinking, and be aware that you are aware of yourself thinking — and the process seems to continue without ending. Johnson–Laird (1983a, p. 505) uses the example of an inclusive map. Suppose a large and quite detailed map of Adelaide were to be laid out on the grounds of the Festival Centre. Then the map should contain a representation of itself in the part that represents the Festival Centre. But then the representation should itself contain a tiny representation, and so on. Such a map is clearly impossible to realise in the physical world because it contains an infinite regress. As Johnson–Laird points out, Leibniz rejected Locke's theory of the mind precisely because it assumed such a regress.

But there is a computational resolution of the paradox. One can devise a recursive procedure for constructing a map that calls itself and continues to do so. Although the physical representation would quickly become too small, the process could, in principle, be carried as far as one wished. Johnson–Laird sees this recursive ability of the mind to model itself as the solution to the paradox of self-awareness. He expresses the resolution of the paradox in terms that combine the images of reflection and recursion:

> At the moment that I am writing this sentence, I know that I am thinking, and that the topic of my thoughts is precisely the ability to think about how the mind makes possible this self-reflective knowledge. Such thoughts begin to make the recursive structure of consciousness almost manifest, like the reflections of a mirror within a mirror, until the recursive loop can no longer be sustained. (1983b, p. 477)

People obviously know something of their own high-level capabilities — their metacognitive knowledge — but in Johnson–Laird's view, they have access to only an incomplete model of their cognition. Consciousness may have emerged in evolution as a processor that moved up in the hierarchy of mind to become the operating system. The operating system has no direct access to what is below it in the hierarchy; it knows only the products of what the lower processors do. Consciousness requires a high degree of parallel processing so that the embedded mental models can be available simultaneously to the operating system. Consciousness, then, is simply 'a property of a certain class of parallel algorithms' (Johnson-Laird, 1983b, p. 477).

Johnson-Laird contends that any scientific theory of the mind has to treat it as a computational device, although he allows that we may have to revise our concept of computation. If we are to understand human behaviour, he contends, we must assume that it depends on the computations of the nervous system. Since the computer is the computational device par excellence, Johnson-Laird views it as not merely the latest metaphor for mind, but the last (1983a, p. 507).

Reflection and Recursion in Teaching

A Model of the Learner

Clearly, there is no one computer metaphor of the mind; there are many. In dealing with practical matters of instruction, a mathematics teacher cannot take time to ask which of several mental models, if any, applies to a student. Teachers do operate, however, with some representation, generally implicit, of their students' minds. The representation may be as crude as that of a container to be filled, or it may be as sophisticated as that of a concrete operational caterpillar metamorphosing into a formal operational butterfly. Can computer metaphors provide a model of students' minds that teachers might find helpful — even as they recognise that such a model will be necessarily incomplete?

One can think of a computer, and a mind, as an information-processing device capable of at least three types of process: assembly, performance, and control. Roughly speaking, assembly processes translate incoming information into a usable form, performance processes use that information, together with information stored in memory to produce some outcome, and control processes manage the sequence and timing of the other processes. Until recently, performance processes have received most of the attention in research on information processing. As Snow (1980) has noted:

> Attention has been turning to the executive functions, but these are thought of mainly as control processes. The primary executive function, however, would appear to be assembly; the computer program analogy has for too long left out the programmer. (p. 33)

Recent work in cognitive science is attempting to redress this imbalance — to give more attention to how executive functions develop, particularly those related to assembly — otherwise known as learning. The picture that is emerging suggests a learner of mathematics who comes into the classroom, at whatever age, with a rich fund of knowledge and a burgeoning set of beliefs. Much of the knowledge is at a completely tacit level; the learner has no awareness of how it is used; it is not available to consciousness. As instruction in mathematics progresses, children gradually organise and integrate separate domains of experiences, and some cognitive processes that initially required attention become relatively automatic — although they can be brought back to consciousness if necessary. An executive structure emerges that allows children to direct their thinking, to reflect consciously on their experiences, and to formulate beliefs about their own thinking and learning. When children can model their own cognitions within their model of the world, the stage is set for self-awareness and for greater control over their cognitive processes.

What Does a Teacher Do?
When the child's mind is viewed from an information-processing perspective, one has a difficult time seeing it as anything like a blank slate. The child comes equipped with wiring already installed and programs already running. Whether one views these programs as microworlds or as domains of subjective experiences, the school-age child is a self-programming being who has already put together many programs for dealing with intellectual tasks. Some of these programs are quite different from the programs that teachers have in mind. Recent work in cognitive science has documented some of the often rather fully worked out misconceptions that learners bring to the study of mathematics and science (R.B. Davis, 1983; McCloskey, 1983). These misconceptions, which Easley (1984) terms 'students' alternative frameworks', are often quite resistant to attempts to change them through direct instruction. Easley suggests an indirect approach in which students take on the responsibility for persuading their peers, and the teacher takes on the role of moderator. Such indirect approaches have been tried at various times and in various places, but, as Easley points out, they have not been investigated very extensively.

The teacher may sometimes need to go beyond being a moderator. Bauersfeld observes that a comparison of two domains of subjective experiences requires that the student have a language from outside both. Part of the teachers' role, then, might be to supply students with a language for reflecting on their own experiences.

A similar active stance for the teacher is proposed by Freudenthal (1978, p. 186) who advocates that teachers exploit discontinuities in children's learning, making them conscious of their learning process. He argues that the attitude of reflecting on one's mental activities should be acquired early. Learning processes, to be successful, should be made conscious, but they too seldom are. In Freudenthal's view, mathematics itself is, to a great extent, reflecting on one's own and others' activity. One of the problems he posed at the Fourth International Congress on Mathematical Education in 1980 was 'How to stimulate reflecting on one's own physical, mental and mathematical activities?' (1981a, p. 142)

It is an excellent question, with no easy answer. Certainly students who have been given no opportunity to reflect on what they have learned are unlikely to develop a reflective attitude. But opportunity alone is unlikely to be sufficient for most students. They need encouragement and probably some explicit instruction in how to look at their own thinking. The view of intelligence as requiring an executive system to assemble and control thinking has been helpful in suggesting some strategies for training children to manage their thinking processes more intelligently. The most helpful strategies appear to be those in which students are not only shown some rules for guiding their thinking but also given some direct training in how to manage and oversee the rules (A.L. Brown and Palincsar, 1982). These strategies seem particularly appropriate for children with learning disabilities (Loper, 1982) and perhaps for children from certain racial groups (Borkowski and Krause, 1983).

Research on training metacognitive awareness (A.L. Brown, 1978; A.L. Brown and Palincsar, 1982; Loper, 1982) suggests that children who are having learning difficulties can sometimes be successfully trained to monitor and regulate their performance, with consequent improvement in learning. Most of the research thus far has dealt with reading instruction, and some of the training programs have been rather mechanical, but there are some useful suggestions embedded in the research literature. An example of an apparently successful training program that takes seriously the idea that metacognitive strategies should be used consciously and purposefully is that of Baird and White (1984), who found that secondary school students could gain more control over how they learned science and thereby understand it better when they adopted a questioning approach to learning.

Some recent work by Anderson and her colleagues (Anderson, Brubaker, Alleman-Brooks and Duffy, 1984) suggests how school practice may inhibit the development of metacognitive skills in some students and encourage it in others. Consider what happens when a teacher assigns students tasks that they find too difficult or confusing and insists that they stay busy to complete their work. For students who have been successful in learning a subject, difficult or confusing assignments are relatively rare. Consequently, they are likely to seek help from the teacher or some other source. They develop skills in learning how to learn even though their formal classroom instruction may seldom have dealt with such skills. For students who have not been successful, however, the story is different. Difficult or confusing tasks are a common occurrence for such students. Under pressure to stay busy, students who have become accustomed to meaningless tasks are unlikely to see themselves as needing to learn how to guide their own learning. In other words, the cognitively rich get metacognitively richer, while the cognitively poor get metacognitively poorer.

Teaching that is aimed at breaking this cycle may fall into the trap of making the learner self-conscious in the negative sense of being distracted by playing simultaneously the roles of actor and observer. One reason Nijinsky pictured himself as though he were in the audience must surely have been that had he paid too careful attention to where he was placing his feet, he might have tripped over them. Too much self-awareness can inhibit performance, yet some self-awareness is needed to make improvement possible. There may be a lesson for mathematics instruction in recent developments in the teaching of athletic skills that allow learners to see what they are doing — to return automated responses to consciousness so that they can be improved upon (von Glasersfeld, 1983). Techniques such as slow-motion photography, accompanied by careful analyses of movement, have been especially helpful to experienced athletes who wish to hone their skills. Such techniques can also be helpful to the novice, provided that the task to be learned has been simplified. The skill of skiing can be learned today in days rather than months in large part because the task has been broken up into 'increasingly complex microworlds' (Fischer, Brown and Burton, 1978, p. 3). The learner begins with simplified versions of the task, using special equipment such as short skis, in a simplified environment such as a packed slope. The task, the equipment and the environment are gradually made more complex, and the learner is given feedback that enables 'bugs' at each level of performance to be corrected. The teacher has a repertoire of exercises that permit the student to debug errors constructively; for example, lifting up the end of the inside ski while executing a turn informs the skier that most of his or her weight is on the outside ski (where it belongs). The student can see both what to do and how to do it. Once performance has been perfected, it becomes more or less automatic, and control can move to a higher level. The mark of an expert is that many aspects of performance are automated, but when a novel situation is encountered, conscious monitoring and control take over. There is much more, of course, to mathematics than skills, but the problem of finding the appropriate level of self-awareness is pervasive.

The portrayal of self-awareness as a product of a recursive process in which the mind constructs a model of itself suggests that teachers may want to give more explicit attention in instruction to how the mind works. Schoenfeld (in press-b) has reported that even when students reach a college calculus course, they may have developed little awareness that they can observe and critique their own thinking. When these students say that in solving a problem they do 'what comes to mind', they may be expressing a view of their minds as passive rather than active agents.

Papert (1975, 1980) has suggested that children can learn to reflect, to think about thinking, by working on a computer. In teaching the computer to think, the child becomes an epistemologist (Papert, 1980, p. 19). Papert may have blurred the distinction between using the language of a discipline and consciously reflecting on that language (S.I. Brown, 1983), but he has drawn attention once more to the ancient dichotomy between education as the acquisition of a fund of knowledge versus education as the development of an attitude toward knowledge and its acquisition (Jahnke, 1983). Teachers should consider ways to help students construct more active models of their minds not simply because an active stance is likely to be more helpful in learning and doing mathematics, but also because the attitude that I am responsible for what I learn and how I learn it is itself a valuable outcome of education.

The teacher who would have his or her students reflect on what they are doing and construct recursively a model of their own cognition should have personal experience in reflection and recursion. Teacher education programs have sometimes provided teachers with mirrors for their teaching through the media of critiques by other teachers and videotapes of lessons, but once teachers get out into the field, they are unlikely to take — or have — much time to reflect on themselves as teachers and learners of mathematics. Some researchers (Thompson, 1984) are beginning to look at how teachers' beliefs are exhibited in their teaching practice, but much remains to be done in examining how encouragement and opportunities to reflect on their teaching might affect how they teach in the future.

Robert Davis has noted that good students, after solving or working hard on a difficult mathematics problem, often seem to be deep in thought, turning the problem over in their minds in an 'after-the-fact analysis' (R. B. Davis and McKnight, 1979, p. 101). When questioned, the students can replay the problem solution, much as an expert game player can replay a game from memory. Davis conjectures that after-the-fact analysis may be a time when students consolidate their knowledge and develop metacognitive knowledge about their procedures. A similar phenomenon can be observed in the practice of outstanding teachers. In 1966 and 1967 George Polya offered a series of seminars for freshmen at Stanford University. After each lesson, which was certainly similar to lessons he had taught many times previously, he would retrace what had happened. He was especially concerned about students who had apparently had difficulty understanding certain points, and he would consider questions he might have asked differently or examples that he might have given. After more than a half century of teaching mathematics, he was still reflecting on his teaching and attempting to improve it.

A challenge to teacher educators and to teachers themselves is to devise ways of encouraging reflection. What good students and good teachers do on their own with respect to looking back at their work ought to be promoted in mathematics learning and teaching.

Reflection and Recursion in Mathematics Education

We have seen, in the case of the learner and the teacher, that there may be some advantage in attempting to get outside the system and look back at it. Let us now step up a level of abstraction and consider mathematics education itself as an enterprise and a field of study. The same metaphors of reflection and recursion can be applied to it. As Bauersfeld (1979, p. 210) has noted, we in mathematics education need to develop our own self-concept.

Reflection on Mathematics Education

How can we in mathematics education hold a mirror up to ourselves? One way, of course, is by examining the activities we engage in — the journals we publish, the meetings we attend, the instructional materials we produce, the research we do. How do these activities reflect the development of our field? Do they suggest where we are headed? Mathematics education as a field is virtually unstudied; we have almost nothing that might be termed 'self-referent research' (Scriven, 1980).

Centres and institutes where mathematics education is taken seriously as a full-time enterprise would be natural places for self-referent research to occur. One of the resolutions of the First International Congress on Mathematical Education in 1969 was that 'the new science [of mathematical education] should be given a place in the mathematical departments of Universities or Research Institutes, with appropriate academic qualifications available' (Editorial Board of *Educational Studies in Mathematics*, 1969, p. 284). Since that time several institutes and centres for the study of mathematics education have been established around the world. More are needed. And it is perhaps time that a little more of their attention is given to reflection on our field. How does mathematics education function in various countries? And is it a learning, developing organism?

Recursion in Mathematics Education

This Congress is the fifth term in a sequence. Each Congress has been a function of its predecessors, but it remains to be seen to what limit, if any, the sequence is converging. A larger question is whether it is possible to maintain over a longer time the network of relationships among people that is formed in a somewhat ad hoc fashion for each Congress. I understand that the Executive Committee of the International Commission on Mathematical

Instruction is considering a plan for the establishment of an international program committee that would lift some of the burden of organising the Congress off the host country. Such a proposal is laudable, but perhaps we should go further. Perhaps the field has reached a point in its development where it needs to set up a permanent executive — a secretariat that would facilitate communication among mathematics educators around the world. The secretariat could at least maintain a central file of people with interests and talents in particular aspects of mathematics education. The International Commission might even wish to sponsor some sort of individual membership organisation, possibly with a newsletter, so that interested persons might maintain contact with one another in the four years between Congresses. I offer these suggestions with no particular agenda in mind but with the conviction that the field is maturing and that it may need a stronger structure.

Some Final Remarks

Two of the most eminent mathematics educators of our time are Ed Begle and George Polya. I had the great privilege of being a student and a colleague of both men. Although they differed on many points, they were equally talented at discerning important questions for mathematics educators to tackle.

Ed Begle ended his last work by observing that a substantial body of knowledge in mathematics education did not appear to be accumulating; at the Second and Third International Congresses he could not see that we had identified knowledge that had not been available previously (Begle, 1979, p. 156). The provocative question that follows from Begle's observations is, 'What do we know about mathematics education in 1984 that we did not know in 1980?' Each of us can formulate his or her own answer to that question as we participate in this Congress. It may be, of course, that knowledge in our field simply does not accumulate — that may be the wrong metaphor. My conviction is that we do know some new things, or at least we see some old things in a new light. As I have tried to suggest, some of this new light comes from a greater appreciation of the value of self- awareness.

George Polya's questions are provocative too. At the Fourth International Congress in 1980, he proposed that we take as our theme for the next Congress: 'What can the mathematics teacher do in order that his teaching improves the mind?' (Polya, 1983, p. 1). That question has not been explicitly taken up for the present Congress, but in the spirit of reflection and recursion, let me propose that we monitor our deliberations metacognitively by asking ourselves Polya's question from time to time. As I have tried to suggest, improvement of the mind depends at least in part on students' and teachers' awareness of the need to turn their cognitions back on themselves.

This paper has been concerned with metaphor because, in my view, all our discussion about how children learn mathematics and teachers teach mathematics ultimately rests on metaphorical constructions, some of which people have attempted to formulate into theories. The metaphor of the human mind as a computer is especially powerful and seductive. To conclude that the computer is the *last* metaphor for the mind requires the assumption, first, that the computer will not change in its nature and, second, that cognitive theory must be computational. To conclude that the computer offers a *complete* metaphor for the mind requires the assumption that all knowledge can be reduced to information and all wisdom to knowledge.

We do not have to make these assumptions, however. We can use the computer metaphor without becoming prisoners of it. We can remind ourselves that in characterising education as information transmission, we run the risk of distorting our task as teachers. We can use the word *information* while at the same time recognising that there are various kinds of it (Freudenthal, 1981b, p. 32), and that something is lost when we define the ends of education in terms of information gained.

Aristotle, in *The Art of Poetry*, saw mastery of metaphor as the mark of genius, for to be good at metaphor is to be intuitively aware of hidden resemblances. Resemblances, but not identities. In using a metaphor, one should heed St. Thomas Aquinas' admonition: 'The more openly it remains a figure of speech, the more it is a dissimilar similitude and not literal, the more a metaphor reveals its truth' (Eco, 1980/1983, p. 295).

Note

This paper was prepared for a plenary session of the Fifth International Congress on Mathematical Education, Adelaide, South Australia, 25 August 1984. For ideas and suggestions, I am grateful to Heinrich Bauersfeld, John Bernard, Al Buccino, Larry Hatfield, Bob Jensen, Jack Lee, Yuang-Tswong Lue, Nik Azis Bin Nik Pa, Jan Nordgreen, Sandy Norman, Neil Pateman, Kim Prichard, Alan Schoenfeld, Ed Silver, Ernst von Glasersfeld, and Jim Wilson.

References

Anderson, L.M., Brubaker, N.L., Alleman–Brooks, J., and Duffy, G.G. (1984). *Making seatwork work* (Research Series No. 142). East Lansing: Michigan State University, Institute for Research on Teaching.

Baird, J.R., and White, R.T. (1984, April). 'Improving learning through enhanced metacognition: A classroom study'. Paper presented at the meeting of the American Educational Research Association, New Orleans.

Bartlett, F.C. (1967). *Remembering: A study in experimental and social psychology*. Cambridge: Cambridge University Press.

Bauersfeld, H. (1979). 'Research related to the mathematical learning process'. In B. Christiansen and H.G. Steiner (Eds.), *New trends in mathematics teaching* (Vol. 4, pp. 199-213). Paris: Unesco.

Bauersfeld, H. (1983). 'Subjektive Erfahrungsbereiche als Grundlage einer Interaktionstheorie des Mathematiklernens und –lehrens' [Domains of subjective experiences as the basis for an interactive theory of mathematics learning and teaching]. In *Untersuchungen zum Mathematikunterricht*: Vol. 6. Lernen und Lehren von Mathematik (pp. 1–56). Koln: Aulis–Verlag Deubner.

Begle, E.G. (1979). *Critical variables in mathematics education: Findings from a survey of the empirical literature*. Washington, DC: Mathematical Association of America and National Council of Teachers of Mathematics.

Bolter, J.D. (1984). *Turing's man: Western culture in the computer age*.

Chapel Hill: University of North Carolina Press.

Borkowski, J.G., and Krause, A. (1983). 'Racial differences in intelligence: The importance of the executive system.' *Intelligence*, Vol. 7, pp. 379-395.

Brown, A.L. (1978). 'Knowing when, where, and how to remember: A problem of metacognition'. In R. Glaser (Ed.), *Advances in instructional psychology* (Vol. 1, pp. 77-165). Hillsdale, NJ: Erlbaum.

Brown, A.L. (in press). 'Metacognition, executive control, self-regulation and other even more mysterious mechanisms'. In F.E. Weinert and R.H. Kluwe (Eds.), *Learning by thinking*. West Germany: Kuhlhammer.

Brown, A.L., and Palincsar, A.S. (1982). 'Inducing strategic learning from texts by means of informed, self-control training'. *Topics in Learning and Learning Disabilities*, Vol. 2 No. 1, pp. 1-17.

Brown, S.I. (1983). [Review of *Mindstorms: Children, computers, and power ful ideas.*] *Problem Solving*, Vol. 5 No. 7/8, pp. 3-6.

Campione, J.C., and Brown, A.L. (1978). 'Toward a theory of intelligence: Contributions from research with retarded children'. *Intelligence*, Vol. 2, pp. 279-304.

Carroll, J.B. (1976). 'Psychometric tests as cognitive tasks: A new 'structure of intellect'. In L.B. Resnick (Ed.), *The nature of intelligence* (pp. 27-56). Hillsdale, NJ: Erlbaum.

Case, R. (1974). 'Structures and strictures: Some functional limitations on the course of cognitive growth'. *Cognitive Psychology*, Vol 6, pp. 544-574.

Chomsky, N. (1980). *Rules and representations*. New York: Columbia University Press.

Cooper, D., and Clancy, M. (1982). *Oh! Pascal!* New York: Norton.

Cutland, N. (1980). *Computability: An introduction to recursive function theory*. Cambridge: Cambridge University Press.

Davis, R.B. (1983). 'Complex mathematical cognition'. In H.P. Ginsberg (Ed.), *The development of mathematical thinking* (pp. 253-290). New York: Academic Press.

Davis, R.B. and McKnight, C.C. (1979). 'Modeling the processes of mathematical thinking'. *Journal of Children's Mathematical Behavior* Vol. 2 No. 2, pp. 91-113.

Davis, R.M. (1977). 'Evolution of computers and computing'. *Science*, No. 195, pp. 1096-1102.

Dewey, J. (1933). *How we think: A restatement of the relation of reflective thinking to the educative process*. Boston: Heath.

Dictionary of scientific and technical terms (2nd ed.). (1978). New York: McGraw-Hill.

Easley, J. (1984). 'Is there educative power in students' alternative frameworks — or else, what's a poor teacher to do?' *Problem Solving*, Vol. 6 No. 2, pp 1-4.

Eco, U. (1983). *The name of the rose* (W. Weaver, Trans.). New York : Warner Books. (Original work published 1980)

Editorial Board of Educational Studies in Mathematics. (Eds.) (1969). *Proceedings of the First International Congress on Mathematical Education, Lyon, 24-30 August, 1969*. Dordrecht, Holland: Reidel.

Fischer, G., Burton, R.E., and Brown, J.S. (1978). *Aspects of a theory of simplification, debugging, and coaching* (BBN Report No. 3912).

Cambridge, MA: Bolt, Beranek and Newman.

Flavell, J.H. (1976). 'Metacognitive aspects of problem solving'. In L.B. Resnick (Ed.), *The nature of intelligence* (pp. 231-235). Hillsdale, NJ: Erlbaum.

Flavell, J.H. (1979). 'Metacognition and cognitive monitoring: A new area of cognitive-developmental inquiry'. *American Psychologist*, Vol. 34, pp. 906-911.

Fodor, J.A. (1983). *The modularity of mind: An essay on faculty psychology.* Cambridge, MA: MIT Press.

Freudenthal, H. (1978). *Weeding and sowing: Preface to a science of mathematical education.* Dordrecht, Holland: Reidel.

Freudenthal, H. (1981a). 'Major problems of mathematics education'. *Educational Studies in Mathematics*, 12, pp. 133-150.

Freudenthal, H. (1981b). 'Should a mathematics teacher know something about the history of mathematics?' *For the Learning of Mathematics*, 2(1), pp. 30-33.

Freudenthal, H. (1983). *Didactical phenomenology of mathematical structures.* Dordrecht, Holland: Reidel.

Fuys, D., Geddes, D., and Tischler, R. (Eds.). (1984). *English translation of selected writings of Dina van Hiele-Geldof and Pierre M. van Hiele.* New York: Brooklyn College, School of Education.

Gardner, H. (1983). *Frames of Mind: The theory of multiple intelligences.* New York: Basic Books.

Garofalo, J., and Lester, F.K., Jr. (1984). *Metacognition, cognitive monitoring and mathematical performance.* Manuscript submitted for publication.

Genkins, E.F. (1975). 'The concept of bilateral symmetry in young children'. In M.F. Rosskopf (Ed.), *Children's mathematical concepts: Six Piagetian studies in mathematics education* (pp. 5-43). New York: Teachers College Press.

Gitomer, D.H., and Glaser, R. (in press). 'If you don't know it, work on it: Knowledge, self-regulation and instruction'. In R.E. Snow and J. Farr (Eds.), *Aptitude, learning and instruction: Vol. 3. Conative and affective process analyses.* Hillsdale, NJ: Erlbaum.

Hawkins, D. (1973). 'Nature, man and mathematics'. In. A.G. Howson (Ed.), *Developments in mathematics education: Proceedings of the Second International Congress on Mathematical Education* (pp. 115-135). Cambridge: Cambridge University Press.

Hayes-Roth, B., and Hayes-Roth, F. (1979). 'A cognitive model of planning'. *Cognitive Science*, Vol. 3 pp. 275-310.

Hilton, P. (1984). 'Current trends in mathematics and future trends in mathematics education'. *For the Learning of Mathematics*, Vol. 4 No. 1, pp. 2-8.

Hofstadter, D.R. (1979). *Godel, Escher, Bach: An eternal golden braid.* New York: Basic Books.

Hofstadter, D.R., and Dennett, D.C. (1981). *The mind's I: Fantasies and reflections on self and soul.* New York: Bantam.

Jahnke, H.N. (1983). 'Technology and education: The example of the computer' [Review of *Mindstorms: Children, computers, and powerful ideas*]. *Educational Studies in Mathematics*, Vol. 14, pp. 87-100.

Jaynes, J. (1976). *The origin of consciousness in the breakdown of the bicameral mind*. Boston: Houghton Mifflin.

Johnson-Laird, P.N. (1983a). 'A computational analysis of consciousness'. *Cognition and Brain Theory*, Vol. 6, pp. 499–508.

Johnson-Laird, P.N. (1983b). *Mental models: Towards a cognitive science of language, inference, and consciousness*. Cambridge, MA: Harvard University Press.

Lawler, R.W. (1981). 'The progressive construction of mind'. *Cognitive Science*, Vol. 5, pp. 1–30.

Levinas, E. (1973). *The theory of intuition in Husserl's phenomenology* (A. Orianne, Trans.). Evanston, IL: Northwestern University Press. (Original work published 1963)

Locke, J. (1965). *An essay concerning human understanding* (Vol. 1). New York: Dutton. (Original work published 1706)

Loper, A.B. (1982). 'Metacognitive training to correct academic deficiency'. *Topics in Learning and Learning Disabilities*, Vol. 2 No. 1, pp. 61–68.

Mandler, G. (1975). 'Consciousness: Repectable, useful, and probably necessary'. In R.L. Solso (Ed.), *Information processing and cognition: The Loyola Symposium* (pp. 229–254). Hillsdale, NJ: Erlbaum.

Maurer, S.B. (1983). The effects of a new college mathematics curriculum on high school mathematics. In A. Ralston and G.S. Young (Eds.), *The future of college mathematics: Proceedings of a conference/workshop on the first two years of college mathematics* (pp. 153–173). New York: Springer-Verlag.

McCarthy, J. (1983). 'Recursion'. In A. Ralston and E.D. Reilly, Jr. (Eds.), *Encyclopedia of computer science and engineering* (2nd ed., pp. 1273–1275). New York: Van Nostrand Reinhold.

McCloskey, M. (1983, April). 'Intuitive physics'. *Scientific American*, pp. 122–130.

Merleau-Ponty, M. (1973). *Consciousness and the acquisition of language* (H.J. Silverman, Trans.). Evanston, IL: Northwestern University Press. (Original work published 1964)

Minsky, M. (1980). 'K-lines: A theory of memory'. *Cognitive Science*, Vol. 4, pp. 117–133.

Möbius, A.F. (1967). *Gesammelte Werke* [Collected works] (Vol. 1). Wiesbaden: Martin Sandig. (Original work published 1885)

Natsoulas, T. (1983). 'A selective review of conceptions of consciousness with special reference to behavioristic contributions'. *Cognition and Brain Theory*, Vol. 6, pp. 417–447.

Newell, A., and Simon, H.A. (1972). *Human problem solving*. Englewood Cliffs, NJ: Prentice-Hall.

Newman, J.H.C. (1903). *An essay in aid of a grammar of assent*. London: Longmans, Green. (Original work published 1870)

Papert, S. (1975). 'Teaching children thinking'. *Journal of Structural Learning*, Vol. 4, pp. 219–229.

Papert, S. (1980). *Mindstorms: Children, computers, and powerful ideas*. New York: Basic Books.

Peter, R. (1967). *Recursive functions* (3rd ed.). New York: Academic Press.

Piaget, J. (1956). 'Les étages du developpement intellectuel de l'enfant et de l'adolescent' [The stages of intellectual development in the child and the adolescent]. In P. Osterrieth et al., *Le probleme des stages en psychologie de l'enfant* (pp. 33–41). Paris: Presses Universitaires de France.

Piaget, J. (1971a). *Science of education and the psychology of the child* (D. Coltman, Trans.). New York: Viking. (Original work published 1969)

Piaget, J. (1971b). 'The theory of stages in cognitive development'. In D.R. Green (Ed.), *Measurement and Piaget* (pp. 1–11). New York: McGraw-Hill.

Piaget, J. (1976). *The grasp of consciousness: Action and concept in the young child* (S. Wedgwood, Trans.). Cambridge, MA: Harvard University Press. (Original work published 1974)

Piaget, J. (1978). *Success and understanding* (A.J. Pomerans, Trans.). Cambridge, MA: Harvard University Press. (Original work published 1974)

Pinard, A., and Laurendeau, M. (1969). ''Stage' in Piaget's cognitive–developmental theory: Exegesis of a concept'. In D. Elkind and J.H. Flavell (Eds.), *Studies in cognitive development: Essays in honor of Jean Piaget* (pp. 121–170). New York: Oxford University Press.

Polya, G. (1983). 'Mathematics promotes the mind'. In M. Zweng, T. Green, J. Kilpatrick, H. Pollak, and M. Suydam (Eds.), *Proceedings of the Fourth International Congress on Mathematical Education* (p. 1). Boston: Birkhauser.

Rappaport, D. (1957). 'Cognitive structures'. In *Contemporary approaches to cognition: A symposium held at the University of Colorado* (pp. 157–200). Cambridge, MA: Harvard University Press.

Rotenstreich, N. (1974). 'Humboldt's prolegomena to philosophy of language'. *Cultural Hermeneutics*, Vol. 2, pp. 211–227.

Schoenfeld, A.H. (1983). 'Episodes and executive decisions in mathematical problem solving'. In R. Lesh & M. Landau (Eds.), *Acquisition of mathematics concepts and processes* (pp. 345–395). New York: Academic Press, 1983.

Schoenfeld, A.H. (in press-a). 'Beyond the purely cognitive: Belief systems, social cognitions and metacognitions as driving forces in intellectual performance'. *Cognitive Science*.

Schoenfeld, A.H. (in press-b). *Mathematical problem solving*. New York: Academic Press.

Scriven, M. (1980). 'Self-referent research'. *Educational Researcher*, Vol. 9, No. 4, pp. 7–11, and Vol. 9, No. 6, pp. 11–18.

Silver, E.A. (1982a). 'Knowledge organization and mathematical problem solving'. In F.K. Lester, Jr., and J. Garofalo (Eds.), *Mathematical problem solving: Issues in research* (pp. 15–25). Philadelphia: Franklin Institute Press, 1982.

Silver, E.A. (1982b). *Thinking about problem solving: Toward an understanding of metacognitive aspects of mathematical problem solving.* Unpublished manuscript, San Diego State University, Department of Mathematical Sciences, San Diego, CA.

Simon, H.A. (1977). 'What computers mean for man and society'. *Science*, 195. 1186–1191.

Skemp, R.R. (1979). *Intelligence, learning, and action*. New York: Wiley.

Snow, R.E. (1980). 'Aptitude processes'. In R.E. Snow, P.A. Federico, and

W.E.Montague (Eds.), *Aptitude, learning, and instruction: Vol. 1. Cognitive process analyses of aptitude* (pp. 27–63). Hillsdale, NJ: Erlbaum.

Thomas, R.M. (1984). 'Mapping meta-territory'. *Educational Researcher*, Vol. 13, No. 4, pp. 16-18.

Thompson, A.G. (1984). 'The relationship of teachers' conceptions of mathematics and mathematics teaching to instructional practice'. *Educational Studies in Mathematics*, Vol. 15, pp. 105–127.

Van Hiele, P.M. and van Hiele–Geldof, D. (1958). 'A method of initiation into geometry at secondary schools'. In H. Fruedenthal (Ed.), *Report on methods of initiation into geometry* (pp. 67–80). Groningen: J.B. Wolters.

Vergnaud, G. (1983). 'Why is an epistemological perspective necessary for research in mathematics education?' In J.C. Bergeron and N. Herscovics (Eds.), *Proceedings of the Fifth Annual Meeting of the North American Chapter of the International Group for the Psychology of Mathematics Education* (Vol. 1, pp. 2–20). Montréal: Université de Montréal, Faculté de Sciences de l'Education.

Von Glasersfeld, E. (1983). 'Learning as a constructive activity'. In J.C. Bergeron and N. Herscovics (Eds.), *Proceedings of the Fifth Annual Meeting of the North American Chapter of the International Group for the Psychology of Mathematics Education* (Vol. 1, pp. 41-69). Montréal: Université de Montréal, Faculté de Sciences de l'Education.

Whitehead, A.N. (1929). 'The rhythm of education'. *The aims of education and other essays*. New York: Macmillan.

DISCRETE MATHEMATICS

Renfrey B. Potts
The University of Adelaide
Australia

Vombatus Platyrhinus

The wombat with the broad hairy nose.
Identification number ICME 5 1234123413.
The peculiar animal peculiar to South Australia.
What has Vombatus Platyrhinus got to do with my talk?
That is a question for you to try to answer.

1. Introduction

I am going to consider two questions:
Q1   What is discrete mathematics?
Q2   What should be the role of discrete mathematics in mathematics education?

I am glad these sorts of questions did not appear in mathematics examinations I have taken. I do not think I would get more than 6 marks out of 10 for my answer to either question — perhaps 7 marks if I were particularly neat.

Among mathematicians there seems to be some consensus that the last decade or two has belonged to discrete mathematics, but there seems to be little consensus about what is discrete mathematics. Discrete to many means not continuous. So I begin my answer to question Q1 with the statement that discrete mathematics is mathematics which is not continuous. I think so far I have earned 1 mark out of 10. I need to say more.

Recent mathematics journals and texts which claim to specialise in discrete mathematics cover a range of topics:

> graph theory, network theory, coding theory, lattice theory, matrix theory, matroid theory, game theory, scheduling theory, discrete probability theory, Boolean algebra, combinatorics, difference equations, discrete optimisation, algorithms.

I suppose this collection of topics is what most would call discrete mathematics. By now my answer to Q1 may be worth 3 out of 10.

To earn more marks I propose to do the following. Firstly, I shall make some general philosophical remarks about continuous versus discrete phenomena; and as I am no philosopher I could be in danger of losing some marks. Secondly, I shall give a rather detailed analysis, pitched at pre-calculus secondary school level, of two examples of applied discrete mathematics. Here I shall be on firmer ground, and here will be my hopes of increasing my marks for my answer to Q1 to 6 or 7 out of 10. How I shall tackle Q2 I shall leave till later.

2. Continuous versus Discrete

Is *matter* continuous or discrete? Obviously the floor I am standing on is continuous. But I have been taught by my teachers and have taught my students that this is obviously wrong. The material of the floor is discrete, with an atomic structure.

Is *energy* continuous or discrete? Obviously the energy of the light shining on me is continuous. But I have been taught by my teachers and have taught my students that this is obviously wrong. The energy of the light is discrete, bundled into discrete elements called photons.

Is the *rational number line* continuous or discrete? Obviously, it is continuous. Between any two rational numbers there is an infinite number of rational numbers. But I have been taught by my teachers and have taught my students that this is obviously wrong. The rational number line has an infinite number of gaps for the irrational numbers.

Is *time* continuous or discrete? This is not so obvious. I wear two watches, a digital watch from Japan and an analogue watch from Switzerland. The digital watch displays the time in discrete units, in seconds. The analogue watch obviously gives the time continuously as the hands rotate. But is it so obvious? If you listen carefully, you will hear my analogue watch ticking. And if you look inside at its works, you would see the unwinding of the spring discretely governed by an escapement mechanism. On closer examination, my analogue watch operates not continuously but discretely. The concept of time has mystified philosophers, scientists and mathematicians throughout the ages. What is meant by the continuity of time? How should Zeno's paradox of Achilles chasing the tortoise be resolved? There are now scientists who believe that time is essentially discrete, and have given the discrete unit the name chronon with a duration of 10^{-20} seconds. Is time continuous or discrete? It surely remains a confusing and unanswered question.

Matter, energy, the rational number line, time. Continuous or discrete? Let me leave this question to the philosophers. There can be no confusion with the mathematical concept of continuity because mathematicians have defined what it means. There is an agreed rigorous definition of what mathematicians mean when they say that a function is continuous. But it must be remembered that the definition is the culmination of hundreds of years of mathematical thought and argument. The definition is man made, a figment of the pure mathematician's imagination, in contrast to discrete mathematical quantities which seem to be part of nature. There seems to be something natural about counting the fingers up to 10, but something unnatural about the definition of continuity. Students find it unnatural. No sooner do we teach the definition than we give examples of functions which appear to be continuous but which, according to the definition, are discontinuous everywhere. This teaching trick, I suggest, can be pulled with continuity but not with discrete concepts.

Matter, energy, the rational number line, time, at first appear to be continuous but this I have questioned. The mathematical concept of continuity is rigorously defined, but is it natural? Have I earned any marks for this philosophical digression? I doubt it.

If I haven't succeeded in confusing you, let me conclude this digression with a comment on a current favourite with some educationists — continuous assessment. Fortunately for teachers and students, continuous assessment is impossible. It must be discrete.

Enough of this confusion. Let me proceed to my first example of discrete mathematics — difference equations, which I shall present at a level suitable for a secondary school curriculum.

3.1 A Linear Difference Equation

A colleague of mine, a Professor of Applied Mathematics, was asked by the mathematics teacher at a secondary school to help a selected group of bright 14 year old students by giving them some enrichment material relevant to current research in mathematics. He was at first apprehensive because the basic tools for his own work involve at least calculus. I suggested difference equations. I will show you why the suggestion worked wonders.

I refer to

$$x(n+1) = a\,x(n) \qquad n = 0,1,2, \ldots \tag{1}$$
$$x(0) = c$$

as a difference equation (ΔE), indeed a *first order linear* ΔE (also known as a recurrence relation). The notation I use is chosen for ease of exposition; teachers would prefer other notation, or perhaps a computer program in BASIC. In fact, the ΔE can be analysed with a hand calculator (or by mental arithmetic) without bothering about notation. By guided experimentation, the student will quickly learn that

if $c = 0$, $x(n) = 0$ for all n
if $a = 1$, $x(n) = c$ for all n
if $c > 0, 0 < a < 1$, $x(n) \to 0$ as $n \to \infty$
if $c > 0, a > 1$, $x(n) \to \infty$ as $n \to \infty$.

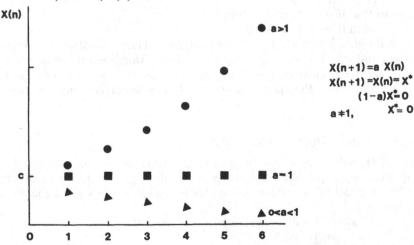

Fig. 1 The first few values of the solution of the linear difference equation for three values of the parameter a.

[I stress again that the notation is not for the student, but for this exposition; $x(n) \to \infty$ may be illustrated by computer 'overflow']. For simplicity the cases $c < 0$ and $a \leqslant 0$ have been excluded. These would not be relevant if (1) were being considered as a model of compound interest, population growth, or radioactive decay, for example.

As is often the case in mathematics, a graphical illustration of an algebraically formulated problem is a help to better understanding and an aid to teaching. In Fig. 1, the values of $x(n)$ are plotted against n, illustrating the different behaviour for $0 < a < 1$, $a = 1$, and $a > 1$.

It is more illuminating and more instructive to plot $x(n+1)$ against $x(n)$ and this is done in Figs. 2 and 3. These figures can be drawn without any calculations. The geometric constructions correspond to solving the ΔE as indicated by the flow chart

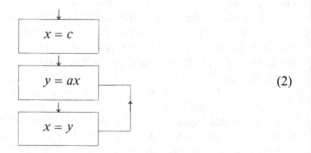

(2)

I shall use this simple example of a linear ΔE, which you will all have recognized as the familiar geometric progression, to introduce the concepts of *stable* and *unstable equilibrium*.

If for any value of n, $x(n+1) = x(n)$, then all subsequent $x(n)$ are equal and such an x is referred to as an *equilibrium point* (or fixed point) and I shall denote it by x^*. For (1) the x^* are given by

$$(1 - a)x^* = 0 \tag{3}$$

so that if $a = 1$, x^* is arbitrary
and if $a \neq 1$, $x^* = 0$.

The stability interpretation of the analysis of (1), as illustrated in Figs. 2 and 3, is as follows. For $0 < a < 1$, $x^* = 0$ is a *stable* equilibrium point in the sense that if c is close to zero, then $x(n) \to 0$ as $n \to \infty$. But for $a > 1$, $x^* = 0$ is *unstable*. For the special case $a = 1$, any point is a neutral equilibrium point.

3.2 A Nonlinear Difference Equation

The ΔE (1) with $a > 1$ may be appropriate to model the increase in the yearly crop from my apple tree in the early years of its growth but a modification to inhibit an eventual 'overflow' of apples is desirable. This gives the motivation for the ΔE

$$x(n+1) = a[b - x(n)]\, x(n) \qquad n = 0, 1, 2, \ldots$$
$$x(0) = c. \tag{4}$$

Comparison with (1) shows that the factor a has been modified by a further $b - x(n)$ which reduces the multiplicative effect as $x(n)$ approaches b. It is no

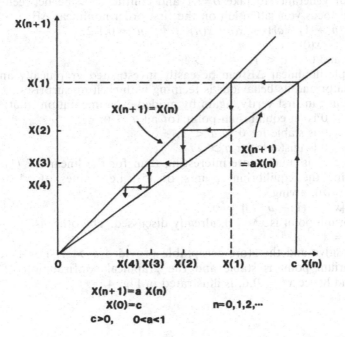

Fig. 2 Geometrical construction of the solution of the linear difference equation, with $0 < a < 1$. Start with $x(0) = c$, then follow the arrow to give $x(1) = ac$, and so on giving $x(2), x(3), x(4)$ ... The equilibrium point $x^* = 0$ is stable.

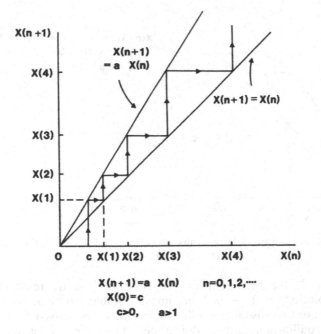

Fig. 3 With $a > 1$, the equilibrium point $x^* = 0$ is unstable.

great loss of generality to take $b = 1$ and confine x to be between 0 and 1.
Thus I shall focus your attention on the first order nonlinear ΔE

$$x(n + 1) = a[1 - x(n)]\, x(n) \qquad n = 0,1,2, \ldots$$
$$x(0) = c. \tag{5}$$

with $a > 0$ and $0 \leqslant c < 1$.

This simple nonlinear ΔE can be easily investigated graphically and with a
hand calculator and its behaviour is teeming with exciting surprises.

The student can first verify, again by guided experimentation, that:

$x^* = 0$ is an equilibrium point for all $a > 0$;

$x^* = 0$ is stable for $0 < a < 1$;

$x^* = 0$ is unstable for $a > 1$.

The case $a > 1$ is much more interesting than for the linear ΔE (1). First let
me determine the equilibrium points of (5), i.e. values of x^* for which
$x(n + 1) = x(n)$, giving

$$x^*\,[x^* - (1 - a^{-1})] = 0. \tag{6}$$

One equilibrium point is $x^* = 0$, already discussed. The other is

$$x^* = 1 - a^{-1} \tag{7}$$

which is positive and therefore acceptable provided $a > 1$. For $1 < a < 3$,
this equilibrium point is stable and the graphical confirmation of this for
$a = 2.5$, and hence $x^* = 0.6$, is illustrated in Fig. 4.

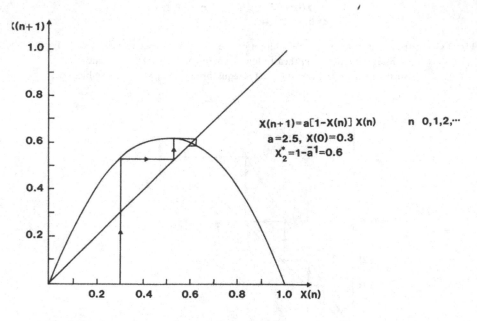

Fig. 4 For the nonlinear difference equation with $a = 2.5$, the equilibrium point
$x^* = 0.6$ is stable.

Beyond $a = 3$, the behaviour of the solution is quite remarkable. The
equilibrium point $x^* = 1 - a^{-1}$ becomes unstable, so that, together with
$x^* = 0$, there are two unstable equilibrium points. As proved from (6) there
are no other equilibrium points. But a new phenomenon occurs — *stable
oscillations*.

Fig. 5 illustrates the numerical results obtained for a value of $a = 19/6$, just greater than 3. The solution soon settles down to an oscillation of period 2. And this happens to be quite a good description of the crop on my mature apple tree. The crop is not the same each year but tends to have a two year period with a heavy crop followed by a light crop, and then back to a heavy crop and so on. In contemporary jargon, the nonlinear difference equation (5) is a good mathematical model for the crop of my apple tree.

The analysis for a 2 period solution is simple in theory but a bit messy in practice. First I use the difference equation (5) twice:

$$x(n+1) = a[1-x(n)]x(n)$$
$$x(n+2) = a[1-x(n+1)]x(n+1)$$

and eliminate $x(n+1)$ to give the second order ΔE

$$x(n+2) = a\{1-a[1-x(n)]x(n)\}\,a[1-x(n)]\,x(n). \qquad (8)$$

For a 2 period solution,

$$x(n+2) = x(n) = x^*$$

and (8) gives now a quartic equation

$$x^{**}[x^{**} - (1-a^{-1})][(x^{**})^2 - (1+a^{-1})x^{**} + a^{-1}(1+a^{-1})] = 0. \qquad (9)$$

For the specific example $a = 19/6$, this quartic factorises easily (which explains my particular choice of a):

$$x^{**}[x^{**} - (13/19)][x^{**} - (10/19)][x^{**} - (15/19)] = 0. \qquad (10)$$

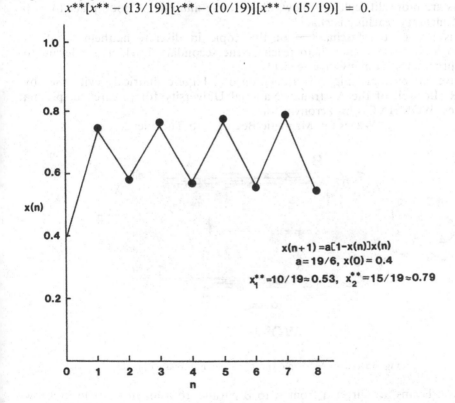

$$x(n+1) = a[1-x(n)]x(n)$$
$$a = 19/6, \ x(0) = 0.4$$
$$x_1^{**} = 10/19 \approx 0.53, \quad x_2^{**} = 15/19 \approx 0.79$$

Fig. 5 For the nonlinear difference equation with $a = 19/6$, there are no stable equilibrium points but a stable oscillation of period 2, between $x^{**} = 10/19$ and $x^{**} = 15/19$ occurs.

Two of the roots $x^{**} = 0$, $x^{**} = 13/19$ are the unstable equilibrium points. They must reappear, since they satisfy $x(n+1) = x(n)$ which certainly forces $x(n+2) = x(n)$. The two other roots correspond to a 2 period oscillation between $x^{**} = 10/19$ and $x^{**} = 15/19$. I am not suggesting this algebraic analysis is suitable for the young student, for whom the numerical results, as illustrated in Fig. 5, are simple to derive and are very convincing.

For larger values of a the behaviour is more exciting. There is something for everyone. For $a > 1+\sqrt{6} \approx 3.449$, the 2 period solution becomes unstable and a 4 period solution arises, then an 8 period solution and so on.

When $a \approx 3.570$ a 'chaotic' region occurs with never-repeating solutions appearing. This result has sent shivers up and down the spines of those addicted to mathematical modelling. It used to be argued that data represented by random never-repeating values must result from some stochastic variable and that the appropriate mathematical model must include some random or statistical term. But the nonlinear difference equation (5) represents a completely deterministic model — yet capable of producing random-like data. Is your spine shivering?

For values of a beyond the chaos region, the behaviour is confused and complicated, with odd numbered period solutions occurring. The numerical results are more difficult to follow, likewise the analysis. A good place to stop for elementary teaching is chaos!

It is one of the fascinations of this topic in discrete mathematics that, although elementary enough to teach at the secondary level, it is relevant to current research in many diverse fields.

I give one example. Fig. 6 is a diagram of a large cylindrical device used by Dr. R. Boswell of the Australian National University for research in plasma physics. WOMBAT is his acronym for

Waves On Magnetic Beams And Turbulence

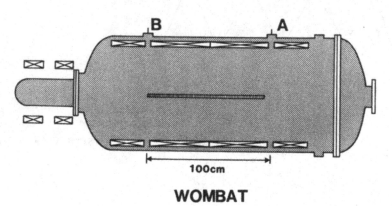

WOMBAT

Fig. 6 Device used for present day research in plasma physics.

Electron beams are directed from A to B parallel to a magnetic field in a low pressure gas. The oscillations detected for increasing beam current give all the features described by the nonlinear ΔE (5). Period doubling occurs, then more period doubling and so on until a chaotic regime is reached. Fig. 7, taken directly from a research paper submitted by Boswell for publication this year

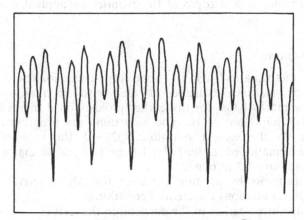

Fig 7 Period three oscillation for the electron beam experiment with WOMBAT.

(1984) illustrates a period three oscillation of the amplitude of the electron beam.

3.3 A System of Difference Equations

I want now to make a very brief extension of the analysis of the single nonlinear ΔE to a system of two ΔE's because it introduces a strange phenomenon which has attracted much interest among applied mathematicians. Those in the know would have recognised the pun — the phenomenon is called a *strange attractor*. Plotting the results on the screen of a micro computer is ideal for observing the analysis which perhaps should be regarded as possible enrichment material for the brighter students.

The specific example I shall consider is the nonlinear system

$$x(n+1) = 1 + y(n) - 1.4\, x(n)^2$$
$$y(n+1) = 0.3\, x(n) \qquad n = 0,1,2,\ldots$$
$$x(0) = c,\ y(0) = d. \tag{11}$$

It is straightforward to calculate, for given values of c and d, the values of $x(1)$, then $x(2)$, $y(2)$, and so on. The equilibrium points, defined by $x(n+1) = x(n)$ and $y(n+1) = y(n)$, are determined from

$$1.4\,(x^*)^2 + 0.7\, x^* - 1 = 0$$
$$y^* = 0.3\, x^*, \tag{12}$$

giving two equilibrium points (0.631.., 0.189..) and (-1.131.., -0.339..).

Numerical experimentation verifies that these two equilibrium points are unstable. For example, starting with $x(0) = c = 0.63$ and $y(0) = d = 0.19$, the successive points wander away and then seem to jump randomly. No periodic oscillations appear but the totality of points start to form a set of curves called a *strange attractor*. Although the equilibrium points are unstable, the solution points do not wander off to infinity but seem to be attracted to the curves. A dynamic feature of the attractor is that successive points jump randomly from one part of the curve to another. The demonstration with a graph plotter recording points in succession is fascinating.

There is considerable current research on strange attractors and some applied mathematicians have proposed them as possible models of turbulence and related phenomena. It is a topic at the frontier of applied mathematics research.

4. First Progress Report

What is discrete mathematics? (Q1) I hope my discussion of linear and nonlinear difference equations has helped my answer to this question. Perhaps I have regained the marks I lost for my earlier philosophical digression and I am back to about 5 marks out of 10. You may think my progress very slow. When will I get to the second question (Q2) of the role of discrete mathematics in mathematics education? But I hope you realise that while I am answering Q1, I also have Q2 in mind.

Let me report progress by arguing the case for ΔE (discrete difference equation) versus DE (continuous differential equation):

(1) ΔEs can be introduced earlier in the curriculum than DEs;
(2)   this allows early introduction of the important concept of non-linearity;
(3)   many of the properties of linear ΔEs also hold for linear DEs;
(4) ΔEs are more appropriate for modelling discrete phenomena;
(5) ΔEs can be analysed numerically with a hand held calculator;
(6) ΔEs are good examples of worthwhile mathematical uses of calculators and micro-computers;
(7)   computing the solution to a DE often requires that it be approximated by a ΔE;
(8)   the strange attractor phenomenon can be illustrated by a simple system of two ΔEs (for DEs, three are required).

Do not get carried away. I am not advocating that ΔE's replace DEs, that discrete mathematics replace continuous. I am suggesting that discrete mathematics should play a more important role in mathematics education. I hope I have made some progress with my answer to Q2.

5.1 Biological Problem

As my second example of discrete mathematics I am going to turn to a problem in network theory, one of the fields of mathematics which has commanded great attention in recent years, one which is acknowledged as a key area of discrete mathematics, one which has no counterpart in continuous mathematics, and one which has sneaked into some school mathematics curricula.

What are my aims? I hope by this second example to give you a better understanding of discrete mathematics. To put it crudely, I hope to end up with 6 or 7 marks out of 10 for my answer to Q1: What is discrete mathematics?

I am going to illustrate the application of network theory to a problem in biology. In Australia, and perhaps elsewhere, the last ten years or so has seen a dramatic increase in the teaching of biology at school. It therefore seems important to me, as an applied mathematician, to give special consideration to the application of mathematics to biological problems. Applications of the calculus are well known; there are applications of difference equations to problems in ecology and populations but these are not so well known. However, I have chosen for my example an application of network theory to what is a central problem in biology: the classification of different species of plants and animals.

One of the approaches to biological classification uses the evolutionary theory relating similar species to construct *ancestral trees* (or more technically phylogenetic trees), of which Fig. 8 is a trivial example. Vertex A represents the common ancestor or root of the ancestral tree, vertices B and S represent two new species obtained by mutation of genes, B is the ancestor of P and C, and C the ancestor of Q and R. The construction of such an ancestral tree depends on fossil data which may be very incomplete. The inclusion of possible extinct ancestors may be quite speculative.

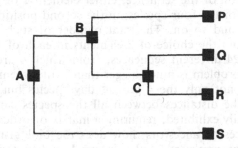

Fig. 8 An ancestral tree. The vertices P,Q,R,S represent the species being classified, the vertices A,B,C represent ancestors which may be extinct or other existing species.

How does the biologist compare species? Is a wombat more similar to the kookaburra than the koala bear or the kangaroo? It is now possible for biologists to answer such a question on a quantitative rather than a qualitative basis, by defining the 'distance' between species. This distance is taken as the minimum number of mutations or changes required in the evolution of one species from another.

A somewhat simplified record of the basic data for the wombat, kookaburra, koala bear and kangaroo is given in Fig. 9. The numbers represent the sequences for a hypothetical protein (ICME 5) common to the four species. In each position of the protein sequence, four alternative elements are allowed, and these are numbered 1, 2, 3 or 4. The protein

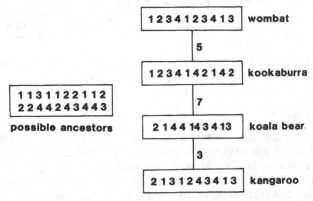

Fig. 9 Protein sequences for four species being classified and all possible ancestors.

sequence 1234123413 is the 'identification number' for the wombat, 1234142142 that for the kookaburra. These two sequences differ in five (the last five) positions. The 'distance' between the wombat and the kookaburra is therefore taken as 5 units. From Fig. 9 it can be checked that the distance between the kookaburra and the koala is 7 units and that between the koala and kangaroo 3 units.

The essence of the classification problem is to construct an ancestral tree connecting the vertices representing the wombat, kookaburra, koala and kangaroo. But first allowance has to be made for all possible ancestors, existing or extinct, and the sequences for these are indicated in the left vertex in Fig. 9. In the first position of the sequence, either element 1 or 2 is allowed to cover the requirements for the four species; in the second position 1 or 2; in the third position 3 or 4, and so on. The total number of such ancestors is 1020. This is calculated from the choice of 2 elements in each of 10 positions giving a total of $2^{10} = 1024$ different sequences, from which 4 are subtracted for the given species. The problem is much bigger than it first seemed. All told it is necessary to consider not only the 4 present day species but in addition 1020 possible ancestors. The distances between all the species can be readily determined but not so readily exhibited, requiring a matrix of order 1024.

Of all the possible choices of ancestors, how does the biologist choose the best ancestral tree? One criterion used, and it is one I shall use, is to choose the tree which, including the four given vertices at least, is of minimum length, representing minimal evolution, the minimal total number of changes.

Fig. 10, an ancestral tree of length 10, gives the required answer. Just three ancestors are required. Their protein sequences and their distances apart are given in the figure.

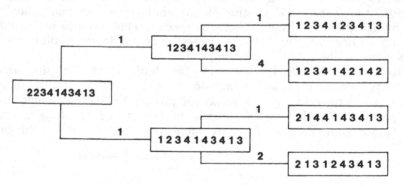

Fig. 10 The minimal spanning tree for the wombat, kookaburra, koala and kangaroo using the protein ICME 5.

5.2 Network Formulation

The precise mathematical formulation, or mathematical model, of the biological problem of constructing ancestral trees requires a brief statement of fundamental definitions and notation of network theory, or graph theory as it is often called. I am forced to do this because of a complete lack of uniformity in the literature. I need to state what I shall use, with the warning that my choice of definitions and terminology will not agree with those of many of the authors of the numerous texts which have been recently published.

A network is a set $\mathscr{V}$ of unordered elements, which I shall call *vertices*, together with a set $\mathscr{E}$ of unordered pairs of vertices called *edges*. Fig. 8 is a geometrical representation of the network with $\mathscr{V} = \{A,B,C,P,Q,R,S\}$ and $\mathscr{E} = \{(A,B),(A,S),(B,C),(B,P),(C,Q),(C,R)\}$. This network has 7 vertices and 6 edges. The sequence A,(A,B),B,(B,C),C,(C,Q),Q represents a *path* from an initial vertex A to a final vertex Q. It is essential that any vertex in a path appears only once. The network illustrated in Fig. 8 is a *spanning tree*. The essential defining property of a spanning tree is that between each pair of vertices of $\mathscr{V}$ there is one and only one path. The tree is said to span the vertices of $\mathscr{V}$.

It is simple to destroy the tree structure by adding another edge. For example, if in Fig. 8 I add the edge (P,Q) then there would be a second path from A to Q, namely A,(A,B),B,(B,P),P,(P,Q),Q. The sequence B,(B,P),P, (P,Q),Q,(Q,C),C,(C,B),B would be what is called a *cycle*. A cycle is a path, except that the initial and final vertices are the same.

Network theory is a fascinating branch of discrete mathematics. From very few definitions, from very little background material, many interesting theorems can be proved. I have already said enough for the following result to be established: a tree with N vertices must have $N-1$ edges. [Hint: proof by induction on N.]

To introduce the concept of distance between vertices, the edges can be prescribed to have certain lengths, as indicated in Fig. 10. These lengths are additive so that the length of a tree is simply the sum of the lengths of its edges, in this case $1 + 1 + 1 + 4 + 1 + 2 = 10$ units. Although I use the terminology length and distance, be careful — the triangle property that the length of one side is less than the sum of the lengths of the other two sides is not necessarily true.

I can now give a precise statement of the so-called *minimal spanning tree* problem. Given a network, with vertices $\mathscr{V}$ and edges $\mathscr{E}$, construct a tree spanning all vertices such that the length of the tree is a minimum. It is unfortunate for the biologists that this is not the minimal ancestral tree problem. Can you see why? Unfortunately, because there is a simple algorithm for constructing a minimal spanning tree. All one needs to do is arrange the edges in ascending order of lengths. Starting with the shortest edge, keep choosing edges in order, but reject an edge if its inclusion would give a cycle. When $N-1$ edges have been chosen so that all vertices are spanned, the result is the required minimal spanning tree. It may not be unique.

The *minimal ancestral tree* problem is different, unfortunately for the biologists but fortunately for the applied mathematicians. Our help is needed. The complicating feature of the minimal ancestral tree problem is that in addition to the set $\mathscr{V}$ of vertices representing the species which must be included, there is another set $\mathscr{A}$ of vertices representing all possible ancestors, some of which may be included in the ancestral tree. The possible ancestral trees must span all vertices of $\mathscr{V}$ and any number of vertices of $\mathscr{A}$. The precise statement of the *minimal ancestral tree problem* is this: given a set $\mathscr{V}$ and a set $\mathscr{A}$ of vertices, and a set $\mathscr{E}$ of edges, construct a tree of minimal length which spans all vertices of $\mathscr{V}$ and any number of vertices of $\mathscr{A}$.

My illustrated example considers just four species, the wombat, kookaburra, koala bear and kangaroo. In practice a biologist would be wishing to classify twenty or more species. Yet the minimal ancestral tree problem for my example is quite large. The set $\mathcal{V}$ consists of 4 vertices, $\mathcal{A}$ consists of 1020 vertices, $\mathcal{E}$ consists of all pairs of the 1024 vertices. The possible ancestral trees to be considered must span the 4 vertices of $\mathcal{V}$ and any number of the 1020 vertices of $\mathcal{A}$. The possibilities are vast.

5.3 Solution?

There is a useless algorithm for solving the minimal ancestral tree problem which may satisfy the pure but not the applied mathematician, and certainly not the biologist. All that is necessary in theory is to construct all possible ancestral trees, calculate their lengths, and choose one with minimal length. The problem is a discrete one, the number of ancestral trees is finite, so go to it. The difficulty is that for an example of biological interest, with about 20 species to classify, the number of possible ancestral trees is finite, yes, but so large, in the many millions, that the biggest of today's computers could not complete the task of constructing all the trees in a reasonable time.

It is a sobering thought that while the masses are intoxicated with stories of the miraculous powers of computers, the applied mathematicians are now very much concerned with distinguishing between problems which can be computed in reasonable time and those which cannot. The minimal spanning tree problem is in the first category, the minimal ancestral tree problem in the second.

Faced with what has been proved to be in practice a non-computable problem, the applied mathematician seeks heuristic algorithms which do not guarantee the optimum solution but yield a good solution in reasonable computing time. This may be quite acceptable to the biologist. To insist on an absolute minimum for the length of the ancestral tree may be quite unjustified. Would other trees having, say, just one more evolutionary change be unacceptable? Various heuristic algorithms for the minimal ancestral tree problem are currently being studied. It is an active field of research, a lively example of discrete mathematics being applied to an important biological problem.

In fact the applied mathematicians claim they are doing much more than help the biologists with their classification of species. They regard their recent work as strong evidence in favour of the theory of evolution, a theory which has been seriously questioned. For 11 species of vertebrates, the ancestral trees were constructed for 5 different independent proteins. The trees obtained were remarkably similar, supporting the evolutionary theory on which the constructions were based. A triumph for discrete mathematics!

6. Second Progress Report

What is discrete mathematics? (Q1) My second example has been taken from network theory, one of the major topics in discrete mathematics. A network is a set of discrete elements, vertices and edges. The minimum problem I have described has nothing to do with the standard maxima and minima problems of school mathematics. Calculus is no help.

My answer to Q1 is complete. My second example has been chosen in order to exemplify essential features of applied discrete mathematics. If I had time and patience, and you and I have run out of both, I could continue with other examples. But I am not aiming for full marks. I shall settle for 6 or 7 out of 10.

And what of the role of discrete mathematics in mathematics education? (Q2) I am not suggesting that my example of ancestral trees is suitable for a school curriculum. But it should help support the case for the inclusion of some network theory. The sort of material in Book 3 of the School Mathematics Project seems suitable.

My first example of difference equations is another matter. I am doing more than suggest that it is suitable for a school curriculum. I recommend, challenge, insist that it be taught in secondary schools — next year!

I am certain I am right. I claim 10/10 for my answer to Q2.

7. Conclusions

Q1 What is discrete mathematics?

I have claimed about 6 or 7 out of 10 for my answer. Perhaps I could have done better with more examples with shorter descriptions. But I chose to be more detailed with just two examples in the hope of giving enough material to help a teacher whose interest may be awakened.

Q2 What should be the role of discrete mathematics in mathematics education?

This is the sort of question which will be asked again and again. It is on a par with some of the questions asked at ICME IV:

>   Is calculus essential?
>   What should be dropped from the secondary school mathematics curriculum to make room for new topics?
>   Where do we go from here?

I have given my answer to Q2 and claimed 10/10. Your answer will have to take many educational factors into account. But I suggest that there are some *facts* you must consider:

- computers have made an enormous impact not only on society but on mathematics;
- mathematics education must respond to the impact of computers;
- computers use discrete, not continuous, mathematics;
- mathematics is an exciting, living, changing discipline;
- most mathematics in school curricula is old mathematics (old = over 40 years);
- linear programming (formulated as continuous, but solved by discrete mathematics) is one of the few young (= not old) mathematics topics in school curricula;
- some discrete problems have been represented by continuous mathematics models which have then been solved using discrete mathematics;
- discrete mathematics, as illustrated by difference equations and network analysis, requires less background material than continuous mathematics, and is simpler conceptually.
- difference equations, by calculators or graphs, is a must!
  I shall watch with interest the work of six American colleges and universities

which have recently been granted funds from the Alfred P. Sloan Foundation "to devise a sequence of courses for the first two undergraduate years in which the roles of continuous and discrete mathematics are balanced". The assessment of their findings should help you in answering Q2.

Another question asked at ICME IV was:

What should have been accomplished in four years to the next ICME? It was partly answered by a research agenda for 1984 which listed:

problem solving
discrete mathematics
applied mathematics

as the three new directions for the 1980s. This gives me confidence that I am heading in the right direction. I shall be interested to see what is accomplished by the next ICME.

References

The references listed below, although confined to the English mathematical literature, should be readily accessible and should prove useful for those wishing to find out more about the material in my talk.

1. May, R.M. (1976) 'Simple mathematical models with very complicated dynamics'. *Nature* 261, pp. 459–467.

   Quoted from the abstract: "First-order difference equations arise in many contexts in the biological, economic and social sciences. Such equations, even though simple and deterministic, can exhibit a surprising array of dynamical behaviour, from stable points to a bifurcating hierarchy of stable cycles, to apparently random fluctuations. There are consequently many problems, some concerned with delicate mathematical aspects of the fine structure of the trajectories, and some concerned with the practical implications and applications".

   This is an excellent review article by one of the world's leading mathematical biologists. I wholeheartedly agree with the author's conclusion that "the most important applications may be pedagogical". As he states: "The elegant body of mathematical theory pertaining to linear systems tends to dominate even moderately advanced University courses. The mathematical intuition so developed ill equips the student to confront the bizarre behaviour exhibited by the simplest of discrete nonlinear systems [such as (5)]. Yet such nonlinear systems are surely the rule, not the exception, outside the physical sciences. I would therefore urge that people be introduced to, say, equation (5) early in their mathematical education".

2. Hofstadter, D.R. (1981) 'Strange attractors: mathematical patterns delicately poised between order and chaos'. *Scientific American* 245, pp. 16–29.

   The author is a regular contributor of metamagical themas to Scientific American, and is well-known for his fascinating Pulitzer Prize winning book "Gödel, Escher, Bach". This popular article on strange attractors, with a touch of romance, provides excellent background material for a study of nonlinear difference equations.

3. Thompson, J.M.T., and Thompson, R.J. (1980) 'Numerical experiments with a strange attractor'. *Bull. Inst. Maths and its Appl.* 16, pp. 150–154.

An interesting father and son paper analysing the system of nonlinear difference equations (11). The computer programming was done by the son during school holidays.

4. Boole, G. *Calculus of finite differences*. 1st edition (1860); 4th edition Chelsea (1970).

   Difference equations is old mathematics!

5. Lighthill, James (ed.) (1980) *Newer uses of mathematics*. Penguin.

   The editor of this book is one of the world's leading applied mathematicians, one who has taken a keen interest in mathematical education, and is a former ICMI President.

   Chapter 4 of the book, written by the present author, gives an introduction to networks with applications to a wide variety of problems in transportation, telecommunications and industry.

6. *School Mathematics Project*, New Book 3, Part 2. CUP 1982.

   An interesting snippet on route matrices of networks (p.76).

7. Penny, D., Foulds, L.R., and Hendy, M.D. (1982) 'Testing the theory of evolution by comparing phylogenetic trees constructed from five different protein sequences.' *Nature* 297, pp. 197–207.

   This paper gives a detailed analysis of the minimal spanning tree problem for 11 species. The example described in the talk for just 4 species is a simplified version. With more species one becomes aware of the rapid growth in the size of the problem and begins to realise that it could become too large for any computer to handle. The authors claim that their results are consistent with the theory of evolution.

8. Zweng, M.J., et al. (ed.) (1983) *Proceedings of the Fourth International Congress on Mathematical Education* (ICME IV). Birkhäuser.

   Is calculus essential? (p. 50)

   What should be dropped from the secondary school mathematics curriculum to make room for new topics? (p. 390)

   Where do we go from here? (p. 434)

   What should have been accomplished in four years, to the next ICME? (p. 435)

WOMBAT

ACTION GROUP 1: EARLY CHILDHOOD YEARS

Organiser: Bob Perry (Australia)

Dedication

This report is dedicated to the memory of Professor E. Glenadine Gibb, whose work in the field of early childhood mathematics education is held in such high esteem. In particular, the Organiser of the ICME 5 Early Childhood Action Group wishes publicly to acknowledge the valuable aid which Glenadine Gibb gave so willingly to the organisation of the group.

Preamble

The Early Childhood group at ICME 5 met for four working sessions. The sessions were attended by approximately 60 delegates including practising teachers, parents, teacher educators, and researchers representing twelve different countries. In a separate session, the Organiser reported to other interested Congress members on the work of the group.

The objectives for the group were:
• to become more aware of current initiatives in mathematics education for children aged 4 to 8 years;
• to identify the needs of mathematics education for children aged 4 to 8 years in a number of countries;
• to discuss strategies for fulfilling those needs;
• to provide advice, based solidly on research and experience, for the implementation of practical curricula for mathematics learning in the early childhood years;
• to develop an international network of early childhood mathematics educators through which the discussions begun at ICME 5 can continue.

For the first session, the group met as a whole to hear presentations on current initiatives in early childhood mathematics education in six countries. Participants then chose the discussion group in which they were to spend sessions 2 and 3. For those sessions, participants met in small groups where they identified and discussed the issues raised by the topics listed in the program. The final session provided an opportunity for the discussion leaders to report to all participants in the action group.

In addition to these programmed sessions, some twelve participants in the group with a particular interest in teacher education met to discuss their teacher education programs and to set up an informal liaison group of teacher educators in early childhood mathematics education. This group was chaired by Evelyn Neufeld (USA).

An International Perspective

In the first working session, a brief overview of early childhood mathematics education in their country was given by George Eshiwani (Kenya), Avelyn Davidson (New Zealand), Jenny Murray (UK) Kathy Richardson (USA), Catherine Berdonneau (France) and Gillian Lewis (Australia). Many important issues were raised in these presentations, the highlights of which are reported here.

Early childhood education in all countries is made up of two facets: 'formal' education (catering for some children) and 'informal' education (catering for all children). Formal early childhood education is a comparatively recent phenomenon in Africa with the earliest official policies not appearing until the seventies. In other countries, such settings have a much longer history. In African countries, preschool settings generally are seen to be elitist, being restricted to children from the middle and upper classes.

In countries where there are written policies on early childhood education, there is rarely any mention of entering behaviour and the ways in which this behaviour might be built upon in the education of the child. The form of early education in many African countries is modelled on that of an English nursery school. This means that the curriculum often consists of materials drawn largely from a culture only faintly comprehended by the children.

Sadly, very few concerted attempts have been made to discover how children learn outside the classroom, and how this might be relevant to school learning. Generally, children do not seem to use their talent for discovery or their curiosity in their mathematics learning.

'Formal' preschool education in New Zealand is provided through playcentres, kindergartens, and child care centres, with children commencing school on their fifth birthday, usually spending two years in the infant classes. Some may spend shorter or longer periods of time according to the position of their birthday in the year and their progress, development and maturity.

Mathematics learning in New Zealand adheres to the premise that learning is based on understanding and mathematics is supremely an activity. Activities and experiences which foster and develop mathematical understanding and language are emphasised. There is a growing awareness of the need to provide for individual pupil differences in mathematics. Pupils should be grouped according to their development in the topic, and program content should be differentiated to meet these identified needs. Appropriate resources need to be collated and used to cater for individual differences among children. Without regular use, mathematical skills and understandings will fade, and maintenance activities are a regular feature of mathematics programs.

In early childhood mathematics education in England, the scene is varied. On the whole, however, teachers take 'learning by doing' as axiomatic. Since its release in February, 1982, the Cockcroft Report has exerted considerable influence on the field.

One of the major achievements in mathematics education in England is in the training of teachers. Primary mathematics diplomas for teachers completing inservice courses have been instituted at many training colleges. The diploma requires teachers to involve themselves with mathematics as well as mathematics education. It now has government support so that teachers receive relief for their classes while they attend college.

Although many examples of excellent teaching and classroom practice have always been part of American education, the majority of children in American elementary schools experience mathematics as arithmetic, and more specifically as arithmetic in textbooks. Teachers see the textbook as the authority and their responsibility as getting the children 'through the text'. Assessment consists of how well the children can complete the workbook pages. In some states, standardised tests emphasising computation are given each year and the results are published in newspapers. This, of course, puts a great deal of pressure on the teachers.

There is a growing recognition that these methods are not working. Children can do arithmetic, but they cannot use it. There are many organisations and individuals who are working towards change. One individual who has had a great influence on many people is Mary Baratta-Lorton. What she offered to teachers through her books and inservice work were real and practical methods to help them teach in new ways. This approach has been successful because it deals with the needs of teachers and children in ways which allow theory to become practical.

In France early childhood education covers two types of schools roughly equivalent to the English nursery and infant schools. Mathematics education in the second of these is quite formal although recent research has affected the awareness of teachers and has resulted in a realisation of the value of discovery and curiosity.

As in other countries, a great deal of research has been undertaken on the ways in which young children learn mathematics. Current research includes work on initial numerical learning, the introduction of computers to young children, and geometry. The work of the Institutes for Research into the Teaching of Mathematics (IREMs) serves as an exemplar of what can be achieved.

In Australia early childhood education involves children of ages 0 to 8, with a major focus on three to eight year olds and their learning at home, in preschools, in kindergartens, and the first three to four years of school. It is difficult to describe early childhood mathematics in Australia as there is no national policy or curriculum, and there are different provisions and different starting ages for the various agencies in each state.

In practice there is an emphasis on the child's language as the link with representation and more formal language. Community languages are taught in some early childhood teacher education courses to facilitate working with those children whose first language is not English. In the state of Western Australia, guidelines have been prepared for the implementation of the curriculum with traditionally oriented communities.

In the light of contemporary theories of cognition, it appears that children's cognitive capacity is frequently underestimated at the preschool level. In preschool centres mathematics is often integrated with other curriculum areas with an emphasis on language and the manipulation of materials. However, formal expository teaching with little or no individualisation still exists in many junior schools. This does not allow for concept and language developments or for alternative modes of representation.

Materials used in schools are often dull, stereotyped, and developmentally inappropriate. At the same time, there is a belief that children should learn through all their senses. It must be kept in mind that the visual and tactile modalities develop earlier and are utilised before the kinesthetic and auditory senses. Children's capacities can be overloaded if too many senses are called upon at one time.

There is a growing concern about the low level of achievement in mathematics of entrants to early childhood teacher education courses. This is usually accompanied by poor attitudes to the subject. As most entrants are female, they naturally provide significant role models for the girls, thus perpetuating current problems of low participation rates and poor attitudes to the subject.

Discussion Group Reports

Potential and performance of mathematics learning in the early childhood years.

Convenors: Tapio Keranto (Finland), Jarkko Leino (Finland)

The group addressed itself broadly to the following questions:
- What have children learned in mathematics before entering 'formal' education?
- What kinds of objectives are realistic at certain stages?
- At which age and in which order is it meaningful to teach mathematical skills?
- At what stage of development is it possible and reasonable to proceed from verbal-actional to mental mathematics to written symbolism?
- What kinds of materials and activities could be used to help learning?

It was felt that the Piagetian stages are often taken too literally by teachers. These stages are not automatic in all children and are not generalisable to all areas. Teachers often do not take into account the previous education and experiences of children (such as 'informal' learning at home). There is very little research into the relationship between Piagetian operational stages and the actual strategies that children use in their school mathematics.

The importance of developing children's mathematical awareness through their interaction with adults was emphasised. Some discussion centred on the conflict between cultural expectations and the differing backgrounds of many children in the same group — a conflict which has considerable implications for mathematics teaching.

The second session began with a presentation of Keranto's study of children's spontaneous use of strategies for solving problems using elementary addition, subtraction, multiplication and division. 'Number-listing' skills are seen to be of great importance in these processes and strategies. Such skills can be improved by instruction and will in turn lead to improvements in other elementary arithmetical tasks and processes. The results of the research give strong support to the view that external aids are significant and in some cases indispensable in the learning and teaching of basic mathematical skills.

Lewis concluded from her research that children move through levels earlier than was previously believed, that children's experiences affect their development of mathematical ideas, and they can be taught certain processes and operations even if, strictly speaking, they are not at the appropriate

Piagetian level.

The development of practical early childhood mathematics curricula.

Convenor: Noel Thomas (Australia)

A child's learning must build upon actual experiences and existing skills and knowledge at the point of entry to formal education. One important objective is to maintain early independence of thought throughout the years of schooling. In the group discussions, three levels of curriculum were explored: the prescribed curriculum, (usually the written documents and associated philosophy), the implemented curriculum (managed by the teacher), and the acquired curriculum (the knowledge, attitudes, and skills developed by the children).

The following important issues were identified.

- Mathematics should not be seen in isolation but in the context of the child's world and his/her total education — in D'Ambrosio's terms, we must consider time, location, and cultural diversity.
- The development of positive attitudes to mathematics among children is crucial.
- Consideration should be given to an integrated curriculum in the early years of schooling. Perhaps integrated guidelines and a thematic approach should be developed.
- Mathematically aware and confident teachers will have an effect on curriculum implementation.
- Mathematics can be made meaningful to children through the development of processes such as patterning, comparing, generating, counting, symbolising, organising, classifying, measuring, representing, estimating, approximating, justifying (or proving), analysing, generalising and inferring.
- Social aspects of curriculum such as communication skills, confidence, and enthusiasm need to be considered.
- Curricula should address learning processes and methodologies such as imagining and making mental representations, manipulating, talking, reflecting, and creating.
- Consideration must be given to the sensitive area of the roles of parents and the rest of the community.
- In the mathematics curriculum, equal weight should be given to all aspects of space, measurement, number, and their applications.

Concern was expressed that maturation and readiness varies so widely in early childhood that the use of national or regional standardised testing of children is inappropriate. In many countries, the continuing dependence of teachers on textbooks as the basis for the mathematics curriculum has resulted in poor implementation of curricula. Another difficulty is the lack of sufficient inservice support to schools. In some countries, incentives are given to encourage teachers to further their professional development.

The format of prescribed curricula was discussed. It was felt that policy documents, guidelines or lines of development are to be preferred so that the teachers can have some freedom and responsibility in determining their own school curricula.

The role of problem solving in early childhood mathematics.

Convenor: Piet Human (RSA)

What is problem solving? It is action taken to overcome an obstacle encountered in pursuing a goal. This may occur outside mathematics when available mathematical knowledge and skills are used to provide a solution, or it may occur outside or inside mathematics when the invention of mathematical knowledge is required.

Strategies to promote problem solving related to the learner:
* Children should be encouraged to tolerate 'not knowing' and yet be comfortable in 'finding out'.
* Each child, both alone and as part of a group of children, should be encouraged to pose problems.
* Children should be placed in positions of cognitive conflict.
* Sometimes children should be allowed to choose from a variety of problems.
* Children should be encouraged to take risks and to explore alternatives.

Strategies to promote problem solving related to the teacher:
* Teachers must be aware of the importance of positive attitudes and the ways in which these might be developed.
* They should encourage and develop the inherent problem solving skills of children.
* Teachers should pursue interesting problems when they arise spontaneously.
* Stimulating materials and environments should be provided for children.
* Teachers need to extend learning situations through questioning to encourage children's responses and representations.
* Teachers should be willing to let a problem situation run according to the children's perceptions even though this may be a different direction from that planned.
* Problems need not always be solved by children. A problem may be used by the teacher as a starting point for new material.

Needed research:
* It is possible that at present children's attitudes towards problem solving and their skills in this area are impeded rather than fostered by schooling. This should be a priority area for research.
* Information is needed on the attention span children bring to their work in problem solving.
* Information is needed on what types of problems are most captivating to young children.
* There is a need to design and construct a developmental model for problem solving ability.
* The relative merits of different strategies for teaching problem solving need to be clarified. In particular, the merits of the two major strategies of the conscious, systematic development of various subskills through exercises that focus on one subskill at a time, and the 'deep-end' approach which encourages development of skills from meaningful problem settings, need to

be investigated.
- There is a need to develop guidelines and suggestions which can help all early childhood educators enhance practical teaching environments through problem solving.

The role of technology in early childhood mathematics.

Convenor: Bob Perry (Australia)

We must think carefully before introducing computers into early childhood education. Computing as a curriculum area and general computer use must be part of the everyday educational program. Schools should have clear priorities for computer use before embarking on expensive purchases. Young children need challenging, stimulating and meaningful activities both with and without a computer. Since the major purchasers of software are parents, the teacher's role must include the education of parents in the role of computers in early childhood learning.

It must always be remembered that a computer should be seen as just one of several tools used in the education of children. Perhaps a computer corner could be set up in a classroom. However, if a computer must be shared with other classes, some kind of schedule is needed. Possible arrangements include leaving it in each class for an extended period, say two weeks; placing it on a mobile cart and booking it for specific times during the day; locating it centrally and taking the children to the machine. In schools where these strategies have been tried, teachers seem to prefer the first suggestion.

Quality software which reflects the philosophy of the curriculum is essential if the computer is to be incorporated into the everyday educational program. We must not forget that we already know about the ways in which children learn. Before they will be able to make proper use of software, all children will require a range of direct learning experiences. We should look for open-ended software which can be used to set up a variety of microworlds. To produce really useful software a team consisting of a learning theorist, an early childhood educator, and a programmer is desirable.

At the early childhood level, current practice is dominated by LOGO, particularly Turtle graphics. To make this more accessible to children, mechanical devices such as a robot turtle or 'Big Trak' are used. There is a realisation that 'off-computer' activities must support the computer work.

The role of language in early childhood mathematics.

Convenor: John Conroy (Australia)

A considerable amount of work has been done in the area of language and mathematics. The relationship between mathematics and language is an interactive one and is dependent on the context and people involved. It is important to start from the children's language and encourage development to standard mathematical language. Teachers need to develop skills which enable them to listen to, monitor, and assess children's language while they are involved with mathematical activities.

Children must be allowed to explore mathematical situations through language. Adequate amounts of such freedom will require teachers to control their enthusiasm for intervention. Hasty intervention is one of the greatest dangers to effective language development in mathematics.

Another problem which teachers must address is ambiguity in their own language. This presents a particular problem for children whose first language is not the language of instruction.

The issue of mathematical symbolism occupied the group for some time. In general it was felt that symbols should arise out of a need felt by the child rather than be imposed by the teacher or the parents. When symbols are introduced they should be accompanied by oral language and the written word.

Work that has been done in integrating mathematics and the language of mathematics should be an integral part of both preservice and inservice teacher education. Communication between researchers and teachers needs to be fostered, and the work of linguists needs to be more readily accessible to mathematics educators. Both researchers and linguists need to communicate the results of their work to teachers. Further research needs to be done on the ways in which children develop an awareness of signs and symbols, particularly where this learning occurs in bilingual situations.

Conclusion.

Only a small part of the work of the Action Group on Early Childhood Mathematics Education can be shared in this brief report. Through the active participation of its members and lively discussions, many important issues in early childhood mathematics were canvassed. However, there were three recurring themes.

Firstly, the distinction between 'formal' and 'informal' early childhood mathematics was made repeatedly. Teachers must take into consideration not only *what* children learn outside 'formal' educational settings, but *how* they learn it. The mathematics education of young children should be a continuum, not a series of gaps joined by a variety of teaching procedures.

A second recurring theme was the importance of parental attitudes and their influence on children's mathematical education. This is possibly a more important issue in the early childhood area than in any other because of the proximity of the children to their 'before school' experiences. Early childhood teachers need to be aware of the importance of communication with parents and of the education of parents.

Thirdly, there was some feeling that early childhood educators in general may not have sufficiently high expectations of the children in their care. In two discussion groups the results of giving children certain types of experiences were demonstrated. We must be careful not to allow ourselves to be restricted unduly by current theories but must strive always to improve the education of all children.

One of the problems with an action group such as this is the maintenance of communication between the members of the group after the Congress. In an attempt to overcome this problem, the group, in its final session, formed an informal study group on early childhood mathematics. The purpose of this group is to continue dialogue between interested workers in the field of early childhood mathematics, to disseminate research and practice to colleagues and, perhaps, in the long term, set up a more formal international study group.

ACTION GROUP 2: ELEMENTARY SCHOOL (AGES 7-12)

Organisers: Claude Comiti (France), Joseph Payne (USA), Wally Green and Alistair McIntosh (Australia)

The organisers of the Elementary School Action Group identified a broad range of topics considered to be important for elementary school mathematics education. Informal and written surveys helped in identifying the topics of greatest concern to teachers, teacher trainers and researchers, and the results helped in formulating the final list of topics for discussion. A list of seven topics comprised the major part of the group's program, with a summary presentation for the final session.

Each section met for one and a half hours for four days. For each section the following questions provided the basis for presentations and discussions:
- What is the present state of knowledge and of practice?
- What are the major problems in teaching?
- What research is needed and who should be involved?
- How can teachers use the knowledge that exists more effectively, being aware of different national and socio-cultural factors?

1. Basic Facts, Computation with Whole Numbers and Computational Estimation.

Convenors: Paul Trafton (USA), Edward Rathmell (USA)

The purposes of this group included:
- discussing the status, trends and forces acting on the curriculum and instructional practice;
- discussing the problems of classroom learning and teaching;
- identifying issues, unresolved questions and direction for needed research;
- identifying promising practices to improve the quality and quantity of learning;
- promoting co-operative international efforts in the field.

Status Reports
Calvin Irons (Australia), Tapio Keranto (Finland), Ping Tung Chang (Taiwan) and Ramakrishnan Menon (Malaysia) summarised the status, trends and forces acting on the curriculum and instruction for these topics in their countries. Their reports indicated that classroom instruction often focusses on the basic skills to the exclusion of higher level processes, including mental computation and estimation. Furthermore, the instruction is often too formal, too soon,

too rote and provides few links to prior knowledge and applications. However, there seem to be forces acting for change in all countries represented. In many cases less demanding goals for computation, due to calculators, and a more developmental instructional program are being suggested by mathematics educators and educational officials. One of the key problems is to get these curriculum changes translated into classroom practice.

Learning Basic Facts. Ed Rathmell (USA)
The presentation in this session focussed on curriculum and learning problems associated with basic fact instruction. The discussion that followed showed a general agreement about the importance of the mastery of basic facts.

There was also general agreement that teachers should facilitate the development of generalisations, number relationships and thinking strategies to help children learn the basic facts. Discussion centred on how to achieve these aims. Representatives from at least five different countries thought that instruction to teach specific strategies is crucial, especially for slower students, while others felt that students should be provided with many opportunities to develop their own strategies. One of the key points involved the links between initial concepts, thinking strategies, extensions in computation such as mental arithmetic and problem solving. There was general agreement that thinking strategies used to teach basic facts should flow naturally from the conceptual bases used for an operation. Furthermore, those thinking strategies should be useful processes for the computational algorithms, mental arithmetic and estimation. The conceptual base and the thinking strategies should also provide a meaningful cognitive structure from which children can easily begin solving verbal problems.

Whole Number Computations. Calvin Irons (Australia) and Bengt Johansson (Sweden)
The first issue that was raised involved the instructional balance between written computation (paper and pencil), electronic computation (calculators), and mental computation. Most agreed that estimation requires greater emphasis.

Another area requiring reappraisal is that of computation with large numbers with the establishment of more reasonable goals for computational proficiency. Several members supported eliminating some aspects of long division, while some others supported the possibility of teaching different paper and pencil algorithms that lead more naturally to approaches used in mental computation and estimation.

The group then considered stages in the development of algorithmic work, with particular emphasis on real world understanding, model understanding, oral understanding and symbolic understanding. While it was agreed that conceptual work is important in teaching algorithms, there needs to be a better fit between conceptual approaches and procedural aspects of carrying out algorithms. Teachers need to take children's thinking strategies into account when teaching procedural rules.

Computational Estimation. Paul Trafton (USA)
Children should become more proficient at computational estimation and its application to the real world through practice, which should emphasise

speed, provide reasonable answers, and be done mentally. Children also need to develop a broad repertoire of strategies allowing them to use estimation flexibly depending on the context.

The decisions children need to make were considered and issues such as when to estimate and the need for 'close' answers were discussed. The importance of imbedding experiences in real world contexts for motivation and application was recognised. Work with estimation and mental computation needs to start early to be thoroughly integrated with all computational work.

One program on estimation was presented using estimation skills including: front-end calculation (for example to add 465 to 275: $400 + 200 = 600$, the rest makes about 100 more, so 700), rounding, grouping to make easier numbers to work with, clustering about assumed means or averages, and using compatible numbers.

It was noted that estimation skills are not easily learned by children and that teachers require curriculum support materials in order to make a sustained instructional effort in this area.

2. Ratio, Decimals and Fractions

Convenors: Friedhelm Padberg (FRG) and Leen Streefland (Netherlands)

The papers and discussions focussed on the important components of concepts of ratio, decimals and fractions, the extent to which these concepts are being learned by students, and the way the concepts and models are used in moving to more formal means of expression with algorithms and/or computational rules. Data from research studies and comments from a wide variety of people from various countries were the basis of most of the conclusions reached in the sessions.

Ratio: Gerard Vergnaud (France) introduced ratio in a broad sense by tracing the development from the beginning idea of 'the more you buy, the more you pay' to the additive properties for age 7 or 8 and more advanced linear concepts at age 10 or 11. The importance of proportions was evident in his opening statement. 'It is not multiplication and division that make young students understand proportion. It is rather proportion that makes multiplication and division meaningful to them.' He noted the long time it takes for children and adolescents to extend the concept of ratio between magnitudes of different kinds, such as distance and time. He asked, 'What does it mean to divide a distance by a duration?'

Stephen Willoughby (USA) gave examples of ratio from his textbooks at the early and later grades using number-cube games. Streefland presented alternative examples for introducing ratio using a picture of a building and having children draw themselves to scale, arguing for less pre-structuring from the point of view of mathematics. Visual reality was modified by showing a similar picture of a scale model of the building. He suggested that visual conflicts help to make the intuitive notions of children more explicit. He called this a qualitative approach to problems with estimation as an intermediate tool on the way to numerical precision, an approach to be incorporated into an elementary school program on ratio. Other examples included density at an early stage, based on the idea of intertwining the separate learning sequences for counting large quantities and measuring area (a small and a large cake with different numbers of ginger pieces and also different ginger taste per bite), ratio tables, multiscale number lines, and stick shadows to illustrate ratio invariance. Ratio should be considered a key notion in elementary school mathematics, as it is a key to applications in physics, chemistry, biology and specific areas of mathematics including probability, similarity and trigonometry. An overall criticism was that the existing curriculum places too much emphasis on the mathematics rather that relating the reality of the child to the mathematical concepts.

Decimals: Diana Wearne (USA) presented a paper prepared with James Hiebert (USA) on the meanings of decimals held by students in grades 4,5,6,7 and 9. They gave written tests to 700 students and interviewed 150 of these. They found that less than half of the students at the end of grade 6 could correctly write a decimal for 3/10, with a fourth of the students making the error of 3.10. About the same number wrote 0/9 or 9/0 when asked to write a fraction for .09. Students tended to treat decimals as whole numbers rather than as quantities, ignoring the decimal point. Only 14% of grade 6 and 37% of grade 7 students thought that .5 was larger than .42. Where 4 of 100 equal parts of a region were shaded, only half of the students were able to write the decimal correctly at the end of grade 9, with 4.100 the common error.

Data show that many students have not connected decimal symbols with either place value or part-to-whole concepts of common fractions. If they are meaningless symbols, then decimal computation turns into a mechanical application of symbol manipulation rules, with an abundance of predictable procedural flaws. The most urgent instructional problem is to help students create meaning for decimal numerals.

Fractions: Friedhelm Padberg (FRG) reported on his study which investigated whether it is better to begin with addition or multiplication of fractions. His results with 28 classes of grade 7 students showed clear superiority for doing addition first. For students who did multiplication first, 19% gave 5/13 for 3/8 + 2/5 while only 9% of those who did addition first made the same error. He argued for addition first because 3 × 4/5 can be done with repeated addition, 4/5 + 4/5 + 4/5 and because addition situations occur more often in daily life than multiplication situations. The most difficult level of exercise on multiplication was 2/11 × 5 where 10/55 was given as the answer by 26% of the students, while 15% of the students gave 15/7 as the answer to 5/7 × 3/7.

Earlier Vergnaud had expressed the view that we need to conceive of fractions as both operators (action of sharing) and quantities (result of sharing). Padberg reported on the changes in German texts where the formal idea of operator is used in multiplication (2/3 means multiply by 2 and divide by 3). Until 1976, both fractions were viewed as operators in a problem such as 2/3 × 4/5. Texts now treat one fraction as a quantity and one as an operator.

Joseph Payne (USA) reported on curriculum problems and issues for rational numbers. He discussed his research results, showing that the set model and number line model are more difficult than models with real objects and plane regions, thus suggesting that sets and the number line be delayed until concepts of objects and regions are firmly learned. Equivalent fractions and equivalent decimals were noted as difficult topics. He emphasised the need for increased instructional time on the meaning of fractions and decimals.

The discussion clearly showed the dominant practice in all countries represented is for students to learn a set of computational rules. Often the rules are mislearned with application of part of a rule for one operation applied to a completely different operation.

The issue of grade placement of computation topics was explored at some length. Suggestions included the delay of multiplication until grade 7, keeping computation relatively simple (small, easy denominators), doing a better job of developing the computation, and delaying the more complex work until grades 7 and 8. The importance of mixed numbers was raised, with the conclusion that operations on mixed numbers are of declining importance. Payne and Milton Behr (USA) suggested that concrete models be related more carefully to the verbal and symbolic rules for computation, illustrating the point using muliplication of fractions.

Summary
Faulty additive reasoning in solving ratio and proportion problems, the persistent peculiarities in the ways students operate with fractions and decimals, and the poor understanding the students possess show that purely numerical treatment and formalisation of rules enter too early in mathematics education. This premature formalisation often destroys the insights and ideas which pupils already have acquired from real life experiences. We should not teach rules too early because of the danger of students performing meaningless operations with meaningless symbols. Rules need to be derived carefully, making sure that the rules and models are related. Algorithms which are often confused need to be contrasted and compared. Above all, we need to develop quantitative feelings for ratio, fractions and decimals, always taking into account the insight and reality that students have. There is a strong need for developmental research to see how these important objectives can be achieved.

3. Verbal Problem Solving.

Convenors: Frank Lester (USA) and Marilyn Zweng (USA)

The principal questions addressed by this action group were:
- How do children solve problems?
- What makes problems difficult to solve?
- Problem-solving instruction: what do we know and what do we need to know?

In addition, presenters responded to the broader question:
- What are the problems in teaching problem solving?

Research reported by Terezinha Carraher (Brazil) showed that many children who are proficient in solving problems outside school (market situations) fail to solve equivalent problems when they are presented in school settings (story problems). Not only did the children perform better in market situations, 98% correct against 73% correct on the same problems given verbally later, they also used different strategies. In market situations the children manipulated quantities, while in school situations they more frequently manipulated symbols. In market situations the children used methods to solve problems that they had not learned in school. These methods were effective and demonstrated that the children had dealt intelligently with the quantities involved in the problem. Carraher contrasted these methods with the 'numeral pushing' taught in school.

Frank Lester (USA) reported on his work showing that children attempting to solve process-type problems within a school setting do not think through problems but rather 'grab' hopefully for answers. The approaches in school settings seem to suppress children's problem solving abilities and strategies. In the discussion general agreement was reached that:
- Teachers must help children make the connections between the language and the quantities involved in verbal problems;
- The emphasis placed on computation algorithms should be reduced and more attention given to the meaning of the operation;
- Key words (clue words such as 'left', 'more', 'altogether') should not be used in teaching verbal problem solving. It was pointed out, however, that verbal problem solving is a linguistic activity, in part. Students need help in interpreting the language of the problems to understand the quantitative relationships instead of making the trival operation – word clue connection.

Charles Thompson (USA) in his presentation gave key components for the initial phases of problem solving. In problem solving it is necessary to analyse the relationships and to model them in order to derive a solution. A key component of problem difficulty is the ease with which a child is able to perform this anlaysis and modelling. Thompson considered these points in relation to verbal addition and subtraction problems. His research indicated that changing the structure of a simple problem affected both the level of difficulty and the method of solution. Following are examples of two of the many types he used:

(a) Bill has 3 red pencils and 2 green pencils. How many pencils does Bill have altogether?

(b) Bill has 2 pencils. Jean has 5. How many more does Jean have than Bill?

(b) was more difficult for the children to solve than (a).

Thompson reported that, especially with young children, difficulty experienced in solving problems will be reduced if physical models are made available, and if students know their basic facts. He recommended using physical models, discussing relationships among quantities, stressing meaning when teaching arithmetic operations, and providing a large variety of problems.

As the child's knowledge of the number system is extended, problems will be more difficult to solve if the meanings given to the operations are not extended at the same time. In his paper Shigeo Katagiri (Japan) gave the following example to illustrate this point.

At first we define multiplication on the integers as repeated addition, but when it is extended to decimals, this definition is not adequate.

Katagiri also stressed the need for teachers to provide students with interesting problems that allow them to explore and develop real mathematical thinking.

Successful problem solving instruction begins with the child's natural language. To this end, Dianne Siemon (Australia) stressed in her paper the value of group discussion and talking about the problem situation in order to facilitate the development of a good feeling for the problem to be solved. Consequently a sound language program is essential.

If instruction is to enable children to solve problems, it is necessary to isolate those factors that help and those that do not. Marilyn Zweng (USA)

reported the results of a study she conducted with children in the age range 8 to 11 years. She found that the children were helped to solve problems if:
- they focussed on the action (putting together, comparing);
- they focussed on the part - part - whole relationships;
- the problem was stated with smaller numbers;
- they made a drawing or used manipulatives;
- they considered the size of the answer relative to the numbers in the problem.

The children were not helped if they were:
- directed to identify the given data;
- directed to identify what the question asked;
- given a less wordy re-statement of the problem.

Instruction should encourage children to use calculators in solving verbal problems so that their attention can be focussed on the process rather than the computation involved.

Finally it was agreed that children need to be presented with a much richer variety of problem types than is usually the case. If teachers are to be successful in helping children become better problem solvers, they must pay less attention in problem-solving instruction to developing a 'bag of tricks' and more to developing mathematical thinking. For example, instruction that encourages children to make and test conjectures is very helpful. In presenting a variety of problems teachers must keep in mind the interests of the children and be careful not to impose adult conceptions of problems on them.

4. Geometry in Elementary School

Convenors: Edith Biggs (UK), Colette Laborde (France)

Colette Laborde presented a summary of the main problems:
- Too little time is alloted to geometry by primary school teachers.
- Geometry seems to stir up a multiplicity of problems much less clearly defined than in the field of numbers.

She showed how the historical conception of geometry has greatly influenced the content of teaching. For example, in France the aim of geometry as taught is not to develop children's mastery of spatial ideas but to introduce deductive reasoning.

Several countries have attempted many innovations and experiments in the classroom, taking into account problems of perception and representation. Research into the properties of geometrical objects, the process of representing three-dimensional (3D) objects in the plane, the understanding and treatment of geometrical figures, and the influence of culture on concept formation show that 'simple' perception activities are not copies of reality but the result of a construction by the child, involving the child's own concepts.

It seems necessary to reveal the conceptions of the various spatial phenomena worked out by the pupils and to find what can be done to facilitate the evolution of these concepts to improve the understanding of space by children.

The following presentations and discussions were based on experiments carried out within a theoretical framework as well as innovations conducted in a more empirical fashion in the classroom.

Seventh Graders' Ideas About Lines and Points. *Helen Mansfield* (Australia)
The purposes of the study were to determine the extent of pupil understanding of what they had already covered from a text book; and whether their understanding was related to their previous experience in the environment.

Pupils, individually interviewed, were first asked to talk about and define parallel lines and to determine whether lines in illustrations were parallel. All but one thought that curved lines might be parallel. Then they were asked several questions to explore their understanding of continuity. For example they were asked what is the size of a point and how many points would fit into a two centimetre length. Some thought the size of a point could vary, one thought it had no size, all gave a finite (often large) number for the number of points, and most of them answered that points could be combined to form a line. The study suggested that mathematical ideas encapsulated in formal definitions may not be seen by the student as abstractions from the real world. It was clear from the discussion which followed that the ideas constructed by the students did not correspond to the work in the text book. The work in the text book should be more comprehensible and more closely related to live situations.

How to Develop Spatial Skills and Abilities by the Use of Materials. Hirokazu Okamori and Kiyoshi Yokochi (Japan)
In a Japanese study, children made objects for themselves which they valued. As an example, mirror transformations were developed in three stages. Children in the lower grades made stamps; each an example of a shape and its reflection. Children in the middle grades made book covers with a kaleidoscopic design. They were given a variety of pictures and had to construct the reflections without folding. In the upper grades, they used a computer and had to check the reflections of oblique axes with a light pen. The study showed that children had made various discoveries about mirror transformation. The discussion which followed stressed the fact that the conflicts the children experienced allowed them gradually to correct their own mistakes.

Shapes With Everything. Edith Biggs (UK)
Some activities conducted with children (age 5 to 9) were presented, using boxes of different shapes to build the tallest tower which will stand up, making sketches of what was seen from above and from the front; using two identical empty containers, cutting the flat net of the second box and painting faces of the net in the same colour as on the solid container. The analysis of these activities showed that they gave children experience of the relationship of a 3D object to its 2D representation.

Examples of paper folding were used to introduce and develop ideas of rotational and mirror symmetry, conservation of area, and the variation of perimeters. Children's work was displayed illustrating the discoveries made in such situations. In the discussions, members returned to the importance of providing elementary students with materials for learning geometry.

Geometry For All. Geoffrey Giles (Scotland)
For the great majority of children who do not expect to take their mathematics any further than school level, the emphasis in their school work

must be on enjoyment and success. A second priority must be the development of basic intuitive skills and understandings that will be retained indefinitely. Thirdly, we have to involve children with geometrical topics which will be used by them because they are helpful in life contexts.

The group worked for a time with materials developed by Giles, including 3D materials and booklets giving 2D representations of 3D shapes to be made with the various pieces provided. This experience demonstrated how the material stimulates children to create their own problems, how they develop problem solving skills, and a variety of problem solving strategies. The discussion which followed stressed that it is essential both for children to use materials and for teachers always to have in mind the concept they would like to help children acquire.

How to Help Teachers to Improve Geometry Activities. Michel Blanc (France) and Jeannine Weber (France)
In one classroom activity, a group G of pupils have a geometrical object (a pyramid) which the other pupils cannot see. The other pupils, who are divided into small groups, ask questions of G in order that they can reconstruct the object. This situation is based on the theoretical framework developed by Guy Brousseau. The discussion showed the interest in this kind of activity and also pointed out the difficulties for teachers in organising such activities in a classroom. Problems of the timing and nature of teacher intervention were raised.

To help teachers in France, books called 'Aides Pedagogiques' (Educational Aids) were written jointly by a group of teachers, teacher trainers and researchers, all of whom are members of the French Commission on Mathematics Teaching at the Elementary School. The books contain background geometry for the teacher and suggestions for designing and organising geometry activities in the classroom, activities which until recently, were often ignored. The discussion was devoted to the place of geometry in the curriculum and to the attitude of almost all teachers towards geometry.

Summary
In conclusion the following points were stressed:
• Attractive material should be provided and 3D material should be handled before 2D.
• Material should be designed so that children can take time to explore and construct models and can correct their own errors without intervention from the teacher.
• The recording of discoveries should also be attractive, as it allows children to interact and learn new language patterns and vocabulary.
• After the exploratory stage, all activities should be carefully structured. In using specific materials the child learns mathematical ideas from the problem given. The solution is a part of the mathematical concept the student is in the process of acquiring.
• Planning activities is not easy. We need to observe children's procedures, to listen to what they say, and to question them to help their learning.
• This has implications for initial training and for inservice training. The teachers themselves need experience with materials they plan to use with children, in order to clarify their own concepts. They need support from the

principal and from each other in planning and preparing materials.

5. Algebra

Convenor: Robert Davis (USA)

What is the meaning of 'algebra' in the elementary school? What are the purposes of algebraic activities and which are appropriate? What are the key problems for research and development? These major questions were addressed in discussions and presented papers.

Algebra has at least three components:
- going from reality to algebraic notation;
- starting with an algebraic expression and seeing possible 'reality' meanings of it;
- going from one algebraic expression to another.

There was general agreement that the third component is least important or appropriate at the elementary stage. The group agreed that 'algebraic thinking' better describes work in the elementary school because it is less formal and less symbolic. However there were conflicting views as to whether elementary school teachers should lead children towards accepted symbols or should concentrate on encouraging children's own problem-solving methods and notation. Reports were given of successful work done with children but not by imitating the approach of traditional secondary school algebra.

In the discussions it was impossible to consider content without considering pedagogy. Questions of what should be taught were highly dependent for their answers on how it was taught.

A distinction was made between two approaches to teaching mathematics:
- The student learning certain 'facts', and being told how to carry out certain procedures (such as 'removing parentheses'). The essence of the whole process is *communication*, with students following clearly-stated explicit instructions.
- The student being helped to build up certain key ideas. An important part of the process is *reflection*. After doing something, the student reflects on what he/she has just done.

The second approach is closer to algebraic thinking. Appropriate algebraic activities for children include looking for, describing and generalising patterns, and representing generalisations in pictorial, graphical and notational forms (both given by the teacher and invented by the child). All these contribute to gradually increasing comfort with the ideas of function and variable. Papers by Larry Hatfield (USA) and Shigeo Katagiri (Japan) explored aspects of such activities.

In elementary school there are many opportunities to begin to consider algebraic ideas (For example: one spider has 8 legs; 2 spiders have 2×8 legs, how many legs do 3 spiders have? 10? 100? 71 spiders?).

What are we to make of these opportunities? A mere collection of nice 'fun' things does not add up to significant learning. If we move immediately from the 'spider' to saying $8s = 1$, however, we go too fast, and many students will not understand what we did. In between the initial spider and the much later use of notation such as $8s = 1$, *we need to insert a wealth of experiences* that will make meaningful the key ideas of algebra. Papers by Chalouh and Nicolas

Herscovics (Canada) documented the dangers of overestimating students' understanding of algebraic notation.

Attention was given to examples of appropriate activities between 'the spider' and 'the notation'. One example adaptable to any grade level is W.W. Sawyer's 'Guessing Functions'. Students make a rule such as 'double and add 3' but do not tell the rule. Other students choose numbers such as 0,1,5,4 and are told the results of using the rule, 3,5,13,11. The game ends when students guess the rule.

In another game from SMP, students are given cards such as $\times 2$, -5, $+1$, an input number such as 5 and a target output such as 11. The students play cards and earn points as in scrabble. One result could be $5 \times 2 + 1 = 11$.

While the group in general opposed manipulation of notation as an elementary school goal, three opposing reactions were of interest:
- Some successes were reported;
- It is important to make symbolic representation functional for children;
- Indeed the powerful purpose of algebra is not merely to represent but to allow you to operate on the representations without getting bogged down in the particulars.

The group reached three main points of agreement:
1. Algebraic activities should follow the same pedagogical path as that advocated for all elementary mathematics:
   - Give students a task or challenge;
   - Do not tell them how to do it;
   - Afterwards, build on this experience by discussing and reflecting on it.
2. Algebraic activities should be planned, progressive and purposeful.
3. Tasks for the future include the following:
   - Observing and charting what children do in algebraic situations;
   - Identifying the representations and symbolisations of ideas that are available at this level;
   - Questioning the extent to which notation and its manipulation is possible with young children (and under what circumstances);
   - Identifying the key ideas of secondary school algebra;
   - Addressing the question: Is elementary school algebra more than preparation for secondary school algebra?
   Many members of the group expressed interest in continuing the dialogue.

6. Using Microcomputers and Calculators

Convenor: Janice Flake (USA)

Microcomputers: The major questions addressed were these:
- How can microcomputers be used to diagnose student errors?
- How do we develop our own software to create mathematics learning and develop concepts?
- What is the current state of computer education?
- How can calculators be used?

George Bright (Canada) reported on work done to use computer methods for error diagnosis, viewing diagnosis as determining gaps in learning. He noted the difficulty in diagnosing some errors because certain inappropriate procedures give correct answers for some exercises. For example, an incorrect rule about zero could give correct answers such as $9 + 0 = 9$ but incorrect answers such as $9 \times 0 = 9$.

Bright stated that the incorporation of diagnosis into regular CAI materials is at a very primitive stage. He listed seven procedures for the creation of diagnostic programs:

* delimit clearly the content to be diagnosed;
* identify curriculum practices for the context;
* list potential errors;
* develop a procedure to specify errors on the basis of student responses;
* keep a record of a student's performance;
* provide documentation on how remediation might be accomplished;
* update the program as needed.

Didier Lavallade (France) gave examples of mathematical concepts involved in simple Turtle-graphics from projects designed by children. Two major types of projects were described; the first a graphic project where the main objective is to draw something; the second, a geometry project where the main objective is to build some knowledge about geometry facts. Geometry projects were shown to arise either from a precise idea about a geometric object, such as drawing a circle, or incidentally by fiddling around with any figure. Lavallade presented examples of the geometric projects developed by children.

Janice Flake (USA) described commercially available software and student programming activities that are helpful in teaching mathematics. She showed how computer graphics can be used to develop concepts of space and number resulting in enhancement of spatial visualisation. Symbolic and visual relationships can be co-ordinated. Activities were presented to enhance logical thinking and problem solving skills, such as isolating and controlling variables, looking for patterns and estimating.

Aurum Weinzweig (USA) demonstrated software that takes advantage of the computer to make rapid displays, to repeat illustrations and to transform figures from one state to another state. One of his programs showed the effective use of colour to teach tesselations. Symbolic and visual relationships can be co-ordinated through electronic blackboards.

Other examples where the microcomputer can help instruction include estimating length and angles, understanding and using variables, learning and using a co-ordinate system, and seeing and understanding transformational geometry. Suggestions were given on using a computer to see relationships between mathematics and other disciplines. For many of the topics, the microcomputer was seen as a supplement to regular classroom instruction in which concrete materials are used and oral discussion takes place.

John Scheding (Australia) gave a report prepared by him and Peter Anderson (Australia) on the status of computers in one state, New South Wales, and the problems they see in schools. There has been more support from elementary teachers than high school teachers. Parents exert pressure on schools to introduce computers so that their children will 'not miss out'. While some elementary schools do not have computers, the number is rapidly decreasing. Some elementary schools have 12 or more machines, one in each classroom and some in a central area. The uses of computers include creative and conference writing, drill and practice (tables and spelling), problem solving, and Turtle graphics and LOGO. Children are learning *with* computers and *about* computers, both important goals for the state. Except for talented students in the upper grades, children are not taught to program. For most teachers, the mystery and apprehension disappears as soon as they can operate a machine, load and run a program.

Considerable discussion was generated about the role of programming in the elementary school. While the participants agreed that students should not be deprived of programming experiences, the purpose of such programming should be to enhance the learning of mathematics. LOGO was currently the most accepted programming language for elementary age students.

Calculators: Donna Carlton (Australia) led the session devoted completely to calculators. Her presentation included a description of a model for introducing the calculator into elementary schools in her region, Western Australia.

She presented data showing that calculators are readily available in many countries. In the USA in 1982, 75% of grade 5 students had access to calculators and in 1983-84 87% of grade 5 students in the UK had calculators available. In a study in 1983 in New South Wales, Australia, 40% of grade 5 students used calculators in the classroom, the highest percent found so far in any study.

As recommended in the Cockcroft Report (UK), mathematics programs must take advantage of the power of calculators and computers at all levels. The calculator is an intrinsically motivating aid and should be used more fully. The effect of the calculator on the curriculum needs more attention, especially as single digit facts and estimation skills are essential in order to evaluate the reasonableness of answers in arithmetic and on calculators. Inservice education is needed to help teachers use calculators effectively in concept formation and problem solving, while parents need to understand how schools use calculators as an instructional aid.

In the discussion it was recommended that calculators be used for tests in school and for employment. The ready availability and portability of calculators compared with computers was also noted.

For both calculators and computers, research is needed to find effective ways to integrate them into the classroom, producing more effective learning of mathematics by students.

7. Means and Modes of Instruction
Convenor: Jorgen Cort Jensen (Denmark)

Discussion was concerned with the interaction of means of instruction (general environment, structured situations, text books) and the modes of instruction (the teacher, language, student interaction). An effort was made to locate significant trends in the range of situations in which children may find themselves. To achieve this aim six papers were presented and discussed.

Mathematics with a starting point in the natural environment of the children, the role and limitations of language. George Eshiwani (Kenya)
In African classrooms there is often a multiplicity of home languages, but a different mother tongue. Language difficulties have significant impact in three main areas:

- Expanding children's mathematical vocabulary by defining terms, inventing new words and borrowing words from other languages.
- Problems of syntax in which logical connectives (such as, and, if ... then, or) have different meanings in every day speech from those in mathematics.
- Transition problems from one language to another.

There was general agreement that these problems exist in most countries, whether developed or not.

The following points arose from the discussion:

- What a child learns and knows of the environment is not reflected in the mathematics classroom, while the mathematics class deals with concepts that are not usually applicable in the home environment.
- 'Ethno-mathematics' (after D'Ambrosio) should be used as a basis for 'learned mathematics', especially in primary school.
- The problem remains of how and when to make the transition from concrete situations to the 'symbolic world' of mathematics.

Mathematics with a starting point in the natural environment of the children, cases of students' interaction in the learning process. Frederik Van Den Brink (Netherlands)

The key idea is that calculating is primarily a 'happening'. A traditional approach uses blocks to represent various arithmetic operations. Although this involves the children in manipulating concrete materials, the role of the blocks is purely illustrative. An example of a living situation could be a bus ride. Your 'play bus' is in the classroom with 'passengers' entering and leaving at stops, the 'driver' and others keeping count of the numbers. In this way calculation means registering the course of events of the performance.

Such a performance may be of at least two kinds: stories around given characters in given situations (bus, train, shop) with the social rules fixed; genuine games (such as, bowling, dice) with rules chosen arbitrarily by the children.

In the discussion that followed many advocated the use of blocks because they are a faster medium to work with compared with relatively slow play-acting. The counter argument was that play-acting situations enable the children to generalise more easily, so that not as much practice is needed.

Mathematics with a starting point in structured situations, with an emphasis on student interaction: the use of games. Ulla Jensen (Denmark)

A game was distributed and played. It served its purpose by forming the basis for some didactical considerations among participants. Many told of their own experiences with a variety of games, and there was general agreement that this topic should receive more systematic attention from research as games are widely used in the classroom. Little is known, however, about the value and effects of this means of mathematical education.

Mathematics with a starting point in structured situations, with the teacher as a main mediating factor. Activities with geometrical patterns. Margaret Kenney (USA)

Geometric models are particularly useful for encouraging visual pattern development and for providing a transition to corresponding numerical patterns and relations. Although pattern recognition, extension and formulation are important skills for young students to acquire, others can be nurtured simultaneously. For example, activities based on model building often call for recording, organising and interpreting data. Consequently students need to learn how to make tables, charts and graphs. They must learn how to relate and compare results, and make predictions. Geometric pattern activities develop problem solving skills. The discussion showed that structured situations range from everyday life to more formalised mathematics.

Mathematics based on the textbook: the use of language. Noelene Reeves (Australia)
Mathematics has a recording system that transcends language and allows mathematics to be universally comprehensible. The problem for children is that too few connections are made between their own oral language for concrete, mathematical experience and other people's language and symbols. The problem is exacerbated by the fact that the special reading and writing demands of mathematics are not being met in current reading and writing programs.

Reeves gave specific examples of reading difficulties such as not being able to deduce words from context, not giving meaning to symbols, misinterpreting pictorial representations, not recognising that mathematics reading is not just left to right, and mismatch between texts and diagrams. Children can be helped by slowing the transition from their oral language to written language, pictorial representations, and finally to symbols. Morever, we need more interesting mathematics books as continuous texts for children.

Mathematics based on the textbook: the teacher as mediator. Jorgen Cort Jensen (Denmark)
The combination of the textbook as the means and the teacher as the mode is very important for the book contains mathematics in a socially and culturally acceptable form, whilst the teacher has the mathematical knowledge and should be able to mediate it. The textbook has also become a major medium employed to promote and push reform. We are now seeing texts that reflect a different view of mathematics education, emphasising the real world, and the open, more or less structured, challenging situation as starting points for mathematics learning.

There is much evidence to show that if teachers are not willing, able and motivated to handle a changed situation, even the best books will not make much difference (and may even be counterproductive). For example, if the purpose of presenting combinatorial ideas is that they be acted out in the classroom or explored through practical manipulation followed by recording in the book, the intention will be lost if the teacher follows the bookwork and omits the activity.

The group agreed on the importance of the textbook both as a source of reference for the children and as a source of activity (or inspiration for activity) for the teacher. Everybody agreed that books were a necessary vehicle for employment and promotion of curriculum change, even though the dangers outlined above were very real and probably unavoidable.

It was suggested that if microcomputers take over the function of drill and practice, and at the same time, initial concept formation is carried out using concrete materials, then the textbook could become superfluous, at least in the first two or three grades. Most believed, however, that the book would still have to be present, firstly as a source of inspiration, secondly as a safety net, and thirdly, as representing the topics of school mathematics.

Summary Comments

Claude Comiti (France) and Joseph Payne (USA)

About 200 participants from 22 countries attended this action group. 20% were school teachers, 40% were teacher trainers, 30% university teachers or researchers and 8% inspectors, principals, and other education officers. Some important phenomena became apparent during the sessions of the group.

By comparison with ICME IV, most papers moved from a statistical counting of children's mistakes to an analysis of children's procedures and an effort to understand them and to relate them to conceptual bases.

More and more activities, research reports and discussions concerned themselves with the improvement of the links between theory, practice, research and teacher-training. Some findings to help improve teaching and some obstacles encountered in the course of teaching provide new themes for research.

There remains one major obstacle in our trend towards significant progress in the teaching of mathematics at the elementary school level, and that is the near absence of good communication between researchers and the great majority of teachers. This lack of dialogue, which would allow teaching, and research on teaching and learning, to progress at the same time, rests primarily upon our failure to understand the didactic phenomena, that is, the actual relationships between the teacher, the students and the mathematical content.

We should keep the following points in mind in making recommendations for change in curriculum and in teaching:

- We should try to give children a better control of the mathematics that is taught by taking into account their out-of-school knowledge.
- Amid many general pedagogical recommendations and educational reforms, we need to remember that mathematics learning is the important objective.
- We should carry on with more and more advanced studies on the understanding of the teaching phenomena and improve preservice and inservice training not only in the definition of training syllabuses but also in the training itself.
- We should encourage dialogue on mathematics education with the public at large, and the understanding of mathematics by the largest number of people.

It is only in this way that we will be able to reduce the number of failures in mathematics and maximise success for the maximum number of children.

ACTION GROUP 3: JUNIOR SECONDARY SCHOOL (Ages 11–16)

Organisers: Alan Osborne (USA), Patrakoon Jariyavidyanont (Thailand), Jack Briggs (Australia)

Recommendations and Concerns

Introduction

Mathematics for junior secondary school students provides a challenge for teachers. Thinking about curriculum has been in terms of completing the primary school experience at one extreme, or the senior secondary school mathematics program at the other. Problems arise because activities and materials are set too high or too low rather than focussing on the characteristics of learners in the 11 to 16 age range. They fail to deal with the transition from child to adult behaviours.

This action group focussed on nine problem domains identified as critical in this transition age. These problem domains were identified in the first session by Alan Osborne (USA). Four presentations directed attention to curriculum problems and learning behaviours at their age level and this served to further sharpen and delineate the problems and issues.

Kathleen Hart (UK) focussed attention on children's expanding capabilities to generalise and their use of variables by examining examples of child-like, incomplete behaviours of young children in algebraic and numerical settings.

Pekka Kupari (Finland) discussed the role of applications in providing a sufficiently rich experience in problem solving. Varieties of problem settings were used to highlight the nature and type of mathematical encounters needed by children. The limitations inherent in exercises and expository presentation modes of instruction were contrasted with real situations that yielded process learning.

Patrakoon Jariyavidyanont (Thailand) examined the performance of Thai students in statistics for evidence of the difficulties inherent in a mis-match of curriculum with school testing and evaluation.

Joe Crosswhite (USA) examined the unique potential of geometry to move children from inductive thinking processes to more systematic deduction within a mathematical structure. Mathematics was highlighted as having two faces, one orientated to the inductive description of reality, and the other orientated to appreciation of deductive argument. Geometry was described in terms of its potential to reveal these aspects of mathematics.

Transition Problem #1: How do we establish the idea of variables?
Participants uniformly agreed that learning to use variables in the process of

generalising mathematical relationships was critical in the transition to mature mathematical behaviour. A common perception across groups was that instruction directed to learning about variables and making generalisations was infrequently realised in many classrooms. This domain of learning was identified as being loaded with pitfalls in management of instruction and in design of learning materials.

Readiness of the learner was identified as crucial. At risk is the immature child not exhibiting the level of cognitive functioning that indicates a readiness to generalise or use variables to describe relationships. Participants were fearful of negative attitudes and frustrations that could arise from too early demand for generalisation.

The corresponding fear of children missing opportunities to learn was pre-eminent in other participants' thinking. These participants were more optimistic concerning the capabilities of learners in dealing with variables. A partial resolution of the quandary was evident in the observations that children needed to encounter opportunities to use variables and to make generalisations in learning activities, but that they should not be held responsible, particularly in the demands of evaluation, to the point of experiencing negative feelings of guilt about non-performance. Teachers should provide opportunities to use the power of variables in expressing relationships, but be thoughtful about expectations for performance for the individual child.

The issue of readiness to learn highlighted the fact that understanding of variables is not an all or nothing affair but concerns complex learning that should be accomplished in small stages. Researchers and curriculum developers need to attend to the incremental character of this learning in order to provide activities that are 'gentle' in character but which fit the needs and cognitive levels of learners in the early stages of learning about variables and making generalisations.

Specific suggestions for dealing with variables in the classroom were organised in terms of a setting of functions. Participants recommended extensive experiences in identifying patterns within a numerical setting. Construction and use of tables, with a premium being placed on identifying and describing patterns across rows or columns, were deemed particularly supportive of learning about variable, whether the tables were based on Burkhardt's categories of curious and dubious problems, or stemmed from real problem situations.

An observer of the discussions would have been struck by the lack of conversation about equation-solving when the topic of learning was that of variable. Discussants apparently felt that function settings and the opportunity to generalise within them were environments more conducive to learning about making generalisations than that associated with acquiring the skills of equation solving. This reflects a change of emphasis in teachers' perceptions; perhaps teachers are now aware that there is a need to develop a comfortable familiarity with the general concept of variable before embarking on an intensive treatment of equation solving.

The commonly advocated approaches to variable and generalisation place a premium on familiarity of settings in initial work. For example, numerical frameworks can provide a setting for numerical problem solving resulting in the use of variables. If real problem situations are used, teachers should take the time for thorough exploration of situations to allow learners to talk easily about the relationships. Premature formal use of language and definitions would severely limit children being able to exploit situations to learn about variables.

Teacher participants were anxious that the settings for learning about variables represent a wide range of contexts and variety in approaches. Activities that were unidimensional, using only numerical settings in table building, were recognised as being substantially strengthened by incorporating graphical work. This can assure consistency of approach across primary, junior and senior secondary schools with greater potential for integrating an algebraic flavour across disciplines.

Transition Problem #2: How do we begin to take better advantage of the characteristics of modern computational tools to support instruction?
Computers and calculators are recognised as legitimate tools for junior secondary school students and for their teachers. Discussion transcended problems of an obvious nature such as the following:
- Where do we find the resources to provide calculators and computers for pupils and teachers to use?
- How do we assure reasonable access to software of proven worth in promoting learning?
- Where do we find the time and resources for in-service education of teachers who lack background, skills and understanding necessary for successful use of the computer and the calculator?

These are real and significant problems widely encountered across the countries represented by the discussants.

Calculators and computers were accepted as tools for investigation, for exploration of patterns, and for problem solving. Both the computer and the calculator should be incorporated into instruction to strengthen the emphases on variables and functions. (See transition problem #1.) The calculator and the computer can each be used in table generation to help learners gain an appreciation of the use of variables as summarising several instances within a problem setting. Use of the calculator and the computer allows learners to encounter real problems, such as population growth or compound interest, that are understandable by learners, but for which the work load of paper and pencil computation interferes with learning the mathematics.

The conservative teacher who resists incorporating calculators into mathematical instruction was a special source of concern. The participants suggested in-service experiences that would focus on:
- research evidence demonstrating the effectiveness of calculators in bolstering computational and problem solving skills;
- activities that would use the special characteristics of the calculator to communicate mathematics;
- addressing problems of building skills in estimation as an adjunct to calculator use;
- providing guidelines for effective use of calculators.

Most participants recommended that all students should have access to calculators for all aspects of their work with mathematics. Students weak in computational skills need access to calculators to deal with the higher cognitive demands of problem solving or modelling so that computation does not provide barriers to understanding and appreciation.

Computers provide different kinds of apprehensions concerning appropriate classroom usage than calculators. One apprehension concerned too easy access to less than desirable software. Participants supported development of good software but insisted that adequate documentation of the software be a necessary part of that software. Documentation should include not only information about the program but also guidelines for its use with learners. Professional groups should encourage the development of cataloguing systems for software that would enhance the ability of teachers to find and access software for designated purposes.

Participants were concerned that computer use and activity in schools were being delegated solely to the mathematics department personnel, thereby creating the impression that computers were only useful in doing mathematics. The action group makes the strong plea that other teachers, not just those responsible for mathematics instruction, be trained in the educational uses of computers. The state of the art for use of computers in mathematics education is that of a field in its early stages of development. Discussants were concerned that too many teachers looked internally to the field of mathematics education for help without realising they were over-looking valuable parental and community resources.

Finally, the action group strongly recommended that all pre-service teacher education include experience in using computers. All prospective teachers should have experience in programming but, more particularly, should have familiarity with appropriate and inappropriate uses of the computer in instruction. Prospective teachers should appreciate the dynamic potential of the computer when the learner controls the computer rather than the computer controlling the learner.

Transition Problem #3 : How do we help learners acquire a mind set toward higher order thinking processes in mathematics?
Discussion groups agreed that higher order cognitive processes needed more emphasis. Participants universally expressed a strong commitment to the development of problem-solving abilities, but were concerned that this may not be the case in all classrooms and at other levels of instruction. Many participants believed that a flaw in curriculum design is the focus on computation that stems from incomplete testing and evaluation efforts. Others voiced their concern in terms of computation being more easily taught by teachers with inadequate backgrounds.

Many discussants believed that by the time students reach the junior secondary school, it is too late to establish the attitudes of valuing higher order cognitive skills. It is difficult to change habituated work patterns and their supporting value structures; hence the recommendation that primary school treatment of problem solving and other higher cognitive functioning in mathematics needs emphasis with correspondingly less attention given to computational algorithms.

The groups recommended that thinking processes would be approached better from a basis of genuine applications of mathematics. Classic problems of textbooks do not have the same potential for influencing thinking processes and developing an appreciation of these processes as do real situations.

A significant potential for encouraging higher order processes was found in mental arithmetic that allowed children an opportunity to use and compare different strategies in computation. A primary point was made that computation in and of itself was not at fault but rather the unrelenting focus on the reproduction of limited algorithmic routines for computation.

The conclusions attained by most groups suggested increased attention needed to be given to:

- curriculum development particularly focussed on problem solving, modelling, and other higher order processes;
- research concerned with how youth attain process skills in problem solving, modelling, and other higher order behaviours.

A strong concern was stated frequently about the interactions between attitudes of children and their willingness to enter problem solving activities with the eagerness and aggressive curiousity needed to sustain the activity. Problem solving activities require concentration. If attitudes interfere with that concentration, the effectiveness of the instructional activity about processes is lost.

Transition Problem #4 : How do we strike an appropriate balance between the need for mathematical talk, a group instruction process, and the need for individualising instruction?

Discussions of this problem concerning the need of learners for both individual and group activities did not address the problem of what is an appropriate balance in the activities. Rather, concerns were more in the area of how to encourage talk that was constructive and productive. The nature and character of tasks for individual student work received little attention; rather it was assumed that students had considerable experience on individual tasks.

Discussions of group processes focussed attention both on the appropriate nature of group activities and on teaching strategies that would encourage active participation by students. Broad scale support was evident for the necessity for students to talk about mathematics, and that the talk should transpire between student and student as well as teacher and student. Among the techniques of encouraging student discussion, the following were recommended (in particular):

- building student's confidence with success; either small or large successes in communicating mathematics should be used to encourage discussion;
- being flexible in arranging seating to be conducive to talk;
- using mathematics of the real world and investigations of that mathematics as stimuli for discussion. Particular attention needs to be given in selecting real world situations that are timely and appropriate for the interests of this age group. The goal of bridging from real world phenomena to mathematics was identified as particularly conducive to talk within the mathematics classroom.

The problem of what is sensible discourse in mathematics for students at these age levels tended to focus on the positive aspects of verbal interaction. For example, participants recommended attention to the relationship between mathematical words and phrases and how symbols interact with or are related to them. Rather than making an issue about precision and correctness head-on and early, it was hoped that the need for precision would develop out of needs discovered in discussion, that students' valuing of correctness and precision would evolve from ambiguities and conflicts in communicating with each other. Standards for mathematical discourse should be developed naturally within the context of communicating mathematics. Errors in communication with words and with symbols provide a natural motivation for precision and correctness.

Participants were concerned about the time and logistics of providing for student discussion of mathematics. Many participants viewed discussion-based approaches to mathematics as less efficient than other instructional approaches. Management of students at these age levels was identified as prone to noise, disruption and misbehaviour if poorly conceived by the teacher.

Transition Problem #5 : How do we help students learn to relate symbols to physical situations?
Major recommendations about this transition problem were concerned with the problem of gradually easing students into effective use of symbols. Generalisation should be approached cautiously with awareness of readiness factors. Problem situations, including those grounded in physical science, were identified as providing a suitable context for instruction. Problem situations that had the potential of revealing important relationships in the learners' environment were viewed as a more appropriate context for instruction than purely mathematical situations. Students should be given time to understand symbols and their use in mathematical modelling.

Discussants should encourage teachers to continue using instructional activities based on concrete activities well into the junior secondary school. The shift to totally symbolic, abstract mathematics was viewed as premature for many learners, not allowing ideas to gestate and grow to fruition. Students should interpret perceptual data and fit it into symbolic constructs. Emphasis on spatial visualisation and understanding of physical relationships has a payoff in more complete learning of the mathematics involved. Students should use symbols in record keeping for manipulations of physical materials and describe verbally the manipulations and the analogous, parallel work with symbols.

Particular attention should be given to graphing as a means of describing phenomena. 'Graphicacy' has a power approaching literacy in developing understanding of symbols for many learners. Constructing and interpreting graphs yields an expanded understanding of both the mathematical ideas and the relationships that the graphs describe. Data interpretation in graphical settings should emphasise functions or descriptions of relationships between dependent and independent variables. Graphs are a means of describing and reporting data and relationships efficiently.

The group also examined the difficulties learners exhibit when dealing with the equals symbol. Participants felt that the equals symbol was among the more difficult for students. The naive understandings, derived from the limited experiences of primary school students, startle many beginning teachers who have little realisation of how their understanding differs from that of their students. Children who view an equals symbol as a direction to compute an answer need a significantly expanded view of equality to cope with equivalence of equations. Attention needs to be given to developing instructional materials and processes throughout the junior secondary school specifically directed to expanding the meanings for equality.

Transition Problem #6 : How can we use junior secondary school mathematics to reduce debilitating stereotypic images of the mathematically inclined?
A surprising feature of the discussions of the transition problems of the junior secondary school was the lack of attention given to stereotyping by learners. In the United States the evidence suggests that the formative, early teen years are those in which invidious classifications of peers are made. The effects of forming these stereotypes carry on into adulthood and hamper young people in realising their own potential. Leder's annotated bibliography concerning girls and mathematics in Australia (MERGA, 1984) provides evidence that the problems are of the same order of importance and relevance in Australia as in the United States. The Cockcroft Report has a special section addressing the problem of females and mathematics. Thus, we conclude that the problems are significant and widespread; hence, the surprise.

Although discussions were hampered by lack of time and even though there were special meetings devoted to the problems of women and mathematics, the organisers of this action group were startled and chagrined that not one of the sixteen subgroups elected to concentrate on this major problem for the junior secondary school. Proportionally fewer women have attained eminence in the field of mathematics or in fields associated with uses of mathematics when compared to men. Teenagers tend to view mathematics as a male domain. Children displaying either high or low deviant behaviour in mathematics are viewed as different by their peers and soon learn to control their behaviour accordingly. We are particularly concerned about children who impose limits on their own behaviour in social contexts, such as the classroom, thereby not realising their own potentials and talents.

The organising group believes that teachers need to be aware of and concerned with this problem. Aspects of the cultural ethos of the junior secondary school need to be better understood in order to have a greater positive impact on young people's self images and their views of others. Even though some features of the classroom and societal context are beyond direct control by teachers, they need to make attempts to help children have a more accurate view of themselves and their peers relative to performance in mathematics. Too much is at stake in stereotyping mathematical performers in this age group for the problem to be ignored.

Transition Problem #7 : What instructional strategies will improve the functioning of memory by students?
Discussion concerning memory and its role in junior secondary school mathematics emphasised the contribution of research to mathematics education. Good sense derived from experience suggested that a key factor in assuring ease of retrieval was assuring that new learning would fit into structured knowledge already possessed by the learner. Teachers must strive to relate new learning to what children know and to their existing frameworks of knowledge.

Discussants viewed memory as requiring more than drill and practice and hope that research will yield more constructive instructional strategies in planning for instruction. Ideas concerning the directions that should be explored by researchers included attention to learning style, the role of individual activities (as opposed to group tasks), and trying to understand more completely how mnemonic devices work.

One discussion group observed that most mathematics teachers have considerable natural ability in remembering mathematics and, partly as a consequence, have difficulty in dealing with students who lack this ability. The teachers do not understand the difficulties of their students and frequently exhibit little sympathy or understanding for these students or their memory problems. Such teachers provide little supportive aid to these students.

Participants pointed out the difficulties inherent in memorising without understanding and wondered whether premature memorisation interfered with learning. The strong opinion was stated that memorisation without understanding was disastrous for most learners. Teacher participants in several groups noted that memorisation did not provide a suitable basis for the encounters with applications and problem solving. A strong concern was expressed for the plight of algebra students who memorise but cannot apply their knowledge to new or different situations.

One of the observations made was the need to distinguish between memory and memorisation. Memory generated in learning situations directed toward building understanding was viewed as different in quality and character than memorised knowledge. The distinction comes from the differences between knowledge that is fitted into an existing framework of knowledge in contrast to memorised material that stands alone as a discrete bit of separate knowledge. The former was viewed as more powerful and permanent. Because it is knowledge related to many other ideas, the knowledge may be accessed through the other elements of the framework rather than only through the idea itself. Thus, it serves the learner better as a base for using and doing mathematics than that memorised as a stand-alone, discrete, separated idea. Memory developed within a framework of related ideas provides a much better basis for learning new mathematics.

A suspicion that research on memory has been focussed on facts, low level skills and algorithmic learning rather than higher cognitive functioning in mathematics, was registered. Discussants were hopeful that researchers would focus more broadly on mathematical behaviour and try to account for differing individual learning styles. Better understanding of the role of memory in problem solving and in applying mathematics, and of the contribution of problem solving activity to remembering, is needed.

Transition Problem #8 : How can we increase the intensity of instruction on important topics of the junior secondary school curriculum?
The transition problem of intensity of instruction was analysed in terms of the selection of content, its scope or depth of treatment, and how it should be sequenced. Issues and concerns were closely related to the traditions and current status of curriculum in the country of participants. Some felt that their junior secondary school students progressed through content of limited scope and that their major need was to provide curricular experiences that would expand students' perceptions of mathematics and its uses. Other participants felt a need to provide more enrichment activities above and beyond that required by the current curriculum and assessment programs. The provision of experiences to extend and enrich the mathematics program was more critical than the issue of intensity of instruction for most participants.

The only significant recommendation shared across several discussion groups concerned applications of mathematics. The opinion expressed most frequently was that one of the most appropriate ways of extending the treatment of a mathematical topic was to incorporate extensive experiences when applying that topic into instruction. This was viewed as a suitable means of extending the learner's experience with the topic while at the same time providing significant opportunities for reinforcing learning.

Transition Problem #9 : How do we resolve the issues surrounding the inclusion of new content in the curriculum?
Discussion of the problem of new content yielded the identification of two major concerns:
* The curriculum is crowded. If new content is identified for inclusion in the curriculum, how are topics identified for exclusion or de-emphasis?
* Some teachers lack experience with and knowledge of new topics. The lack of preparation affects not only their competence but often extends to their confidence and perceptions of their own worth as teachers. If new content is to be introduced, how do we provide for in-service education for such teachers?

Another pressure identified by participants was in terms of the senior secondary school and assessment programs dictating what new content should be included with little cognisance of the current content of the junior secondary school curriculum, and with little familiarity with learners of this age span or their interests. A further concern was expressed in terms of schools and/or mathematics departments creating alternative options for election by students but at the same time not creating suitable processes for counselling students about those electives.

Discussants were concerned about the sources of ideas for new topics in the curriculum. If the needs of school leavers are to be considered, it is incumbent on the designers of curriculum to consider current uses of applications of mathematics in commerce and industry. Having the potential of improving the reality base of the curriculum, examining local applications of mathematics, and using local business and professional people as resources, strengthens community involvement with the schools in productive ways as well as providing school leavers with useful tools.

Content alternatives that were identified as new but of sufficient importance to warrant inclusion in the required junior secondary school curriculum were the following:
* probability;
* statistics with an emphasis on data collection, interpretation and reporting;
* mathematics integrated with other disciplines;
* applications of mathematics to real-life situations;
* career options that use mathematics;
* estimation.

Participants viewed new content topics as a mechanism to provide excitement in the study of mathematics for junior secondary school students and were pleased with the opportunity that new content would provide to stimulate, enchant, and enthuse colleagues who perceive mathematics teaching as dreary.

Participants were quite suspicious of attempts to install new curricula without adequate materials and support for implementation. They were concerned that if new content is incorporated in the curriculum, then evaluation and assessment processes and instruments should be adjusted to correspond to these changes. This adjustment should transpire in a timely fashion with mathematics teachers contributing to the process to assure that more than low-level skills are assessed.

Few topics were identified for exclusion or de-emphasis, but of these, favourite nominees included: logarithms, simplification of complex rational expressions, computational practice on tasks better accomplished by use of a calculator, repetitive drill work not designed to extend learning or memory, and fractions.

Conclusion

The majority of participants in the junior secondary school action group were classroom teachers. Discussions demonstrated the realities and the recency of their experiences with learners. The professional commitment and enthusiasm displayed in addressing problems unique to pupils of this age span were both heartening and reassuring.

Examination of the reports of each of the 32 discussion sessions revealed that five major themes or belief structures about teaching at this level pervaded deliberations about the transition problems. Whether assumptions about the nature of the learning process or beliefs developed through the wisdom of accumulated experience, these themes provide a significant basis for decisions that must be made by mathematics teachers.

Theme 1. Learners need to develop a thorough understanding of the mathematical concepts that extend doing mathematics to higher cognitive levels than would the study of facts or skills alone.

Theme 2. Learners need extensive experiences with problem solving that must include applications of mathematics and experiences that allow learners to organise and understand the perceptions of their environment.

Theme 3. Teachers must use knowledge of what children know and can do in order to design instructional activities of appropriate cognitive demand and expectation.

Theme 4. Instruction about computation, for its own sake, particularly in areas where calculating devices are part of the environment, is not as important as respecting the role of computation in developing expanded understanding of other aspects of mathematical thinking.

Theme 5. Inclusion of new content and/or instructional emphasis must be carefully planned in terms of the capabilities and needs of children, but must also respect the school's capability to implement them.

Finally, the junior secondary action group was profoundly concerned about the state of the knowledge, experience and commitment of colleagues and peers. Several groups argued vehemently that new approaches, content and tools of instruction were not the most critical need in addressing problems of the junior secondary school program in mathematics. They identified the knowledge and professionalism of teachers as the boundary to progress in junior secondary school mathematics.

The following recommendations were offered by Betty Lichtenberg (USA) in her summary of discussion section proceedings for consideration by the total junior secondary school action group. Participants requested treating them as resolutions subject to vote. Each was passed with no nay votes and no abstentions. The resolutions are that:

National and international professional groups in mathematics education work for:

- the establishment of policy to increase the supply of qualified mathematics teachers;
- the provision of adequate availability of in-service training and retraining in the teaching of all areas of mathematics.

ACTION GROUP 4: SENIOR SECONDARY SCHOOL (AGES 15–19)

Organisers: John Egsgard (Canada), Trevor Fletcher (UK),

The Mathematics Curriculum, 15–19

Prior to the Congress, an analysis of current practice identified a number of key problems in the design of the mathematics curriculum for pupils aged 15–19:

- There is a great increase in the number of students remaining in secondary education and continuing to senior level. In former times courses catered for a group of moderately elite students; now the group is much wider, and in some countries the 'average' students, for whom a fresh style of course is needed, constitute a majority.
- In most countries the established courses, developed for students bound for tertiary education, lead to calculus and linear algebra. What should be central topics in courses for the majority? Should computing and the applications of mathematics receive more consideration?
- Since the students now display wider variations in attainment, the curriculum needs to be more flexible. This is especially so where the curriculum is a national one. Is it a solution to the problem to provide a core with options? If so, what should be in the core and what in the options? Curricula are often overcrowded, so what should go out?
- What new approaches are there to topics such as calculus and statistics?

The following conclusions emerged from the Congress:

- There is general agreement that it is necessary to develop fresh styles of courses for students participating in senior secondary education, but the approaches adopted in different countries vary greatly. Some countries are attempting to accommodate variations in attainment by designing differentiated courses — that is, students are grouped by attainment and follow courses of different levels of difficulty. In other countries this is unacceptable and it is policy to develop modules providing a core and options.
- There is general agreement on the present place of calculus and linear algebra in courses for the elite group of students but no agreement on the position of application oriented topics such as statistics, mechanics and computer mathematics. However, it is clear that statistics and probability should ideally be given a more prominent place, and some participants were very concerned about the unsatisfactory place of geometry.

- Hand calculators should be used to their full potential and microcomputers should become a major teaching tool.
- The curriculum will continue to evolve. Continuing re-examination will be necessary in order to reduce overloading and to update the topics contained therein.
- The influence of university entrance requirements needs to be studied.

National Reports

In some countries national reports have been prepared as a basis for discussion of the above matters and to develop national policies. A group investigated such reports, but rarely found any that detailed problems of the curriculum and suggested reforms, or directions for reform for the entire country. Those from the United States and from England and Wales are very unusual. There are documents in many other countries but these take a great variety of forms. In some cases these are only lists of prescribed topics, without any inquiry into the reasons for policies or any attempt to derive the curriculum from the social context. Where there is a national curriculum with a detailed syllabus, as in Japan, the document describes the ideal or intended curriculum. In Bulgaria the corresponding document also describes a set of materials and so it is an approximation to the implemented curriculum. A deeper study, which is concerned also with the achieved curriculum, is to be found in the Second International Study of Mathematics Achievement (SIMS). In SIMS there is an underlying commonality enabling comparisons to be made among practices in different countries.

Some national reports, such as the Australian Council for Educational Research numeracy survey (1982), or the National Assessment of Educational Progress in the USA (NAEP, 1983) or the reports of the Assessment of Performance Unit (APU) in England and Wales, are concerned, like the SIMS report, with performance. These report on the achieved curriculum without necessarily going on to deduce any implications for curriculum design.

To complicate matters further, the 'nation' covered by a 'national' report is not always a country, but may often be a state, region or other governmental subdivision. The topic group considering national reports looked through the entire ICME 5 program for sessions that could contribute to an understanding of the senior secondary mathematics curriculum as seen from such reports.

The extent to which a report focusses on problems should not be viewed as a measure of the seriousness of those problems. In fact, the correlation may be negative, for, just as it is the higher scoring student who is most bothered by the grade on a test, so we find Japan as self-critical as any other country despite very high standards of performance on attainment tests.

International reports confirm the findings of the curriculum group noted earlier. Increasing numbers of students are proceeding to the twelfth grade, and the mathematics needed for future engineers, scientists and mathematicians is universally felt not to be appropriate for all, and in some places it is felt not to be attainable by all. As a result, there is a tendency to increase the options available in courses, to insert applications, and to give less emphasis to set theory and the structural ideas from abstract algebra and logic which were introduced twenty years ago.

'Integrated' Curriculum

In many countries it is long-established practice to teach mathematics courses in a way which interrelates the various areas of the subject. By contrast, in some countries (such as the USA) it is normal to teach algebra and geometry in separate courses, perhaps even in different years. One group was concerned with the advantages and disadvantages of an integrated curriculum. Their discussions centred in turn on the students, the teachers and the curriculum itself.

There was substantial agreement that it is possible for students to see various topics of mathematics in new relations in an integrated curriculum, and less opportunity for students to be bored by over-exposure to one aspect of the subject. An integrated curriculum is more flexible, and topics can be given varying emphases, depending on the students' needs. Topics can be delayed also to accommodate the maturing intellectual processes of the students.

Some considered it a serious disadvantage of the integrated approach that students can fail to retain skills and concepts because there is insufficient concentration on any one topic. Others argued that spiral development in an integrated curriculum can compensate for this and so assist retention. A further problem arises in some countries from the mobility of students. Students are at a disadvantage if they change schools and find topics missing or treated in different sequence.

For experienced and mathematically competent teachers an integrated curriculum provides an excellent opportunity for creative, imaginative teaching. On the other hand, an integrated curriculum can be frustrating for teachers who lack expertise in some of the topics which it contains. Such teachers can omit these topics or teach them inadequately. If some topics are over-emphasised and others minimised by the teacher, then this causes problems for teachers later in the program who presume that the foundation for a spiral approach has been taught previously.

In countries where algebra and geometry, for example, are taught separately teachers may teach one of the subjects adequately but be unable to teach the other. If an integrated program is introduced, these teachers will need to be retrained or be pooled in a team with other teachers so that all topics can be handled.

It is a definite advantage of an integrated curriculum that it has potential to display the full nature, even the beauty of mathematics, its usefulness and its applicability to real world problems.

There was almost universal agreement in the group that in some countries suitable text material is lacking.

Geometry

The deliberations of a number of groups confirmed that geometry remains an enigma. Its place in senior secondary mathematics has been debated for many years, but there is still little agreement on what should be done. At senior secondary level it is necessary to provide for three types of student — those who will not take mathematics through to the end of this stage, those who will take a general course in mathematics to the end of secondary school, and those who take mathematics as a precursor to tertiary mathematics.

There are great differences in the geometry curriculum for senior secondary school between countries, and even between educational jurisdictions in the same country. The differences occur in *why*, *what*, *how* and *when* geometry is taught. Some differences concern the amount of geometry taught, the content and variety of topics, the approach, the integration of geometry with the rest of the course or its concentration in separate units, and the clientele to whom the courses are taught.

Notwithstanding these differences, the group discussing geometry felt that the main strands of analytic (coordinate) geometry, Euclidean geometry, transformation geometry and vector geometry should be woven together to different degrees and at different times to produce the *how* and the *what*. This weaving together is fully compatible with the aims of those who emphasise problem solving in mathematics, for the student can draw on a multiplicity of approaches to find solutions.

When considering geometry in the senior secondary school, it should be assumed that considerable work with transformations has been done in the preceding years. The first notions of transformations are best learnt through the use of manipulatives that are natural in a pre-secondary school setting. This preliminary work should result in good intuitive notions of the properties of transformations and the properties of some basic geometric figures that can be derived from transformations. The properties can then be used as the axioms for geometric proofs at the senior secondary level. This approach is consistent with the van Hiele notions that students remember and use those facts that they have learned through experience. The transformations studied should be restricted to isometries, similarities, shears, one and two way stretches. Great insights can be gained by studying linear and quadratic functions through their graphs and transformations of them. Matrices, which are ideally developed in an algebraic context, can be used to describe transformations and to consolidate and develop further the transformations of figures.

The basic concepts of coordinates and coordinate graphs should be prerequisite for any work at senior secondary level. Although much work in analytic geometry is associated with algebra, the implications for geometry are significant. For example, the linear function is related to straight lines and the quadratic function to parabolas. Analytic geometry should certainly lead to an initial, informal study of conic sections.

Some teachers feel that there is little value in children studying Euclidean geometry in a modern mathematics program. In addition, as a result of the recent trends away from the teaching of Euclidean geometry, there are now teachers with little or no experience of teaching the subject who need to be convinced of its value. But Euclidean geometry provides a good model of our space and it is important for future study in mathematics. Students should certainly have some exposure to Euclidean geometry at senior secondary level.

Students should learn to write clear and logically sound arguments as solutions to problems; but the traditional 'two-column proof' or a rigorous adherence to Euclidean axioms does not accomplish this. The use of less elegant or local sets of axioms is recommended.

Most students should undertake some work with vectors, preferably based on their natural occurrence in physical situations. For the more able students an extended study is appropriate. Vector ideas are a natural outcome of a study of translations. When translations are described as ordered pairs of numbers, the basic idea of the free vector has been established.

Topology contains some good intuitive problems and ideas, but the role of these is in enriching the curriculum. Spherical geometry, associated with spherical trigonometry, is a suitable extension to the course for a gifted few.

A participant from France did not consider the preceding survey appropriate to the situation in his country and raised the need for an algorithmic and dynamic approach to geometry, as in the 'turtle geometry' of Papert. This, he said, is made increasingly necessary in technical education by developments in the numerical control of machines and in computer graphics.

The non-academically inclined (slow learners)

The curriculum has to be considered not only in relation to the various subdivisions of mathematics but also in relation to the needs of students of differing mathematical attainment. The mathematical attainment of many of the extra students studying mathematics after the age of 15 does not equip them to take the courses prepared for mathematically able students. Also, the methods used to teach mathematically able students usually fail in classes of non-academically inclined students who may even have difficulty reading their mathematics texts.

As an initial step in solving the problem of helping the non-academically inclined student to learn mathematics, the group endorsed the use of the calculator at all times. Such access should help restore the students' confidence in their ability to make accurate calculations.

But success in calculation alone will not be sufficient for these students. There must be changes in course content that emphasise real life structures and the mathematical skills which students will need in the world of work. Lists of skills expected by employers have been compiled around the world. Some clearing house could be established for these lists so as to make them readily accessible to teachers in all countries.

Special teaching methods must be developed that will motivate these students and build their self-confidence. Video tapes and computer software should be considered as possible sources in this regard.

Many of these students have a history of failure. A variety of assessment formats must be utilised that will provide these students with the success they need. Teachers must be careful to teach and assess only what these students can learn. As well as written tests that can be prepared for by continual practice, the group suggested the following:

- students could be allowed to take tests when they are ready;
- students could give a three to four line assessment of their own achievement;
- the teacher could use individualised programs of testing including oral questioning;
- project work and group work could be used in assessment;
- multiple choice tests could be used for syllabus coverage, with longer answers to check higher level skills.

Computers

Both the curriculum and methods of teaching will be affected greatly by the microcomputer. One group was particularly concerned with such problems and based its discussions on the following assumptions:
* the computer will gain universal acceptance and will not go away;
* the computer, used appropriately, will make possible a resurgence in mathematical problem solving skills;
* the computer, to be used effectively, will require different teaching styles and suitably prepared teachers.

The following propositions gained the support of this group:
* Introduction to the computer as an object of study must occur early in the educational life of a student. Computer science, or communication technologies, should be recognised as a discipline apart from mathematics related to all disciplines and should be taught, if possible, by computer science specialists. A solution must be found to the problem of losing mathematics teachers to the computer sciences education program.
* The use of the computer as a teaching tool should provide student interaction with the computer and should be designed to improve student problem solving skills. Complete teaching packages should include text materials, pencil and paper exercises, demonstration computer use for extension of concepts, individual student follow up on a computer, and higher order evaluation. As a result, teachers should have available both classroom demonstration computers and a laboratory situation suitable for classes of students. Care should be taken to enable certain students to continue to develop understanding and facility with mathematics. Further, the school-home interface should be encouraged.
* Developers of software and manufacturers of hardware should be urged to move toward standardisation of format and equipment as has occured with audio and video tapes and tape machines. Users now find themselves locked into particular hardware and their specific software, rather than having access to a wider range of both. It may well be that it is still too early in the development of micros to standardise. This could be a topic of discussion for ICME 6.
* Local, state and national educational policy makers should be aware of the need to retrain present teachers and educate new teachers to take advantage of the new technologies including, but not limited to, the computer. Policy makers should be urged to provide the opportunity and the incentives needed to upgrade the mathematics teaching profession to understand and use the new technologies.
* ICMI, or some other body, should build and extend a computer users and developers network.

Variations in teaching methods among various countries

Representatives from 11 different countries met to discuss variations in teaching methods. They learned that most countries have a national or state curriculum, with an external examination at the end of secondary school. Within this curriculum, algebra and geometry are usually integrated. On the average, mathematics classes meet 5 times weekly for 50 minutes. Most countries attempt some form of grouping by ability and achievement, especially after the end of compulsory education at age 15 or 16.

A survey of teaching methods among the various countries led to the conclusion that most teachers begin their classes with a review of the work that will be needed to enable the students to understand the new material of the lesson. This is followed by a check of homework assignments with the teacher or a student working out the more troublesome problems for the class. The major part of the lesson is given to the introduction of new material. The lesson ends with the students starting the homework assignment in the atmosphere of a supervised study. The group agreed that frequent deviations from this basic method were essential for good teaching.

Teachers working with a set curriculum and/or preparing their pupils for an external examination reported feeling stifled by the demands of the system because they believed that their primary task was to cover the syllabus adequately and to prepare pupils to be successful in the examinations. In most cases the length of the curriculum left no time for any form of creative teaching. There was some feeling that a change to the length of the curriculum, and the addition to the examination of areas such as problem solving, investigational work, applications in nature and the physical world, and projects, have the potential to improve the quality of teaching in most classrooms.

The group expressed an awareness of the need to increase student interest in mathematics and to reduce anxiety so as to enable the students to learn mathematics more effectively. The most successful teachers were identified as those who:

- find a way to link enrichment activities with the standard courses;
- enable the students to see that mathematics is useful to them through its presence in and its effects on our world;
- use novel and stimulating ways of teaching standard topics;
- show the link between mathematics and other subject areas such as physics, chemistry, biology, geography and economics;
- encourage their students to investigate mathematics either singly or in a group;
- assist their students to apply, in a thoughtful way, the routine formulae they may have learned by rote.

The group agreed that internal assessment by the teacher should carry equal weight with external examinations, and should take a variety of forms such as written tests, oral questioning, homework, group tasks, individual projects, investigations, field work, and open book examinations.

Assessment

Two aspects of assessment were discussed by a group particularly concerned with the topic: assessment, as part of the school's teaching methods, and assessment for certification at the end of the course. It was agreed that the internal assessment of students' work should indicate to the student how well the work covered in the classroom is being mastered. Such assessment requires that teachers have clearly defined programs of work, coordinated within their departments.

It is important that teachers assess the ability to use the mathematics being studied to full competence. This includes the accurate performance of technical skills, but this should not be all. The student's ability to apply the mathematics the student knows to various situations is also important (although difficult), as is the ability to understand and develop mathematical ideas.

The form of reporting on student progress also requires careful consideration. Parents and potential employers sometimes ask only for the bare essentials of the student's overall performance, in particular the final grade. More detail is required if the student is to be able to use the information for guidance for future action. There is growing interest in *profile-reporting*, that is, the assessment by teachers of the student's achievement in each of the objectives for a course. Any system using the assessments of individual teachers requires careful moderation among teachers and schools.

Certification occurs in many countries at a level of secondary education corresponding to the 16 + age group, but the common concern of all members of the group discussing certification was assessment for certification at the conclusion of secondary education. The countries represented, however, showed many differences in the type of assessment used, particularly the use of internal or external assessment or a mixture of both.

A final assessment program is constrained in varying degrees by the purposes for which it is used for certification. In most countries, final assessment not only marks the conclusion of secondary education, but also contributes to the selection process for higher education and employment. Where the results of the assessment are seen as only one piece of information supplemented by school reports, principals' recommendations, interviews, and other information, then the constraints on the assessment are not unreasonable. Where the selection of students for tertiary education rests entirely on a single, final assessment, then the pressures on students, teachers, schools and their curricula are more obvious and often undesirable.

Whilst external assessments provide public credibility, too much can depend on them, in which case the assessments become threatening for students. Even so, many teachers are able to help students meet the challenge of the external examination. Where internal assessments are required, teachers may find it difficult to establish the different relationships needed for their two roles: supporting their students whilst judging them as an assessor.

Internal assessments may also place students and teachers under continual pressure because of an increased workload for both, and so the professional responsibilities of the teacher appear to be greater. In spite of the difficulties, internal assessment permits greater curriculum flexibility and a wider range of types of assessment. It encourages the teaching of aspects of mathematics which do not lend themselves to external examination. It also encourages the learning of mathematics as a co-operative endeavour rather than as a competitive one. The greatest impact of central control of curriculum, particularly when it is associated with external assessment, is the pressure for curriculum planning to be 'top-down', rather than building up from the bottom on the basis of child development.

The development of assessment instruments based on clear statements of objectives and criteria, together with more descriptive, informative modes of reporting, seems to hold out the promise for improvements in the future.

These seem to be more useful for both students and parents and to offer the potential for encouraging the better effects of assessment on the teaching and learning of mathematics whilst discouraging some of the less desirable effects of some present practices.

The all too familiar comment made to the teacher who was teaching an unusual topic: 'Why teach that? — It's not examined!' — is a sad comment on examinations. It underlines the need to employ wider criteria for assessment.

Secondary mathematics teachers in developing countries

Students in all countries of the world do not have the same opportunities in senior secondary mathematics. One topic group discussed what other countries can do for senior secondary mathematics teachers in developing countries. The group decided that support should be given to efforts within the country to redesign curricula so they are related to local needs and conditions. It is important that this be done with strong local direction. Financial support is crucial, to train and upgrade local experts in curriculum and methodology and to offer continuing professional support. These leaders will usually be in mid-career and have an indispensible role to play in relating mathematics to the cultural, technological, linguistic and social setting in which it is taught. Whilst this is a role which few can fulfil, the need to train administrators, principals and others to strengthen the educational infrastructure is particularly important.

The group also felt that assistance in improving both the quantity and quality of teacher training is required. In particular, the training of local educators to take responsibility for teacher training and curriculum design should have priority. Expatriates participating in this work require a high degree of professionalism and should preferably make a long-term commitment to the task. Visits by foreign experts without a strong local counterpart identified to continue their work are of dubious value.

The group acknowledged that the shortage of materials, books and journals contributes seriously to the isolation of educators in developing countries. Schemes for subsidising publications, reduced rates for journals, and so on, should be supported. The task of finding means to produce local text books and materials is urgent. The establishment of professional development centres needs financial assistance. The issue of making proprietary software for microcomputers available at low cost should be explored by the governments of those countries where the software is being developed, and manufacturers should be encouraged in this area.

Co-operative and exchange programs have an important role to play. It is necessary to have communication outside as well as from within developing countries. An exchange of information and materials, as well as the movement of people can be encouraged. The design and initiation of such exchanges should rest as much as possible with the developing countries as full participants, although support will have to come mainly from developed nations. Individuals must be carefully selected for such exchanges on the basis of their adaptability and sensitivity to the cross-cultural nature of the task, as well as their professional qualifications.

The group emphasised the need for a clearing house in mathematics education for information on development programs, international co-operation and exchange programs. While UNESCO was suggested as an appropriate agency, it was anticipated that the ICMI would need to use its influence to achieve such a task.

Finally, the group strongly recommends to the program committee for ICME 6 that a full action or working group be devoted to 'international co-operation and assistance in mathematics education'. The planning for such a group should begin as soon as possible and involve invited representatives from as many developing countries as possible.

The shortage of mathematics teachers

One problem that is common to many countries, both developed and developing, is that of finding sufficient teachers to teach mathematics. A discussion group devoted to this problem was able to report on the experience of three countries: Australia, France and the United States of America.

A major factor contributing to the shortage of teachers of mathematics in Australia is a misunderstanding of the actual situation. Young people are discouraged from becoming teachers by the popular impression and constant reporting that there is an oversupply of teachers. Without official acknowledgement of the real situation, it is impossible to communicate that this oversupply does not apply to mathematics teachers. Other factors which may contribute to disinterest in mathematics teaching in Australia are discipline problems in the schools, and the fact that a begining teacher has little control over the location of the first teaching assignment. A first appointment is often to the outback or a region many hundreds of kilometres away from that requested. Also, the rapidly developing fields such as computer technology and engineering are more attractive because they are considered more interesting with better working conditions.

The shortage of mathematics teachers in France is of relatively recent origin. It is a problem confined mainly to the northern half of the country. Again, the attractiveness of other fields and working conditions in the schools are significant factors. Another factor, apparently, is the dogmatic teaching style that has been prevalent in French schools, and which has tended to turn students against mathematics.

One of the steps taken by the French government to lessen the shortage of mathematics teachers is similar to that taken in South Australia: the placing of auxiliary teachers in permanent positions. Also, the government has lowered the passing scores for the licensing examination. Neither of these solutions is considered satisfactory. It is hoped that changing the curriculum so as to veer away from the dogmatic approach will have favourable long-range consequences.

There are several reasons for the gap between the supply of mathematics teachers in the United States and the demand. First is the unattractiveness of the teaching profession in general as in Australia and France. Low pay compared to other professions, discipline problems in the schools, and low prestige are major factors contributing to this unattractiveness.

A second major reason for the shortage of mathematics teachers in the United States is the loss of personnel to the rapidly expanding field of computer technology. This field not only attracts many young people who at one time might have been teachers, it also siphons off teachers with several years of experience. Making this effect more pronounced is the fact that computer technology, engineering and other fields requiring mathematical skills are today as open to females as they are to males. A generation ago, teaching attracted many well qualified females. We are not as fortunate today.

Policies adopted by states and local school systems in the United States to help alleviate the mathematics teacher shortage involve one or more of the following provisions:
* free courses in mathematics for teachers at State-supported colleges or universities;
* loans to students who are working toward certification in mathematics;
* salary supplements for mathematics teachers.

The payment of salary supplements cannot be used in all localities as strong teacher unions usually defeat any attempts to introduce them in fields where there is a shortage of teachers.

It seems clear that improving working conditions for teachers, increasing salaries and raising the prestige of the profession would do most to improve the prospects for an adequate supply of teachers.

This report is based on the work of the following sub-groups. The name of the Convenor for each sub-group follows the group's topic for discussions:
'Advantages and disadvantages of an integrated curriculum in the senior secondary school'. *Stanley Bezuszka* (USA)
'Assessment in the senior secondary school'. *Win McDonell* (Australia)
'Variations in teaching methods in the senior secondary school among various countries in the world'. *LeRoy Dalton* (USA)
'The computer in the senior secondary school'. *Jack Price* (USA)
'The response of various countries to the shortage of mathematics teachers at the secondary level'. *Donovan Lichtenberg* (USA)
'The secondary mathematics curriculum as seen from recent national reports'. *Zalman Usiskin* (USA)
'What can other countries do for third world senior secondary mathematics teachers?' *Jean-Phillipe Labrousse* (France), *John Berry* (Canada)
'Mathematics in the senior school for the non-academically inclined'. *Albert Schulte* (USA)
'Geometry for the senior secondary school'. *John del Grande* (Canada)
'The curriculum: What is in it now? What changes should be made?' *Hiroshi Fujita* (Japan), *Harold Taylor* (USA).

ACTION GROUP 5: TERTIARY (POST-SECONDARY) ACADEMIC INSTITUTIONS (18+)

Organisers: Brian Griffiths (UK), Jack van Lint (Netherlands), Jane Pitman (Australia)

I. Introduction

After consulting with our international list of advisers, the Organisers decided to concentrate on six broad areas which are detailed in the reports of the Splinter Groups. It was agreed that other topics would not be neglected, as they would occur in other parts of the Congress Program.

In dealing with the six areas for group work, the conventional lecture form was avoided except for four short lectures to the whole group on the following topics:

- The Content/Activity Dilemma in Tertiary Mathematics Teaching. *Brian Griffiths* (UK)
- The Teaching of Statistical Literacy to Students in the Agricultural Sciences. *Andi Nasoetion* (Indonesia)
- The Role of Third World University Mathematics Institutions in Promoting Mathematics. *Mohammed El Tom* (Sudan)
- The Education of Mathematical Engineers. *Jack van Lint* (Netherlands)

As the first lecture was largely concerned with explaining the structure of the Action Group to the participants, a summary is included here. The other three lectures will be published in the International Journal of Mathematics Education in Science and Technology.

In the first lecture Griffiths pointed out that workers in tertiary level institutions normally pursue the three activities of teaching, administration and research in various mixtures. The last has greatest prestige, and consists of the creative activity of producing output in the form of theorems, algorithms, models, *etc*. This 'fabricated mathematics' evolves, eventually stabilises, and forms a body of 'content'. Administrative activity involves, among other things, formulating curricula, assessment and selection procedures, in response to pressures from the surrounding society, and meeting constraints both imposed and implied. Teaching activity, which may well include reflection and scholarship, is normally discussed in terms of content, via syllabuses that contain content-words like matrices, equations, and so on. Syllabuses in turn lead to conventional-mode examinations that test only low-level skills, and cause the dilemma of the title. Furthermore it was observed that mathematicians tend to discuss these matters at a crude level, without the care for definition, analysis, and qualified hypotheses that are usual in the writing of mathematics. Indeed, as they had little practice in such discussion, the

Organisers decided to adopt the mode of the annual conferences on undergraduate mathematics teaching held in the United Kingdom as the most appropriate.

These conferences were initiated in 1976 when Michael Atiyah, as President of the London Mathematical Society, felt that it was necessary to respond to societal pressures concerning numbers of students and potential changes in secondary curricula. The Society still sponsors the conferences, with administrative help from the Shell Centre. They are attended by one or more representatives from each tertiary mathematics department in Britain, plus other interested people. Plenary lectures are given, but usually only to present factual material. All other sessions are held in small groups whose members stay together for three and a half days to prepare a written reply to a task assigned by that year's organising committee. The reply is then submitted to a second group that must criticise the draft in writing, and the final versions are published in the proceedings of that year's conference. Early conferences worked in terms of content or 'product' for syllabus construction, but eventually moved in the direction of activity or 'process'; and it is the 'process' that is found more valuable than the hastily produced product of the working groups. The purpose is to impart a more professional attitude to the non-research side of a lecturer's work, and also to ensure that replies to Government pressures can be more effective in explaining the academic point of view.

On the basis of this experience, it was thought appropriate to adopt a similar mode for Action Group 5, even though most participants would come from very different backgrounds. Six Splinter Groups were formed, which could work for four one-hour sessions. With a total of well over 150 participants, these groups were much larger than those at Nottingham, and were advised to split if required. To expedite their work, the ICME 5 microfiche contained some preliminary thought-provoking material and reading lists. As well, each Splinter Group was assigned in advance a convenor and two leading contributors to provide short papers as bases for discussion. Finally, participants were warned that groups might have silences, that nobody should speak for too long, and that time was probably too short for them to produce an impressive final product. It was emphasised that it was the process that mattered, and they should try it out at greater leisure once they returned home.

In the event, some of the final products were very good, and the Organisers had great difficulty in doing justice to them in the compressed reports that follow. It seemed from observation that many participants enjoyed the process that they were undergoing.

Finally, there were two summary sessions, reported separately below, after the reports of the Splinter Groups.

II. Reports of Splinter Groups

Mathematics for Non-Specialist Users.

Convenor: Geoff Ball (Australia)

This group considered service courses for students whose main interest is in a user area such as the physical sciences and engineering, the biological sciences,

or the social sciences, rather than in the mathematical sciences themselves. Typical examples are:
- the multi-purpose and special purpose or remedial service courses offered by mathematics departments; and
- courses in mathematics, especially statistics, taught within user departments, sometimes by their own regular teaching staff and sometimes via special arrangements such as joint appointments or resident mathematicians.

The group members represented experience of service teaching in a wide variety of user areas, types of institution, and modes of instruction. Despite this variety, they shared concern over several major issues which appear to be relatively independent of particular institutions. The starting point was John Taffe's thought-provoking talk based on his contact with students in service courses at the Australian National University, Canberra.

Students. There is often inadequate consultation between mathematics instructors and students in service courses, especially in large departments. Many students have difficulty in relating the mathematics they have been taught to their other courses. This creates doubts as to the relevance of their mathematics courses and highlights the need for careful selection of illustrative examples. Even when examples are chosen from the literature of user departments, there is evidence of a significant gap between the mathematics taught and the mathematics actually used, and this gap is generally even wider in multi-purpose service courses (even though such courses have advantages from other points of view).

Methods of instruction. There has been little attempt at innovation in the more traditional service courses, though the increasing impact of computers is beginning to cause changes in content and emphasis. It is in more recently developed courses, particularly where the course is presented to a homogeneous group of students, that the mode of instruction has been most influenced by the nature of the course. Jean Roussel (France) and Lilia Del Riego (Mexico) described examples of team teaching, both of which were devised for life-science courses over the last twenty years. (See ICME IV Proceedings p. 213>) Two areas warranting further study are the role of technology in service courses and the techniques of team teaching.

Consultation with user departments. Development of a relevant syllabus and appropriate methods of instruction requires consultation between the user and provider departments and compromise from both, a process which involves significant difficulties. Adequate consultation, especially in the creation of new courses, consumes great amounts of time, often over many years. Unfortunately, involvement of teaching staff in this work is often neither encouraged nor even recognised for promotion purposes. In some institutions, very detailed requirements by user departments have led to unduly narrow and fragmented courses with little mathematical appeal. Furthermore, there is little opportunity in such courses for the introduction of mathematical content which may arouse developments within the user departments. If the user departments are conservative, there is a further danger that the only mathematics being taught is that which was relevant to the user departments in previous decades. (This is a good reason for involving active mathematicians in the relevant teaching.) Ideally, both departments need to be represented by staff who possess a broad knowledge of both disciplines else relevance and innovation may be overlooked. Similarly, the presentation of realistic material is enhanced if these abilities exist. Team teaching and resident mathematicians may be ways of overcoming difficulties, but the development of such strategies takes many years.

Recommendations.
- That ICMI investigate the existence of good examples of procedures to enhance communication between mathematics and user departments, with the object of disseminating this information as widely as possible;
- That further study should be made to identify those areas where a gap exists between the service material taught and the needs and expectations of the user departments and their students;
- That organisers of local and international conferences follow the lead of ICME 5 in recognising the need for closer attention to the difficulties associated with service courses. Sessions should be arranged to encourage the interchange of ideas on the aims, problems and successes in this area. (The survival of some mathematics departments may well depend upon the excellence of this aspect of their activities.)

Theoretical Mathematics.

Convenors: Ramesh Kapadia (UK), Anthony Briginshaw (UK)

The Splinter Group divided into two subgroups with 24 and 14 members respectively. Although the topic is very wide, the discussions tended to concentrate on two things: rigour and abstraction, and the extent to which undergraduates should be exposed to them. It was clear that the words 'rigour' and 'abstraction' meant different things to different people, and there was interesting and animated discussion, often provoked by strategic questions inserted by the chairpersons of the subgroups. A considerable amount of written material was produced, much of it in the form of working papers, and difficult to summarise in a limited space. Consequently, this report simply highlights particular aspects of the work.

Rigour. To several, 'rigour' refers to 'rigorous proof', but there was agreement that there is no absolute standard for it since attitudes to proof change in time, and depend on who is talking to whom. It seems to concern 'convincing one's peers', and therefore its appearance in an undergraduate course will depend on the type and interests of the student. Most engineering students probably need very little analytical rigour (and are unlikely to see the point). It was remarked that 'One cannot teach rigour, but only teach rigorously'. In the latter spirit, instances were given of teaching with 'local' rather than 'global' rigour (for example, to derive the Mean Value Theorem algebraically from an intuitively based version of Rolle's Theorem), and of the introduction of counter-examples, new hypotheses, and 'monsters' in the spirit of Lakatos (see References).

The question of rigour and meaning was also raised, through an example that evaluates $\sum_0^\infty 2^{-n}$ first by algebraic manipulation (a trick?) and second by dividing a rectangle into an infinite number of pieces. (See also the article by Page.) Rigorous analysis can be related to the process of testing a model; after we teach the 'rigorous' definition of limit or continuity, we check 'obvious' statements like $x^2 \to 4$ as $x \to 2$ to show that the definition models aspects of our intuition. Students find this hard to understand, but one group spent some time on the Schema: Experience $\to$ Intuition $\to$ Rigour $\to$ Experience. It was also suggested that analysis can be treated as a set of algorithms for providing estimates. A further aspect of rigour is its potential for clarifying concepts and arguments, provided the need for such clarification is felt by those who are in communication.

Abstraction. One use of abstraction was thought to be to show how to organise a collection of isolated results or techniques. It was related to a scale of student needs and understanding ranging from cookery book exposition to refined conceptual understanding and organisation. Consequently, the goals of a course should be matched to the type of student attending it. Some argued that abstraction occurs in subjects other than mathematics (Philosophy, Law, for example), and so the difficulties of teaching abstraction are not necessarily confined to mathematics. For one meeting, the larger subgroup split into very small groups of 'students' and 'teachers' in order to scrutinise proposals for teaching a specific topic like unique factorisation or Rolle's Theorem. Their experience was then shared in the larger group.

References. Many details of the points made can be pursued through the list of references at the end of this document which group members identified as useful. The loss of detail might, however, be partially justified by the following paragraph from the material produced by one group: 'We have not sought ... to reach any earth-shattering conclusions in our meetings. Rather, we are happy that our exchange of views has stimulated each of us more clearly to formulate his or her understanding of their own views (now automatically modified by that exchange)'.

Pioneering New Courses.

Convenor: Mohammed El Tom (Sudan)

Some specific examples of 'pioneering' were first considered, such as the introduction, against conventional examination constraints, of a problem solving course to second-year students at Birmingham University. The group then concentrated on the theme of how to produce change in academic institutions.

It was agreed that academic institutions often develop like biological organisms, essentially governed by environmental factors, together with internal mechanisms and inherited traditions. Adequate understanding of their developmental processes should help enhance our ability to manage change more successfully. Such understanding requires an investigation of a variety of experiences in order to unravel the factors responsible for inducing an awareness of the need for change, the types of change that have occurred, and the forces at work that inhibit or promote the process of change.

Types of change. Changes in undergraduate curricula may usefully be grouped into two major types:

Type A: Large-scale change involving the introduction of a new undergraduate program. Such large scale change is relatively rare and appears to be the outcome of rather special circumstances. It may involve a major change either in the philosophy underlying previous programs of study or in institutional structure policy itself.

Type B: Small-scale change directed towards a particular course, or unit, or, at most, a specific section of the undergraduate program such as statistics. An example of such change might involve complete or partial revision of an individual existing course, or the introduction of a new course, or mode of assessment or style of presentation of an individual course. This type of change leaves more or less intact the basic orientation of the existing program.

Forces at work. Changes of types A or B arise from the interplay of several forces, the nature of which determines the type of change. For example, a type-A change is more likely to occur after the establishment of a new institution (perhaps following a large expansion in student population), or the marriage/divorce of two or more interest groups responsible for some aspect of mathematics teaching. A type B change may come about as a result of any one, or a combination of several, of the following forces (which would be unlikely to lead to a type A change):

- the appearance of a new textbook (such as Birkhoff and Maclane's *Survey of Modern Algebra*);
- dissatisfaction of an individual as a result of teaching a course;
- dissatisfaction of one or more colleagues with an apparently stagnant institution, or caused by their research activity;
- advent of new disciplines (e.g., impact on traditional assessment procedures of statistics and computer science);
- realisation that other disciplines are attracting potential mathematics students in significant numbers (computer science for example);
- persistently poor performance of students in a given course;
- changes in the level of knowledge of newly admitted students;
- demands of other institutions serviced by one's own institution (the task (say) of designing a course to suit students of architecture);
- demands of societal institutions (such as industry);
- advent of technological aids;
- deliberate appointment of new staff to create change.

Constraints and Mechanisms. Constraints usually inhibit change, while mechanisms exist for effecting change. As one might expect, one major constraint is of an institutional nature, namely the lack of adequate and timely understanding by colleagues and insufficient administrative support. Two examples where change is inhibited by constraints are the advocacy of reorienting the curriculum in view of developments in informatics, and the creation of a post in an emerging discipline, each a type B change. Examples of other significant constraints (but not pertaining to the institution) are lack of appropriate support material, student pressure (perhaps arising out of insecurity regarding their exam performance, especially when the new course is unfamiliar), and large class size. To effect desirable change requires careful application of management techniques. For example, one must secure the co-operation of one's colleagues and the support of one's institution. When a new course is introduced, it is also important to convince students that the assessment procedure will make allowance for the fact that the course is new.

Some questions. Two particular questions seem worthy of serious reflection:

- How can we implement transition from the awareness of a need for change to relevant practical activity?
- How can a mathematics department or faculty ensure 'rolling reform' by having some permanent mechanisms for creating awareness of desirable

change and introducing and evaluating such change? Are there relevant insights from sociology?

The art of applying mathematics.

Convenor: Ron McLone (UK)

The aim of the group was to make proposals (that is, to set down general principles and guidelines) for the undergraduate curriculum whose principal objective is to instill in students the skills specifically designed to foster the art of applying mathematics to real problems.

Preliminary material was presented by the convenor and Dilip Sinha (India). The latter stressed the need to motivate traditional mathematics, and to provide time for project work within the curriculum. A possible course structure was proposed. The convenor suggested that it is the acquisition of specific skills relevant to the applying of mathematics which is more important than the detailed content of any particular course. The following list of abilities, considered important for a practising mathematician, was suggested as a basis for the discussion:

- ability to identify situations and to formulate problems which are suitable for mathematical treatment;
- ability to handle and make sense of natural or experimental data, especially the representation and interpretation of data;
- ability to determine the variables and parameters with which to describe observation. Understanding the nature of these variables, (as random or deterministic, logical or numerical, discrete or continuous);
- ability to select significant variables;
- ability to translate information (such as data, verbal description) into pictorial form (such as a graph, diagram, flow chart);
- ability to recognise patterns in data and in processes;
- ability to generate mathematical expressions to summarise observations whether directly or through the paths suggested above;
- ability to set up a model representing the system and relating its significant variables. Use of previous experience of a variety of models to select good models;
- technical ability to manipulate the mathematical expressions of the model to achieve desired objectives;
- ability to consult books, journals, and other resources for additional mathematical techniques or extra non-mathematical information as required;
- understanding of when to change a model, method or objective in discussing a problem. Recognition that numerical or computing techniques may be essential at some stage;
- ability to recognise what constitutes a solution and how this depends on the problem being considered and the objectives. Evaluation of the success of models (for example, examining the significance of differences between solution and observation) and making decisions about further refinements;
- ability to work effectively in a group;
- ability to communicate clearly, especially in writing;
- ability to co-operate with non-mathematicians at all stages of model building, including interpretation of final results.

Possible Strategies. What can be done to develop these abilities in undergraduates? The majority of the students, especially in the early years, cannot handle situations involving several of these abilities. One strategy is to present situations involving one or two of them. Indeed, experience suggests that many students find great difficulty in coping with the first item on the list alone. Successive development will lead to problems for which a larger number of these abilities are necessary. Some possible strategies are:

- present illustrations of formulated models (as motivation);
- introduce necessary mathematics (to give confidence);
- 'mathematise' situations lying within the experience of students (for first year students only the elementary mathematics of their secondary schooling should be involved);
- confront students with problems illustrating the inadequacy of their existing 'problem solving' schema;
- confront students with problems showing inadequacy of the mathematics they already know;
- insist that students communicate, defend and validate their own models and solutions;
- use small group work with criticism and communication (related to above);
- introduce project-oriented courses (including industrial or commercial experience) or a detailed case study;
- provide lectures by specialists in the field of a project;
- provide lectures from industrial speakers to demonstrate modelling carried out in a practical environment.

Generally the progression would be from the simple and accessible to the more complex and complete 'project-type' exercises in the final year. Thus: (Convenor's introduction) 'the aim of a first year applications course could be to demonstrate all the thirteen abilities listed above, and to provide examples which enable students to cultivate most of them in turn; in subsequent years the students might, with greater maturity both in mathematical technique and critical judgement, be encouraged to tackle work of a project or modelling workshop nature which allows them to develop these skills into the fully professional approach expected of the mathematics graduate'. This was endorsed by the whole group.

Where should the necessary time come from? From most units offered for a mathematics degree some material can be dropped without seriously decreasing the quality. This assertion was based on the grounds that the students would acquire other skills, equally demanding and much needed, and that many graduates have shown capability in furthering their mathematical knowledge, when it is necessary to do so, in employment. (A better solution may be to take more time for the program. After all, a better mathematician is the goal.)

Assessment. It needs to be emphasised that most of the abilities listed above are not assessable by conventional, timed, written examinations. A variety of assessment modes is required, with the appropriate mode dependent on the skill being developed and assessed. We cannot lay down rules, but there is clear evidence from existing courses (such as those at the Technological University Eindhoven in the Netherlands and Southampton University in England) that project work will figure significantly in the assessment. Besides traditional modes such as course work, seminar presentations, individual oral interviews, a critical review of work carried out by others, further possibilities include 'practical classes', similar to those in physics, chemistry or engineering, with associated writing of a report, and the involvement of industrial managers for projects in industry. There is a need for those involved in modelling courses to write up their experiences, particularly in relation to assessment methods, and for a bank of these reports to be created. (The Convenor volunteered to consider the setting up of such a bank.)

Attitudes. The pursuit of these suggested strategies will require changes in traditional attitudes on the part of teachers (and also students; but discussion of these is here omitted). University mathematics teachers need to rethink their attitudes both to mathematics and to undergraduate education. Their own philosophy of mathematics is based on their own education and continued work in a closed academic environment. Much of the present curriculum is dominated by a proof-oriented approach to mathematics. A more judicious mix of this approach and algorithm-oriented approaches is required. Mathematics education should contain both content-based and activity-based material. Indeed, what we seek is a restoration of our age-old balance of application and theory in the curriculum, which has been distorted over the past century. However, colleagues should be reassured that the resulting education will still produce graduates capable of employing the most sophisticated mathematics when needed.

Computers and Tertiary Mathematics.

Convenor: John Hunter (U.K.)

At the first meeting of the group papers were presented by Jim Richardson (Australia) and Lim Chong Kiang (Malaysia). The ICMI Discussion Document, 'The Influence of Computers and Informatics on Mathematics and its Teaching' [Ens. Math. 30 (1983) 161-172] and a paper by Sim Hitotumatu (Japan) were distributed. The group split into three subgroups dealing with Modelling and Symbolic Manipulation; The Effect on Teaching and Learning; and Management of Computer Resources for Mathematics Education. The following presents the main points discussed, views expressed, agreed beliefs and some of the flavour of the debates.

Modelling and Symbolic Manipulation. There is a great need for development of software for teaching purposes. Great care must be employed in using 'black boxes' for introductory teaching. The use of computers for presenting material which is difficult or impossible to present by other methods should be investigated; for example, graphic displays for non-linear problems. New kinds of thinking may develop in students who have had early contact with LOGO or symbolic manipulation software. The dangers of 'naive' use of packages point to a need to develop estimation techniques even for symbolic manipulation, analogous to those for numerical work. The experimental and discovery approach to model building can be made more attractive and realistic by use of computers since real data can be used.

Effect on Teaching and Learning. There is a growing availability of easily used packages for graphics displays of three-dimensional relations and these will have a major effect on the development of the geometric intuition of students. For classroom purposes a pre-recorded video cassette or computer display provides an alternative to 'live' display (this avoids bugs). Without changing syllabus content, one can use a micro with a large screen in the classroom for demonstration purposes, for example, graphics in calculus, transformation geometry, convergence of sequences. Available packages were discussed. Some of these will lead to changes in the mathematics syllabus (perhaps less drill in calculus). A shift from traditional theory to problem solving was expected. Teachers should provide some understanding of the processes being used in a package, and avoid the 'black box' approach. There is a need for more investigation into Computer Assisted Learning (CAL) presentations of activities such as proving results, solving problems with several solutions, dealing with abstract work.

Management of Computer Resources for Mathematics Education
Hardware
It is usually preferable to decide what sort of software will be used and then consider what hardware is capable of running it. Students seem to have less fear of micros than other computers. At present, the major choice is between terminals connected to a time-sharing mini or mainframe computer and individual micros, possibly linked by a network or file-server.

Some drawbacks associated with micros are:
- micros may not have enough power for a given problem and may be too slow;
- physical security of floppy disks, 'mice' and the micros themselves present problems;
- the logistics of distribution of course material is more difficult with individual micros. Possible solutions include a library-like loan system for disks, purchase of disks by students, a whole system of material on one floppy which is sealed into the machine, hard disks, shared file-servers, or local networks;
- keeping track of student progress is more difficult on micros;
- many of the existing symbolic manipulation packages are not suitable for use on existing micros.

Some drawbacks of time-sharing terminals are:
- response time on a heavily loaded system can be poor. This can destroy student enthusiasm. Some form of rationing of connect time may be necessary. More computing power is preferable;
- graphics seem at present more expensive and less convenient to deliver through mainframes than through micros. Although the resolution of microcomputer graphics is poorer, it is adequate for most teaching purposes;
- there is some evidence that the friendliness of the user interface on time-sharing systems is lagging behind that provided by micros, for example, for editing the command line, windows, mice and so on;
- if the computer crashes, all work ceases;
- problems can arise when the computer is not close to the terminal laboratory, as with modems, landlines and networks;
- a central computer may be partially or completely outside the political and/or technical control of the mathematics department using it.

Since any chosen hardware is likely to be video, ways of providing students with printouts of their work, should this be helpful, will have to be considered. For example, a session on a video terminal may be recorded in a file which can optionally be printed at the end.

To allow for mechanical failure of hardware, deadlines for any compulsory assignments on a computer should be flexible.

Laboratory Organisation

If enough staff are available, running scheduled supervised sessions in the laboratory is an option. This is certainly desirable when students first use the laboratory. Access to a laboratory on a 'cafeteria' system can be successful; if demand is great, a booking system may be needed.

If line printers are in the laboratory and not under operator supervision, rules of procedure to ensure a tidy distribution of printed output will have to be devised.

Feedback from students should be encouraged; comments, corrections, and other responses should be entered on the machine or in a suggestions book. Comments should be read, acted upon if appropriate, and answered.

Physical security of the laboratory has to be considered very seriously and suitable procedures adopted.

Development and Maintenance of Computer Teaching Materials. We consider this a most important item and make the following recommendations:
- start by purchasing outside material, and when developing one's own, avoid 'reinventing the wheel';
- start with a small senior class. Try materials for large lower-level classes on a pilot group first;
- plan a new system in the abstract before it is put onto the computer. Test planned material with students;
- update computer teaching materials regularly;
- make student use of the system easy, friendly and informed.

Personnel. Several problems concerning the necessary support staff were discussed. We mention:
- programming and technical assistance (possibly retraining existing staff);
- co-operation of the subject author (working in a suitable author language) and the person mounting the materials;
- familiarity of tutors and supervisors with the system;
- administration.

Politics and Finance. Here the following problems should be considered:
- low budgets of mathematics departments;
- maintenance costs;
- the new kind of staff, such as technical and laboratory assistants (salaries);
- representation on the controlling committee of a central service system.

A Recommendation. The splinter group strongly urges ICMI to consider the problem of trying to obtain some degree of standardisation in the use of computers in tertiary mathematics (languages, packages, software exchange).

The secondary/tertiary interface

Convenor: Milton Fuller (Australia)

Two short talks provided the catalyst for the discussion. Helen Whippy (PNG) described the severe problems of transition of students in Papua New Guinea and the approaches used at Unitech in Lae. Francine Cnop-Grandsard (Belgium) described individualised computer-managed instruction systems for remedial courses for first year students at two levels in Brussels. Other members provided information on work on the secondary/tertiary interface in other countries. The Convenor is preparing a register of members and their work in this area; this work formed the basis of the discussion. The group considered ways of helping the transition to tertiary studies both at secondary level and within normal tertiary studies, and ways of helping students with special problems. There was concensus on the following points:

Liaison. Constructive liaison between tertiary institutions, secondary teachers and career advisers is essential. In some places, good liaison is hampered by strains arising from conflict over changes at late secondary level. Some suggestions for helping good liaison include: co-operation of both secondary and tertiary teachers in promoting mathematics among school students (via clubs, seminars on tertiary studies, competitions and journals); greater involvement of tertiary teachers in local mathematical associations; study leave or other opportunities for secondary teachers to up-date their knowledge of the use of mathematics in the real world and in tertiary studies.

Secondary level. Good liaison is particularly needed to provide information and guidance for secondary students on the mathematical content of various tertiary courses, especially on the mathematical demands of courses which may appear to be non-mathematical, such as business studies. In secondary mathematics teaching, some topics should be treated to greater depth. There should be greater emphasis on developing the ability to 'think mathematically' and to solve problems by combining and applying a variety of ideas, not necessarily from the same part of the course. Students also need help in developing independent learning skills before they attempt tertiary studies.

Tertiary level. The transition to tertiary mathematical studies poses serious difficulties to large numbers of students, partly because of recent easing of entrance requirements. Suggested ways of helping to smooth this transition include the following:

- entrance requirements for particular courses should be correctly related to the content of the course;
- early in the first year, there should be a workshop run in conjunction with the first year mathematics course and covering study skills specific to learning mathematics in the particular institution;
- there should be some form of regular clinic or consultation to help individuals and small groups cope with first year mathematical studies. This must be provided at the times when the students are available and need help. The helpers can be fellow-students, tutors or more senior teaching staff. The emphasis should be on getting students to work on their individual area of difficulty, and feed-back can be given to class instructors;
- the total work load of first year students, particularly those with special problems, should be carefully monitored.

Remedial help for special problems. Tertiary institutions should provide well organised remedial help in mathematics, perhaps through special learning/ remedial units. Remedial programs should be preceded by well researched diagnostic tests to identify areas of weakness. Help should be available immediately on the basis of test results. Such programs should help to overcome specific background deficiencies and also develop confidence in learning mathematics. They should be available both prior to, and concurrently with, regular first year tertiary courses, and, in general, no credit should be given for them. (There should then be little danger that provision of remedial courses would discourage adequate secondary preparation.) Remedial programs should be carefully planned to avoid merely re-hashing schoolwork.

At least in the initial contact, some novelty of approach is vital.

Conclusion. The problems considered are world-wide and must be seriously faced by the mathematical community. At ICME 6 special attention should be devoted to this topic.

III. Computers and Tertiary Mathematics

Presenters: Anthony Ralston (USA): *The Importance of Discrete Mathematics in the Tertiary Curriculum*; and John Crossley (Australia): *The Problem of Displacing Old Mathematics to make way for Mathematics for Computing*

This session provided a discussion of the impact of computer science on mathematics, and both presentations would have fitted equally well into a general discussion of the mathematics curriculum. The main theme in both presentations was that in undergraduate mathematics a shift is necessary from 'calculus only' to 'calculus and discrete mathematics'. To introduce present-day students to mathematics almost entirely through calculus is to deprive them of some perspective of what mathematics is about. The historical reason for the amount of attention paid to calculus is that, since the first industrial revolution, the majority of applications of mathematics have depended on calculus since it is concerned with answering questions about the natural world (and, in fact, has answered many of them). This is no longer true since the computer revolution because the questions of applied mathematics now often concern computing, information, economics and management, and so on. Certainly the calculus courses offered today (often heavily emphasising mechanics) are not in accord with this trend, and they definitely do not serve the needs of students in computing science.

So, the first recommendation was to teach less calculus and to compress parts of it, using such developing symbolic mathematical systems as MUMATH, MACSYMA. These can replace a lot of drill and practice in skills which are no longer useful. Furthermore, calculus can be made more algorithmic. Indeed, algorithms should play a central role in much of mathematics. A shift from proof-oriented and existential mathematics to constructive mathematics could gradually take place by making proofs themselves more like programs.

Both speakers stressed the importance of a course in discrete mathematics in the first two years, equivalent in weight to the calculus course (eventually the two could be integrated). The leading themes of such a course would be algorithms (their construction, analysis, and verification) and mathematical induction. Here, the discovery aspect of mathematics should be emphasised, stating problems and letting the students compute small cases, make conjectures, and then prove them. Other topics suggested for a course included recursion, difference equations, linear programming, discrete probability, combinatorics, mathematical logic, and also some of the usual (linear) algebra. The level would be more demanding (being for mathematics majors) than that in the classic books of Kemeny, Snell et al.

Crossley pointed out that whereas in mathematics education most of us have a fixed idea what the basic knowledge is, this is not true for computer science, which is still evolving. Already the emphasis has shifted from numerical analysis to symbol handling, information processing, and so on. Similarly, discrete mathematics is very diverse and has, as yet, no well-defined body of standard knowledge. In the coming years the needs of computer science may keep changing.

Ralston observed that the ideas expressed above were no longer the opinions of only a few mathematicians but were receiving widespread recognition in many countries. Strong opinions in this direction were often ignored until recently, but his recent article on discrete mathematics and calculus (to appear in College Math. Journal in November 1984) attracted quite strong opposing reactions which he considered a sign that discrete mathematics is being taken seriously by the mathematical community.

A more important fact is the program, subsidised by the Sloan Foundation, in which six institutions in the USA are redesigning their curriculum for the first two years in the manner described above. They are Montclair State, Colby, St. Olaf (Colleges) and (Universities) Florida State, Delaware and Denver. It is also significant that several publishers are desperately trying to publish books on discrete mathematics for undergraduates.

Crossley issued a warning not to make the mistake of 'new math' again. We should ensure that teachers teach what they can teach and build on the skills they have. Thus, the change to more discrete mathematics should be an evolution. Nevertheless, both speakers felt that it was desirable to effect a serious change before 1990. One of the subjects which might suffer (once again!) is geometry. On the other hand, topics from discrete mathematics, such as graph theory and orderings, are often geometrical in nature. During the discussion several members of the audience stressed the importance of teaching more geometrical insight at the secondary level and of using geometry for representing data, computer graphics, and so on.

IV. Tertiary Mathematics Around the World

In this simultaneously interpreted session, five short talks presented a sample of some current ideas and activities in tertiary mathematics education around the world.

Ilpo Laine (Finland) described recent trends in the university teaching of mathematics in Finland. The official policy of decentralising university teaching and research has led to the establishment of several new university-type institutions which have had to build the teaching of mathematics up to Ph.D. level using limited resources. The decision to form a reasonable research environment implied some limitation on the range of advanced mathematics that could be taught, and at Joensuu, where Laine himself was involved throughout, the problem was solved by a clear concentration on analysis. Another major trend, particularly in the two older universities, has been the increasing role of applied mathematics. This is illustrated by the growing popularity at Helsinki of two options (applied analysis and stochastics) among the four now available for final studies at masters level.

Lilia Del Riego (Mexico) described a recent mathematics program developed at the Universidad Autonoma Metropolitana–Iztapalapa in Mexico City. The aim was a more applicable program which would optimise resource use and would suit the real needs of mathematics students who would major in various areas, including social and biological sciences as well as physical, engineering and mathematical sciences. The program was developed after considerable experience and extensive consultation with experts in the respective fields. The first half of the program consists of compulsory courses in both pure and applied mathematics. In the second half, students choose one of the following options: traditional (pure) mathematics; computer science (logic, programming, optimisation, simulation); applications in engineering and biological sciences (numerical methods, differential equations, mathematical models, stochastic processes); economic-administrative option (statistics, simulation, modelling in economics). All students take a minimum of four courses in their option after consulting the coordinator.

Lucilla Bolletto-Cannizzaro (Italy) described some work on the training of mathematics educators in Italy. She and two colleagues have been involved for three years in a didactics option taken in the third year by prospective mathematics educators who then complete a dissertation in their (final) fourth year. The aims include introducing students to the use of personal computers for mathematics education purposes, involving them in open, unstructured situations involving both mathematics and mathematics education at school level, and encouraging some of them to pursue further research in the area. Although the students produced some excellent work (including implementation using computer graphics), there were some serious problems due partly to the heavy work-load by comparison with other more traditional options, and partly to the lack of any experience of open-ended study and investigation in the first two years of the students' studies.

Sim Hitotumatu (Japan) spoke on computer education in university mathematics departments in Japan. In 1983, there was a detailed survey of the forty national and (major) private universities and a symposium on this topic was organised by the Mathematical Society of Japan. The major concern was not the (very considerable) problems of teaching computer science as such, but rather how to use computers in mathematical research and in teaching college level mathematics, how to overcome the prejudice and ignorance of computers among mathematicians, and, perhaps above all, how to get mathematicians to really 'play' with computers. As in other countries, shortage of experts and funding problems add to the difficulties. The speaker gave some examples and suggestions, and reported that, despite the difficulties, far more progress than expected has already been made in mathematics departments in Japan.

Zbigniew Semadeni (Poland) drew on his experience in three different university systems (in Poland, North America and Australia) to pose some questions facing tertiary mathematics today. The fact that reputable institutions answer these questions in opposite ways should encourage us to examine assumptions about teaching which we often take for granted. One should, for example, inquire about when and how students are subdivided (mathematics majors and others, pass and honours); methods of assessment; teaching arrangements (optimal use of staff with large or small groups); use or not of prescriptive syllabuses and of textbooks; length of study-units; methods of teaching (particularly whether or not the structure should be 'logical', building up with minimal repetition, or 'spiral', returning to the same topic at a deeper level, both for a single course and for the total course for mathematics majors).

Discussion: The discussion following the five talks concentrated on the value of spiral teaching, and on ways of teaching students to work independently. There is clearly a need for serious research on such problems of mathematics education at tertiary level.

References for II (b)

Davis, P., Hersh, R. (1982) *The Mathematical Experience*, London, Penguin-Harvester.

Lakatos, I. (1975) *Proofs and Refutations*, Cambridge, C.U.P.

Mason, J. and Pimm, D. (1984) 'Generic Examples: Seeing the General in the Particular', *Ed.Studies in Maths*, 15 (pp. 277-289).

Hana, G. (1983) *Rigorous proof in mathematics education*, Toronto, Canada O.I.S.E. Press.

Two-Year College Mathematics Journal (1981). A complete issue devoted to 'proof', 12 (2).

Michener, E., (1978) 'Understanding Understanding Mathematics', *Cognitive Science* 1.

Solow, D. (1981) *How to Read and Do Proofs*, Wiley.

Grabiner, J. (1974) 'Is Mathematical Truth Time-Dependent', *Amer.Math.Monthly*, 81 (pp. 354-365).

Polya, G., (1973) *How to solve it*, Princeton Uni. Press.

Leapfrogs, *Learning and Doing Mathematics*, Hutchinson.

Leapfrogs, *Complex Numbers*, Hutchinson.

Toeplitz, O. (1949) *The Calculus: a Genetic Approach*, Grundlehren no. 56, Springer.

Courant, R. and John, F. (1970) 'Calculus', *Interscience*.

Page, W. (1984) 'Knowledge acquisition and transmission: Cognitive and Affective Considerations' in *New Directions in 2-year College Mathematics*, Springer 1984.

ACTION GROUP 6: PRE-SERVICE TEACHER EDUCATION

Chief Organisers: Willibald Dorfler (Austria), Claude Gaulin (Canada),
Hilary Shuard (UK), Graham Jones (Australia)

For the first of the four sessions devoted to Action Groups, an opening presentation was commented on by two reactors. For the second and third sessions, some 150 participants divided into eleven working groups, some of which focussed on presentations while the remaining four were discussion groups. In the last session, reports were given by the leaders of each group. The report that follows provides a summary of the presentations made in the first session, together with summaries based on the reports prepared by the leaders of the working groups.

For the first session Geoffrey Howson (UK) highlighted the work carried out by the international group BACOMET (Basic Components of Mathematics Education for Teachers) whose work began in 1980 and will result in the production of a book addressing major components identified in the project.

Although the BACOMET group recognised that there are a variety of contexts and traditions in teacher education, Howson reported that it was possible, through a process of extensive international collaboration, to identify certain basic components of the didactics of mathematics which should be given high priority in teacher education. According to their criteria, a basic component must be 'fundamental' (playing a decisive role in the way mathematics teachers function), 'elementary' (in the sense that it must be of immediate interest to intending teachers) and 'exemplary' (illuminating important didactical or practical functions of the teacher).

Howson outlined the basic components proposed by BACOMET which will be amplified in the forthcoming publication as, social norms and external evaluation, mathematics as a school subject, teachers' cognitive abilities, texts and their use, observation of students, tasks and activities, classroom organisation and dynamics.

He also identified the following pervasive and continuing issues for teacher education:
- the often contradictory demands of short-term and long-term perspectives in the training process;
- the conflict with respect to the provision and integration of theoretical and practical knowledge;
- the tension created in teacher education and in teaching resulting from societal expectations and norms in mathematics education;
- the knowledge and view of mathematics that teacher educators want students to acquire;

- the role of tasks, materials and textual materials in mathematics teaching;
- the development of a balanced approach in teacher education to the provision of 'domain specific information' and 'overall orientation'.

The first reactor to Howson's address was Bent Christiansen (Denmark) who stressed that teacher education must provide the teacher with the 'know-how' to establish classroom organisation, based on routines and practical knowledge. However, in order to avoid 'freezing', the training must not focus solely on one methodology. Hence it is important during training to emphasise theoretical knowledge about teaching and learning processes, to generate an orientation towards long-term planning and evolutionary development. He added that mathematics education in schools is largely organised through tasks. Consequently, tasks and the pupils' activities in performing them are of central interest for didactical research and teacher education. Christiansen briefly described Leont'ev's theoretical conception of activity, but noted that a didactical theory has yet to be produced which takes into account the total system, incorporating tasks, pupils, teachers and teacher educators.

The second reactor, Graham Jones (Australia), addressed the issue of social norms and stressed the need for pre-service teachers to understand the expectations of society and pupils towards mathematics and mathematics education. The demanding task for teacher educators is to inculcate in student teachers powerful inner convictions that contemporary methods and new approaches to curriculum will be more successful and more rewarding for children than those often espoused within the community and given acquiescence in the schools.

With respect to the knowledge and view of mathematics needed by pre-service teachers, Jones argued that it was not sufficient to increase the number of mathematics courses required during training. There must be an over-riding emphasis on 'relational' mathematics (Skemp) in teacher education programs if the predisposition towards 'instrumental' mathematics in schools is to be reversed.

Jones also noted that pupils are acutely conscious of their mathematics teachers as persons rather than simply as purveyors of mathematical knowledge, and research findings suggest that mathematics teachers are not rated highly for a personalised approach. He advocated a 'humanising' orientation to mathematics teaching as an additional basic component in teacher education.

Pre-Service Education of Primary and Middle Grade Teachers

Convenors: (1) Erich Wittmann (FRG), Renee Berrill (UK)
            (2) Fred Goffree (Netherlands), Rolf Schwarzenberger (UK)

Two groups worked on this topic. The first group based its work on a series of presentations while the other operated as a discussion group. The following major theme dominated presentations and discussion in both groups. All participants wished to move, in their own countries, towards a structure in which primary student teachers would follow integrated courses combining the study of mathematics, didactics and psychology, and which would include work with children as individuals and in small groups. Fred Goffree (Netherlands) presented an example of such an integrated course, in which student teachers were able to do mathematical problem solving at their own level within a didactical framework, to reflect on and analyse their own and children's learning processes, and finally to carry out similar work with children.

Erich Wittmann (FRG) gave examples of 'teaching units' which provide student teachers with opportunities to do mathematics, study learning processes and observe how they themselves learn; to create different forms of social interaction; and to plan and carry out teaching. The need for practical work was a major theme of his approach, as it was in the presentation by Chich Thornton (Australia) who stressed that student teachers themselves needed to handle manipulatives when studying number, and to analyse what children, say, do, and write when using these materials. Renee Berrill (UK) illustrated how low-cost manipulative materials for practical mathematics could be constructed by student teachers in developing countries from environmental materials, so that those teachers whose schools were not fully equipped would not be barred from a practical approach to primary mathematics.

Kiyoshi Yokochi, Yoshimichi Kanemoto, and Hirokazu Okamori (Japan) suggested how primary mathematics education, and teacher education, should be re-structured for present-day needs. Real problems and problem solving should form a basis for the curriculum, and primary teachers need to be able to construct problems which are both suitable for and of interest to children.

Another recurrent theme was that of attitudes to mathematics, and the problem of anxiety among primary student teachers. Lea Spiro (Israel) described the vicious circle of student dislike and fear of mathematics in their own childhood, a dislike that was often strong enough to prevent real changes of attitude at college level, so that teachers returned to the classroom to pass on a dislike of mathematics. Many discussants placed high priority on attitude changes, and believed that integrated courses such as those described above could lead to more positive attitudes. Ronald Welsh (Australia) is investigating the attitudes of teachers to mathematics and science. Further work is needed in this area.

The group chaired by Goffree explored means of providing unifying threads which might run through integrated courses, and thus produce truly unified programs. Possible approaches included an historical approach, merging context and pedagogy, and showing the individual and collective development of man's mathematical thinking; a developmental approach, proceeding through 'student as learner', 'child as learner', 'student as teacher', to finally focus on 'teacher as educator'; a chronological approach in which the course moved through the mathematics of the lower primary years to the upper primary years.

Alternatively, courses could be based on a particular focus, such as:
- problem solving or investigations related to children's work;
- work with individuals and groups of children emphasising knowledge of children's mathematical strategies;
- use of children's solutions, errors, and reactions to problems, leading to the study of children's thinking, and to mathematics at the students' own level;
- interweaving of workshops with lecture sessions;
- a planned inter-relationship between teaching practice and the college program, involving mathematics educators in the supervision of teaching

practice.

Discussion throughout was vigorous, but was hampered by lack of time to explore fully the major issue of integrating the components needed in a unified course.

Pre-Service Education of Secondary Mathematics Teachers

Convenors: (1) Keith Selkirk (UK), Peter Bender (FRG), Fritz Schweiger (Austria)

(2) Joop van Dormolen (Netherlands), Kevan Swinson (Australia)

Two groups were concerned with this topic. The first concentrated on the mathematics component of initial training, and included three presentations, while the second was a discussion group on the mathematics education component. In both groups, consideration of the topic actually turned on the preparation of prospective mathematics teachers for practical teaching in secondary schools.

Mathematics itself was seen as an important component of teacher preparation. In planning mathematics courses it is important that the depth and abstractness should be such as to give genuine insight into school mathematics. In addition to knowledge of mathematics, student teachers also need knowledge about mathematics, such as the nature of mathematical activity, the methods and historical development of mathematics, and applications of mathematics. Fritz Schweiger (Austria), in his presentation, stressed the role of fundamental ideas of mathematics (in the sense used by Bruner), the pervasiveness of these ideas, and the possibility of organising the mathematics curriculum around them. Peter Bender (FRG) made explicit, for all members of society, another aspect of knowledge about mathematics, namely its importance which is derived from the intimate relationship between social and physical structures and mathematical structures. Prospective mathematics teachers need a strong conviction about the necessity for all pupils to learn some mathematics. This conviction is gained both from working for a time outside the educational system, and from observing and analysing the work of individual pupils and classes. If students are to benefit, these experiences need to be accompanied by courses which encourage reflection upon experience. Rational reflection and theoretical modelling of personal experience appears to be a basic method of developing the professional knowledge of teachers in a way which makes that knowledge available for effective classroom use.

In the 'consecutive' model of teacher education, the mathematics component is completed before the professional education component begins. The group supported the alternative concurrent integrated model by reference to the professional activity of teachers in school. This activity is always holistic, demanding integration between the components. Student teachers should not be expected to integrate separate components into a whole without some support. In his presentation, Keith Selkirk (UK) illustrated how this integration can be enhanced by appropriate textual materials, even when the course follows the consecutive model. His materials consist of tasks to be carried out during teaching practice. They demand the gradual merging of mathematical and didactical knowledge, and are intended to increase the autonomy and independence of the student teacher, reduce his anxiety during teaching practice, and increase the mutual trust between teacher-educators and students and between teachers and pupils. The group strongly supported the view that teacher education should be seen as a joint venture, in which teacher educators, student teachers, teachers and pupils are all active as full participants, rather than some being seen as passive recipients.

Although the discussion in both groups ranged widely, some central inter-related and independent issues emerged. Observation (using a range of methods) of individual pupils and classes at work was seen as an important way of developing students' sensitivity to psychological and social processes, and of providing motivation for the study of educational and didactical theory. Observation techniques should not concentrate only on deficiencies, but encourage students to see positive aspects. Students also need to observe their own behaviour when interacting with others. If observation is to be well focussed, it needs to be based upon some conceptual framework which concentrates the students' attentions. Observation was also considered as an important tool for encouraging mental flexibility and a willingness to change opinions and attitudes.

A major goal of teacher education is to reduce the influence on the student teachers of the way they themselves were taught, of the available textbooks, and of other pressures of the school system on classroom teaching, so that they can develop a self-confident, but self-critical, expertise as a teacher of mathematics. A more confident professionalism and the exhibition of a body of professional knowledge exclusively held by teachers of mathematics, might go some way towards improving the social status of teaching, and encourage more able students to become teachers.

Use of Observation and Assessment

Convenor: Klaus Hasemann (FRG)

In pre-service teacher education, greater emphasis is now being placed on the preparation of student teachers to observe children in order to understand their mathematical behaviour. Observation procedures are also used to assist student teachers in evaluating their own teaching behaviour. Four presentations were made in this group, all reflecting developments in observation techniques.

Error analysis was the starting point for Klaus Hasemann's (FRG) presentation on the analysis and description of children's mathematical errors in terms of cognitive models. Taking illustrations gleaned from clinical interviews, it was proposed that even if different children are dealing with the same examples, following the same teaching, their mathematical thinking is often different. These differences seem to result from differences in the children's internal representation of their knowledge. As well as introducing student teachers to children's mathematical behaviour, observation also provided a basis for cognitive explanations in terms of the theories of van Hiele, Skemp and Vergnaud. The models developed by Minsky and Davis serve best to provide cognitive explanations of children's observed behaviour.

Thomas Cooper, Michael Redden and Rod Nason (Australia) presented a clinical approach to mathematics teaching which has been used in pre-service teacher education in the United States and Australia. The program involves observation of a child solving problems, comparison of the child's solutions with the student teacher's, teaching a concept to a small group [following Wilson's five stage activity cycle (1976)] and one-to-one follow-up with a child who is experiencing difficulties. Based on similar work carried out in Papua–New Guinea, Philip Clarkson (PNG) suggested that student teachers should hear pupils talk. He proposed three kinds of activities for student teachers to meet this goal: teaching a small group of pupils, conducting one-to-one interviews with pupils using the error categories of Newman (1977), and preparing a type of orientation module incorporating measurement and spatial activities. These activities can also lead to further work on diagnosis and remediation.

John McQualter (Australia) described an observation and assessment technique based on Personal Construct Theory, which has been used to analyse the beliefs of student teachers in relation to mathematics teachers at large. The procedures can also be used to illustrate the variety of decisions to be made when teaching, to guide the student in the change from student to teacher and to facilitate the counselling process between supervisor and student teacher.

In the discussion that followed, two major conclusions were reached by the group. The first was an acknowledgement of the central importance of diagnosis of children's level of understanding of specific mathematical concepts as part of teacher preparation. The revised notions of frame seem to have potential in explaining the relationships between teachers' conceptions, children's 'concept images' and mathematical concepts. The second was that simplification of the Wilson model would be valuable in order to extend its use in pre-service teacher education and in other mathematics teaching contexts.

Contribution of Research to Pre-Service Teacher Education

Convenors: Nicholas Balacheff (France), Daphne Kerslake (UK)

This working group examined new ways of relating research in mathematics education to pre-service teacher education. Four presentations served as a focus for the work of the group.

Nicholas Balacheff (France) described a pre-service course which aims to produce teachers who will be able to determine their own answers to problems they encounter in the teaching-learning process. The teachers are expected to evolve and adapt their practices without becoming prisoners of one model. This research-oriented course is based on theoretical content including analysis of the relations between teaching and learning in mathematics, and epistemological analysis of the construction of mathematical knowledge by the use of theoretical constructs such as epistemological obstacle, didactical transposition and didactical contract; but it also includes practical activities such as clinical interviews with pupils, and design and observation of classroom activities. The course culminates in the presentation of a report incorporating the student's critical reflections on the teaching and learning of a mathematical concept. In a related West German study outlined by Haussmann, results on children's acquisition of geometrical concepts under different conditions (haptic versus visual experiences) were presented to student teachers. Their task was to interpret the results and generate a teaching plan based upon the results. Haussmann noted that the process involved the lecturer in providing theoretical background to the results, the student teacher in developing ideas for teaching and the classroom teacher in evaluating the applicability of the student teachers' plans. Daphne Kerslake (UK) described research activity incorporated in the final year of some pre-service courses in the United Kingdom. In her program, students undertake a supervised mini research project throughout the year. The project is intended to give students insight into research methods, rather than to produce important results, and must focus on some aspect of teaching, such as differences between girls' and boys' problem solving approaches.

Heinz Steinbring (FRG) described the West German EPAS project in which materials were developed for teachers of lower secondary pupils as part of teacher education. The basis for the construction of these materials centres on the concept of task system as a means of enabling teachers to comprehend the multi-faceted nature of concepts such as proportion and probability. The task system approach is also used as a curriculum framework for developing mathematical concepts in the teaching program.

In discussion, the following reasons were proposed for making student teachers more informed and aware of research in mathematics education:

• teaching is more complex in all its dimensions (mathematical, psychological ...) than teachers usually perceive;
• theoretical and empirical results shed light on this complexity;
• research findings point to areas where change is needed and provide means to bring about new developments;
• scientific conceptualisation helps teachers to achieve a balance between long term planning and short term survival measures;
• familiarity with research enables teachers to perceive and to formulate didactical problems.

To attain these goals, it is important that student teachers are involved in research activity. However, they are not being educated as researchers. On the other hand, teacher educators need to be actively engaged in research if they are to use it effectively in pre-service teacher education.

Research on Teacher Education and Evaluation of Pre-Service Teacher Education

Convenors: Alfred Vermandel (Belgium), Steven Nisbet (Australia)

In this working group, sessions focussed on research associated with didactic models, attitude changes, and aspects of the evaluation of pre-service teacher education.

Jan Maarschalk (RSA) proposed a didactic model of heurostentics as a means of enabling pre-service teachers to promote creativity in pupil learning. According to the model, teachers usually blend two extreme modes of teaching: the ostensive (expository) and the heuristic (discovery) and activate a search design (SD) (a way of thinking to acquire knowledge) which is dependent on the mode of teaching. It is hypothesised that a wider SD associated with a more heuristic mode enhances creativity while a narrow SD, ostensive mode, dampens creativity. In a microteaching program based on the model, student teachers in South Africa not only showed a significant increase in the use of more heuristic modes but also used a wider SD and consequently increased pupil creativity. Steven Nisbet (Australia) outlined a study in which a mathematics curriculum course for secondary pre-service teachers in Australia was specifically designed to foster more favourable attitudes to mathematics and mathematics teaching. Class sessions included games, puzzles, and hands-on experience with resource materials, focussing on practical ways of teaching mathematics. The results showed significant improvement in attitude towards mathematics (interest, fun, stimulation patterns) and towards aspects of teaching mathematics (looking forward to it, a less mechanical style of teaching).

David Clarke and Dudley Blane (Australia) reported an evaluation of the effectiveness of a mathematics curriculum course for secondary pre-service teachers. The realities of the situation which beginning teachers would enter were identified by surveying graduates after their first two terms of teaching. Additional information was obtained from mathematics coordinators at the graduates' schools and from supervising teachers associated with the teacher education program. The data provided insight into the likely teaching responsibilities and problems of beginning teachers and produced suggestions for improving the present pre-service course. Alfred Vermandel (Belgium) reported on an evaluation of a new pre-service course for upper secondary teachers which has been operating in Belgium. The course program originated from an analysis of the professional tasks of a teacher, and components are structured around these tasks. Central to the research is the assumption that it is possible to predict the degree of effectiveness of future teachers by their adherence during training to either a traditional or a mastery learning instructional paradigm. The new program which emphasised the mastery learning approach was compared with a program oriented to the traditional approach. Results indicated that graduates of the new program have greater confidence in the effectiveness of their teaching, are more inclined to view themselves as good teachers and have a more optimistic view of learners' capabilities.

Discussion following the presentations raised the following points:
- the need for further research on classroom processes focussing on pupils, teachers, mathematical actions and associated interactions;
- the need for improved methodology in the construction of self-appraisal and self-observation techniques for student teachers;
- the value of attitude questionnaires as instruments in teacher education research;
- the degree to which special mathematics courses should be designed for teacher education;
- the need to define effectiveness in teacher education evaluations. It was proposed that the effectiveness of a teacher education program is determined by the extent to which the program's objectives are realised;
- the need, in evaluating effectiveness, to assess the objectives themselves;
- the need to examine congruence between pre-service teacher education goals and the needs of first-year teachers;
- the extent to which diversity in teacher training should be developed to accommodate diversity in the graduates' subsequent teaching assignments;

- the extent to which instructional paradigms are evident in the thought processes and attitudes of future teachers;
- the validity of evaluating the effectiveness of programs on the basis of student teachers' adherence to one or other instructional paradigm.

Use of Computers and Calculators

Convenors: Joan Wilcox, Jack Wrigley (Australia)

Four presentations by Bernard Cornu (France), Klaus Graf (FRG), Larry Hatfield (USA), Tony Ralston (USA), and some shorter contributions, served to stimulate the work of this group. As stressed by all presenters, there is an urgent necessity for preparing all prospective teachers to use computers and calculators to facilitate the learning of mathematics. The first reason is that computers are likely to have a dramatic effect not only on mathematics itself, but also on curricula and teaching strategies in school mathematics. Secondly, in the next few years, many secondary mathematics teachers will probably find themselves teaching courses on computers or computer science. Thirdly, with the advent of computers in schools, the role of the teacher as the provider of information is likely to make way for a new cooperative type of relationship between the teacher, the learners and the computers, which calls for appropriate pre-service education.

As Hatfield says: 'The choice is not between the computer *or* the teacher. The only acceptable, and more promising, paradigm is the computer *and* the teacher whose didactical knowledge permits effective usages of the machine to enhance learning and teaching mathematics.'

Several presenters thought that the preparation of all primary and secondary teachers should include an introduction to computer science (understanding of computers as powerful but limited tools, insight into algorithms and their connection to programming, interactive and recursive procedures, introduction to more than one programming language), and the use of computers in the classroom (various ways of using computers with pupils, types of software, software evaluation, etc.). Some also suggested that analysis of algorithms (design, analysis and verification of algorithms, algorithmic approach to mathematics), and discrete mathematics (for which a list of possible topics was presented by Ralston) should be included in the preparation of secondary mathematics teachers.

The use of calculators was only briefly discussed. There was general agreement that efforts should be intensified to give student teachers an adequate preparation for using simple calculators to stimulate learning. This applies regardless of the fact that the interest of researchers and teachers seems recently to have shifted to computers and that some believe that simple calculators might well be replaced in the future by pocket microcomputers.

Use of Applications

Convenors: Ian Putt, Francis Morgan (Australia)

At all levels of mathematics teaching, there is a growing emphasis on the use of applications and on the process of mathematical modelling. This group examined the implications for pre-service teacher education of such a change. Two presentations served to stimulate discussion. In the first, J.N. Kapur (India) outlined the contents of a module he was recently commissioned to write by UNESCO. The book, which is intended for both pre-service and in-service education, contains some one hundred examples of situations for mathematical modelling, all requiring only pre-calculus mathematics and taken from areas as diverse as geography, physics and chemistry, astronomy, human physiology, business and commerce, epidemics, environment and pollution.

In the second presentation, Ian Lowe and Charles Lovitt (Australia) described material produced by the Reality in Mathematics Education Teacher Development Project (RIME). This material includes detailed lesson plans for using applications in junior high school mathematics. Although it was designed primarily as an in-service vehicle, RIME has been found useful in pre-service education, linking college sessions and teaching practice through the use of a video and a handbook and through the lesson plans.

In discussion the following points were emphasised:
* The time at present allotted to applications and modelling in pre-service teacher education varies considerably between institutions and countries, but it is generally brief. This situation needs urgent improvement, in cooperation with those responsible for the mathematics courses.
* All student teachers should experience a variety of mathematical applications and modelling situations. However, before being able to apply some mathematical ideas, student teachers themselves need confidence and proficiency in using them. Hence, the applications and modelling activities used need to fit the mathematical background of the students involved.
* Student teachers should have opportunities, both in method courses and in teaching practice, to work with and discuss many applications and modelling situations appropriate for the level at which they are going to teach.
* Prospective secondary teachers should develop good understanding of and critical attitudes towards such questions as: 'Why, when and how should applications and modelling experiences be used in teaching mathematics?' What is the distinction between 'doing applications' and 'problem solving' at school level? What is the relation between applications and the process of modelling and mathematising? What characterises a 'realistic' application, as opposed to the artificial psuedo-applications often found in textbooks?
* Initial teacher training can benefit from existing materials on applications and modelling designed for in-service teachers.

Influence of the Cultural, Social and Economic Context

Convenors: Geoffrey Wain, Marilyn Nickson (UK)

This discussion group concentrated on the theme of problems posed for initial teacher education by the developing information and technological culture. Children in many societies are rapidly becoming part of a new culture: schools, on the other hand, often represent a conservative trend in society to which many teachers and parents subscribe. Pupils are now beginning to bring to school informal mathematics which their teachers do not share, or which they do not recognise. This mathematics is drawn from experience of computers, television and other aspects of the new culture. School mathematics needs to make use of this 'hidden mathematics of daily experience', and teachers must recognise the growth of an alternative informal education system which is eroding the monopoly of education formerly held by the school system. The group felt that a coherent partnership between the formal and informal systems is needed if the school, and its mathematics, is to maintain credibility with the majority of children.

In many countries school systems evolved during the change from an agricultural to an industrial society. In the future most people who find work will find it in handling information and in service of their fellows, rather than in manufacturing industry, while some developing countries may greatly compress the industrial phase. However, much of the present content of school mathematics is rooted in the industrial phase of society. Some group members believed that, especially in developing countries, mathematics education should take note of the real needs of society, and seek ways to enable knowledge to serve the community in a direct way. Student teachers and teacher educators therefore need to maintain roots in their communities, and to be aware of the possibilities. In developed societies, it is difficult to foresee what mathematics children should learn in the future, and it is therefore essential that student teachers should not be 'frozen' in inflexible attitudes which will resist change.

The group felt that pre-service teacher education should ensure that students become aware of the changes taking place in society; however, it is not clear how far these issues are specific to mathematics education, or how they are to be incorporated in general teacher education. Student teachers certainly need to become aware of their own perceptions about mathematics, of their attitudes to it, and of how these attitudes interact with the perceptions and views of pupils.

Only a few of the issues involved in a single aspect of this complex theme were identified during the discussions. Other aspects, such as the social setting of the classroom, the influence of cultural, social and economic factors on individual learning, of how far the preparation of student teachers should fit them to act as agents of change, or to teach only the present mathematics curriculum, were barely touched upon. The group therefore recommended that at future meetings, ICME should address in greater detail the theme of the influence of context on mathematics education.

Training of Mathematics Teacher Educators
Convenors: Shirley Hill, Johnny Lott (USA)

Members of this discussion group were drawn from Australasia, the United States and the West Indies. There was wide divergence between their situations and views, and consensus was not always reached. Discussion was illuminated by Kilpatrick's view of the importance of making mathematics education itself a field of study, and thus giving teacher educators some means of reflecting upon their own activity. The group also discussed Christiansen's model of teacher education:

Teacher Educator ↔ Teacher ↔ Pupils ↔ Tasks

and interpreted each arrow to mean that a successful interaction was indicated. It was also agreed that teacher educators need to remain close to pupils, so that as far as possible the school may be the laboratory for mathematics teacher education. Participants therefore felt that an additional arrow should be added to the model to show interaction between teacher educators and pupils.

The group focussed first on the job of the mathematics teacher educator and the ideal preparation for it. The job varies greatly in different countries. It may include content or methods or both, it may or may not be carried out in a school setting, it may include preparation of student teachers for primary or secondary teaching or both. Teacher educators often also have to undertake other tasks such as in-service education, research, or curriculum development. In most circumstances, not only do teacher educators need a suitable background in mathematics, but also awareness of links between mathematics education and areas such as psychology, computer science and language, together with knowledge and experience of evaluation strategies, research methods and problem solving techniques.

It was recognised that the wide-ranging nature of the background needed made it essential for teacher educators to continue to broaden their knowledge, and to keep up-to-date while on the job. Developments in technology and other areas, as well as new trends in mathematics education itself, make this task urgent and difficult. The teacher educator's ongoing self-evaluation must determine directions for his or her own professional development, and this self-evaluation needs to be informed by observation of teachers and pupils at work. Although in the last analysis, professional development is the responsibility of each individual, the group stressed the importance of institutional support for all aspects of professional development, including the availability of time and funding.

Finally, the group examined the question of how far a teacher educator's own teaching could, or should, be a model for students' teaching behaviours. Teacher educators, although they operate in very different conditions from those of school mathematics classrooms, can provide models, both through their own effective use of the variety of teaching methods and strategies which they recommend, and through the style of their interaction with students. Teacher educators need to engage in continuous self-analysis and growth if they are to be effective. Moreover, it is important for teacher educators, as for other teachers, to avoid 'freezing' their teaching styles in the balance of strategies which they have established at a particular time. They should always be ready to reassess that balance in the light of changing circumstances.

Conclusion

Several important issues pervaded the thinking of the entire Action Group and gave pointers for future investigation.

- It was considered vital that the education of a mathematics teacher should be seen as a whole, in which all the parts support one another. This manifested itself in the demand for the development of integrated courses at primary level and concurrent programs at secondary level.
- The inclusion of focussed observation of individual children doing

mathematics was shown to bring varied benefits in teacher education. Moreover, the development of practical theories which will assist teachers to identify cognitive processes would enhance observation activities. A closer relationship of teacher education to research and curriculum development was also advocated and was seen to be compatible with the increased emphasis on observation.

It was accepted that society is changing rapidly and consequently pre-service teacher education should enable teachers to avoid freezing in the teaching process and in their orientation to merging developments in mathematics education, such as computing, modelling, applications, and the hidden mathematics of everyday experience. Thus teachers should adopt a proactive rather than a reactive approach to teaching and learning.

ACTION GROUP 7: MATHEMATICS IN ADULT, TECHNICAL AND VOCATIONAL EDUCATION

Organisers: Rudolf Straesser (FRG), Jeannette Thiering (Australia)

1. The Topic of the Action Group

Prior to ICME 5, mathematics education in technical, vocational and adult education had not been included as a major area for discussion and debate. There were however, reflections and presentations referring to these subjects in the previous Congresses but under different headings. As ICME 5 is the first Congress to tackle these subjects as a whole, an attempt to describe them seems necessary.

1.1 Mathematics in technical and vocational education

Technical and Vocational Education is described by UNESCO as 'the educational process ... (which) involves, in addition to general education, the study of technologies and related sciences and the acquisition of practical skills and knowledge relating to occupations in various sectors of economic and social life'.

Technical and vocational education normally takes place after compulsory education. In different nations the institutional system of technical and vocational education varies from full-time technical colleges (as in France) to isolated and/or part-time activities in private enterprises or colleges (as in England for some levels of qualification). Some nations do not have any institutions which could be named technical or vocational colleges.

UNESCO formulates two major problems with this part of education. First there is a widespread disapproval of technical and vocational education compared to 'general' education. Second there seems to be a problem with the adaptation of the social and economic process of production and distribution of goods and services and the preparation for this process in technical and vocational education. This problem materialises in the shortage of competent teachers and in the lack of adequate curricula and teaching/learning materials.

As technical and vocational education is closely related to 'occupations in various sectors of economic and social life', it seems easy to decide what mathematics should be taught in this part of the educational system: just look for the mathematics used in the various sectors and condense it into a curriculum! This approach was tried and led to interesting results. The mathematics used on the job centres around topics such as basic arithmetic, percentages and proportions. In some vocations, these techniques already seem to be all the mathematics needed.

Unfortunately this approach of a direct transcription of the professional use of mathematics into a curriculum did not work for the reason that the 'real' professional use of mathematics is not as easily discernible as claimed. 'Considerable differences ... were found to exist even within occupations which might be assumed ... to be similar. It is therefore not possible to produce definitive lists of the mathematical topics of which a knowledge will be needed in order to carry out jobs with a particular title' (Cockcroft). Another problem seems to be the identification of the uses of mathematics on the job which partly explains the difficulties of employers, but does not ease the task of the curriculum developer. In addition to these problems, the curriculum developer has to bear in mind that there are 'important differences between the ways in which mathematics is used in employment and the ways in which the same mathematics is often encountered in the classroom' (Cockcroft).

The Cockcroft Report itself takes another approach. In its recommendations it offers, on the one hand, some broad topic areas (calculation, calculators, fractions, algebra, estimation, measurement), while on the other hand, heavy emphasis is put on three more general aspects of the problem:

- 'It is of fundamental importance ... to appreciate the fact that all mathematics which is used at work is related directly to specific and often limited tasks which soon become familiar.'
- ' ... it is important that the mathematical foundation ... should be such as to enable competence in particular applications to develop within a reasonably short time.'
- ' ... it is possible to summarise a very large part of the mathematical needs of employment as a feeling for measurement' (Cockcroft).

The paramount difficulty in the use of mathematics on the job can be seen in the close relationship between mathematics and the vocatiohal situation — sometimes making it impossible to distinguish between these two aspects of a given situation. So the question of 'application' of mathematics (as it is discussed in general education) presents a new facet. Finding the appropriate formula and/or technique to cope with a given vocational situation seems to be the most widespread problem for young employees or apprentices. Most of the mathematics needed in technical and vocational situations is already taught in compulsory education; the problem is the 'application' of mathematics to the professional situation. Technical and vocational education (especially when organised along classroom lines) seems to overlook the need for learning how to *use* mathematics in professional situations and puts too much stress on the topics the students, for the larger part, already know.

Mathematics in adult education

Adult Education comprises all activities where people try to broaden their knowledge and/or skills while they are already full, mature members of a society, that is, responsible for themselves and their future. Such activities are not necessarily related to earning a living. They are undertaken for various reasons as, for example, the possibility of applying for a job, the search for promotion in a job, enrichment of leisure time or philosophical interest. Institutions, goals, learning/teaching strategies, learners, teachers and curricula in adult education are even more heterogeneous than in technical and vocational education — and so are differences between nations all over the world. One major classification of adult education is of college-type activities in classrooms as opposed to activities of 'distance education', that is, 'spatial separation of student and teacher ... use of teaching material in permanently recorded form, ... teacher control and goal directedness' (Pengelly).

As for mathematics in adult education, the current state of the literature is unsatisfactory. This statement from ICME 3 still holds true — with some exceptions.

First a certain classification of adult education activities concerning mathematics seems to emerge: besides the classification along institutional lines (college-type activities vs. distance education), one can distinguish activities aimed at fostering adults' every-day use of mathematics as opposed to courses leading to formal certificates. As for the subjects taught, an approach comparable to the 'direct-transformation-of-needs-into-curricula' manner in technical and vocational education could be tried for adult education. The research study for the Cockcroft Report on adult education seems to follow these lines in trying to identify the 'mathematical needs of adult life'. The report gives a list of needs, namely 'ability to read numbers and to count, to tell the time, to pay for purchases and to give change, to weigh and measure, to understand straightforward timetables and simple graphs and charts, and to carry out any necessary calculations associated with these ... Most important of all is the need to have sufficient confidence to make effective use of whatever mathematical skill and understanding is possessed, whether this be little or much' (Cockcroft).

Besides curricular questions, adult mathematics education often has to cope with the fact that its learner has already failed in mathematics or has the feeling of being a failure. Fear and feeling of failure tend to hinder learning in adult education and render even more difficult adult learning of mathematics, often already difficult because of the isolation of learning in distance education. This widespread feeling can be further illustrated by the reluctance of adults to be interviewed about mathematics which was expressed by half of the people approached for an interview for the Cockcroft Report.

In contrast to this statement one has to bear in mind that adult education in mathematics is not only confronted with people being 'handicapped' mathematically, but with many adult learners who are really interested in and competent to do mathematics.

Most of the activities in adult mathematics education seem to be taught in close integration with other subjects, such as reading and science. The different subjects are even more interrelated than in technical and vocational education, which is another difficulty for a discipline-oriented approach to adult mathematics education.

2. Key Issues

2.1 Key issues in technological and organisational change
Before examining separate statements for technical/vocational education and adult education, one key issue common to both areas must be identified, namely the ongoing changes in technology and the organisation of work. These changes not only affect the everyday life of a vast majority of persons, but have a special influence on mathematics education.

There are direct effects which must be taken into consideration, for example, the widespread use of 'mathematical technology', hand-held calculators and computers, mainly in industrialised countries. Computer-Aided-Design technology (CAD) and numerically controlled machine tools (CNC-machines) obviously have immediate implications for the mathematical competence required on the job by an operator of these machines as well as for vocational education in preparation for such jobs. The introduction of computers to occupations in business and administration is another example of the direct effects of technological and organisational changes on mathematics education. There are also indirect effects on mathematics education. A growing specialisation of work, for instance, leads to different requirements in employment — lowering the level of competence for a lot of employees while concentrating the planning and control of the work in the hands and minds of a few experts. These developments affect the mathematics required on the job. Some descriptions of working situations, especially in industrialised countries, illustrate this. In developing countries the growing need for technical and vocational education — and the growing need for basic knowledge in arithmetic and measurement — can be taken as another illustration of this phenomenon.

2.2 Key issues in technical and vocational education

Even if it is difficult to know what mathematical procedures and knowledge are really used in the different sectors of production, distribution and administration, it is important to know what mathematics is used on the job, and how it is used. For industrialised countries, with some reservation, the curricula of technical and vocational education may be taken as an indicator of the mathematics that is used.

In addition to the 'core' present in all vocations, there are specific topics added to technical and vocational education for future metal-workers, electrical and electronics apprentices, and some jobs in the building sector. These topics include basic algebra (especially manipulation of formulae), reading and interpreting tables, diagrams and graphs, and some more advanced techniques. Commerce apprentices are trained for business calculations and some basic descriptive statistics. Geometry and technical drawing play an important role in the construction area (metal and wood workers, bricklayers). For developing countries there seems to be less information on what is 'really needed' and there are few such indicators.

Analysis of the use of mathematics also has to take into consideration the worker's practical knowledge of the job in question. What comes out of such analysis may even be different from the question of what is and should be taught and learnt in mathematics in technical and vocational education.

Besides content, ways of teaching and learning can be analysed. The Cockcroft Report's references to the differences between learning in a classroom and using mathematics on the job can be transformed into a search for this difference and for its theoretical and constructive consequences. Are there special opportunities and difficulties when the learning takes place in the more realistic environment of a workshop or an office, and how does this difference affect the learning/teaching of mathematics?

Ways of teaching/learning are important for classroom activities too. This raises the question of an alternative structure for including mathematics in the total curriculum in technical and vocational education. The integration of technical/vocational knowledge and techniques with mathematics can lead to different learning/teaching strategies from those in traditional mathematics instruction. This can even lead to a total omission of mathematics, for instance, as a consequence of technological or organisational change (introduction of machine-algorithms, prefabrication). Such changes also challenge the teacher training institutions which are responsible for a continuous up-dating of the teachers' competencies before and during service.

Other problems which have been discussed only rarely are closely related to the learners. What is the role of mathematics in the way learners look at their jobs? What is the role of mathematics in the way learners react to the requirements of their (future) jobs? What is the subjective, individual role mathematics plays for the learner?

Last but not least, problems with unemployment must not be forgotten in this context. What is the role of mathematics in entrance tests, in criteria for selection, in consequences for the learners' attitudes to mathematics? Does mathematics or failure in mathematics really matter when applying for a job? What about the role of mathematics in losing a job?

2.3 Key issues in adult education
Issues in adult mathematics education can be divided into two aspects, namely curriculum problems and problems with methods of teaching and learning.

A lot of learners in adult education are interested first in mathematics for survival and real-life applications. The list from the Cockcroft Report is an example of such a curriculum. What is the mathematics curricula for leisure and for rehabilitation, or for an eventual study of mathematics for its own sake? What of the mathematics requirements of entrance tests for jobs or further education access? There is a need for remedial schemes to prepare students for further education — taking into account the learners' attitudes to mathematics described earlier — and to support them during their courses.

Adults are said to learn differently from children, even if the subject is the same. What effect does the participation of the learner in the selection of content, the student-centred and negotiated curriculum, and pressures from educational, political and social institutions have on adult learning?

What research is there into the learning processes of adults? How should teachers cope with atypical adult students: the extremely gifted, the mildly mentally retarded, or speakers of languages not used in the society they live and learn in? Is the learning process in distance education altered by the isolation of the learner and the slow feed-back inherent in distance education?

3. Report of Congress Discussions

3.1 General debate: technological and organisational change
The first session raised three major issues:
- Students in technical courses only want the mathematics which they will meet at the workplace the next day. They want to get rid of school mathematics and restrict the mathematics they learn to what is 'really used'. This approach to mathematics in technical and vocational education cannot be followed for several reasons:
  - Employers often use mathematics in entrance tests only as a way

of selecting applicants. The tests are not adjusted to the needs of work — if these are known. Technical and adult education has to react to this.

- Besides employers, there are many other influences. Departments such as engineering have their own mathematics syllabus and want just some service mathematics from a mathematics course. Unions have their ideas of training schemes in order to keep control over the development of the area they are organising. Teachers like the way they have always taught and sometimes do not like changing.

- In the production and distribution of goods and in administration, whole job categories are being phased out because of the introduction of information processing machines. As a result, certain courses and teachers are becoming redundant.

- Technical education cannot be restricted to mere training for the jobs available at present. It should aim at training students to the highest possible level in order to give them a professional identity as well as the knowledge and flexibility to cope with technological and organisational developments.

- Widely available hand-held calculators have a growing impact on mathematics education. At least in college-type situations a developed sense of estimation is needed to check the 'answers' given by a calculator. On the other hand, some users of calculators begin to ask for an explanation of what is done by the calculator, so a new interest in understanding basic arithmetic arises.

- Technological change is often taken as the only active part in ongoing changes at work. In contrast to this view, technology, the organisation of work, technical and adult education and mathematics itself — besides other agents mentioned above — seem to interact in a complicated way when decisions on mathematics curricula have to be made. The introduction of the computer-assisted-design technology (CAD) in West Germany can be taken as an illustration of the fact that there is no one-way-street from technological change to the mathematics 'needed' and the design of adequate mathematics curricula.

3.2 Technical and Vocational Education

A central issue for this subgroup was the relationship of mathematics proper to other topics in technical and vocational education and to everyday practice in the workplace. A contribution from the United States of America showed some interesting examples of how to narrow the gap between college and work by establishing special links between single schools or technical institutions and interested companies ('adopt-a-school' program), or having teachers working in private enterprises on a part-time basis. Sometimes teachers of technical education are not even paid for their summer leave, and are thus forced to earn their living by having work experiences outside formal education.

In other countries it is quite difficult to develop a curriculum which meets the needs of work. Experts from the professions — such as engineers or teachers from engineering departments — seem to have trouble with teaching mathematics. Therefore, they often only reluctantly take part in curriculum development for mathematics by identifying the mathematics needs of their courses. The only source available in this case is often the technical textbook where the mathematics curriculum developer hopes to find the mathematics needed for special jobs. An answer to this problem was seen in deliberately involving as many teachers as possible in curriculum development.

A contribution from Australia pointed to the fact that, besides the mathematics explicitly represented in textbooks, one should not overlook implicit use of mathematics in, for instance, table-reading and technical drawing. Mathematics is sometimes used at the workplace and in technical education in such limited contexts that the generalisation and transfer to other situations must be taken as very difficult and not trivial for the student or apprentice.

Curricula can integrate mathematics and, for example, the knowledge of science needed by future metal-workers. This can illuminate the applicability of mathematics, but does not necessarily lead to easy transfer to other fields.

Another major issue was the separation of curriculum developers and teachers. Many technical teachers have to teach mathematics without being trained for this subject, hence they teach curricula which they have not really absorbed themselves. The Research-Development-Dissemination (RDD) model of curriculum development seems to be most inefficient in teacher training for technical education.

Another contribution from Australia demonstrated a different pattern, that of integrating curriculum development and teacher training through workshops bringing together resource people and teachers from only one occupational family at a time, for example, carpentry and joinery. Students are offered material on decimals, estimation, teaching concepts, formulas, and solving word problems, and work on this together in small groups with the curriculum developers. The material prepared will soon be published.

A contribution from the United Kingdom describing a training scheme for unemployed youngsters aged 16 to 17 brought up another issue. In order to offer some work experience to the unemployed, this government-initiated program tries to give participants a taste of whole occupational families (such as office practice, retail trade or motor vehicle). After two years of experience with separate numeracy courses which were not really successful, from September 1984 integrated courses will be trialled, thus giving students an opportunity to learn vocational mathematics different from that taught at school.

Unemployment in general sometimes affects mathematics teachers in rather direct ways. In order to hold the staff of departments which formerly asked for service courses from the mathematics department, many service courses are now taught by the department itself. Reduced demand for mathematics courses taught by mathematicians may threaten the future employment of mathematics teachers in technical education. In other countries the rising number of unemployed youngsters leads to the contrary situation where technical colleges are used as a way out of unemployment, hence they need an even larger teaching staff.

The subgroup on technical and vocational education came back to the central issue of the discussion: how to teach mathematics and who should teach mathematics so that it is relevant to the students and their future jobs?

3.3 Subgroup on adult education

This subgroup first considered curriculum questions. The early mathematics curriculum grows out of the need for clear communication and informed responses to the demands of adult life. Apart from simple concepts and calculations, many mathematics conventions must be learnt, such as measuring from left to right. After this stage, courses become more mathematical and the content is affected by the use of calculators in almost every case. As their use increases, students have to learn efficient techniques, including the memory buttons, and match this with better estimation skills, which also must be taught. When adults are familiar with the context of a problem, they usually check the reasonableness of an answer by interpretation and commonsense; they also have to learn to check the answer on the calculator by numerical estimation skills like rounding. The importance of decimals is increasing, and there are new ways of teaching them by reference to counting of parts without using fraction notation. The four operations with fractions are now less important as they are hardly ever used in adult life, but the concept of fraction and equivalent fractions are still needed. Adults also need to understand the meaning and application of percentages because of their frequent use in everyday life.

There always seems to be insufficient time given to spatial mathematics even though it is relevant to many practical situations. Teachers find that it engenders less fear in anxious adults than number work. The computer language, LOGO, is a valuable new way to teach geometry. Measurement skills are essential and develop well with learning the dimensions of one's own body, with active measuring experiences, and in close relationship to the study of nature as an introduction to the study of science.

Sometimes teachers are forced to teach content they would not choose, such as operations with fractions, because students seek reassurance about their ability to cope with school topics and because of entrance tests for employment or study. In the development of curricula many teachers wish to emphasise more of the processes of mathematics rather than specific content. In general, curriculum decisions are made by negotiation between the teacher and the students.

The other main area for discussion was teaching and learning in adult mathematics education. Many contributors presented examples of small interactive groups where the emphasis was on concept formation, pattern and structure, usually with the use of concrete materials. Through the group process, students' anxieties were allayed. Others stressed the need for an individualised approach in classes or centres set up for self-paced self-instruction with teachers available. Most agreed that students found individualised systems threatening and that as much group work as possible should be included. Other examples included self-help groups in the suburbs using personal computers; tutors teaching well-motivated adults quite extensive and difficult mathematics, for instance, the mathematics required to get a pilot's licence; and classes for older students where teachers must take into account the memory difficulties of the aging. In all adult classes essential goals were to develop self-confidence and willingness to take risks in problem solving.

The search for good printed materials usually starts with school textbooks where much of the content is unsuitable. Project CAM (Careers and Mathematics) books have recently been published in Australia, based on the study of mathematics in use in industry and services, and, although designed for the secondary schools, they are a useful source for adult education.

Teachers expressed confidence in most materials prepared for distance education because the writers have imagined and anticipated students' difficulties, and their explanatory style and self-instructional design make them particularly helpful. Teachers preferred to use concrete materials wherever possible for concept formation, including both commercial aids, such as Dienes multi-base arithmetic blocks and simple objects like small ceramic tiles.

Adults, it was agreed, learn differently from children. The development of abstraction, generalisation and similar skills appears to occur at times in rapid leaps. There is a need for research into how adults learn and how best to teach them.

4. A Final Remark

All participants in the Action Group agreed on the recommendation to include topics related to adult, technical and vocational education in future ICMEs.

5. References

Braun, Hans—Georg, (1982) 'Mathematischer Unterricht in Berufsschullen — Didaktische Konzepte' in Straesser, R. (ed.), *Mathematischer Unterricht in Berufsschulen — Analysen und Daten*, Bielefeld.

Costello, P., Jones, P., Phillips, B., (1984) *Careers and Mathematics*, Vols. 1–3, The Institution of Engineers, Australia, Melbourne.

Cockcroft Report, (1982) *Mathematics Counts — Report of the Committee*, London (HMSO).

Dawes, W.G., Jesson, D.St., (1979) 'Is there a Basis for Specifying the Mathematical Requirements of the 16 Year Old Entering Employment?', *Int.J.Math.Educ.Sci.Technol.*, Vol. 10, p. 391–400.

ICME 3 Proceedings, (1979) *New Trends in Mathematics Teaching*, Vol. IV. Paris (UNESCO).

ICME IV Proceedings (ed. by M. Zweng et al.), (1983) Boston.

Knox, C., (1977) *Numeracy and School Leavers — a survey of employers' needs*. Sheffield (Sheffield Regional Centre for Science and Technology).

McLeod, J.W., Thiering, J.V., Hatherly, S., (1985) *Trade Mathematics Handbook,* TAFE National Centre for Research and Development, Adelaide.

Mathematics in Employment (16–18). Report. (1981) Bath.

Pengelly, R.M., (1979) *Adult and Continuing Education in Mathematics*. Paris (UNESCO), p. 85–106.

Ploghaus, G., (1967) 'Die Fehlerformen im metallgewerblichen Fachrechnen und unterrichtliche Massnahmen zur Bekampfung der Fehler', *Die berufsbildende Schule,* Vol. 19, p. 519–531.

Straesser, R., (1980) 'Mathematik in der Teilzeitberufsschule', *Zentralblatt fur Didaktik der Mathematik (ZDM),* Vol. 12, p. 76–84.

UNESCO 1978: *Classification of information about technical and vocational education.* Paris (UNESCO).

UNESCO 1978a: *L'evolution de l'enseignement technique et professionnel — Etude comparative.* Paris (UNESCO).

THEME GROUP 1: MATHEMATICS FOR ALL
Problems of cultural selectivity and unequal distribution of mathematical
education and future perspectives on mathematics teaching for the majority

*Organisers: Peter Damerow (FRG), Bienvenido Nebres (Philippines),
Mervyn Dunkley (Australia), Bevan Werry (NZ)*

Introduction

Many factors have brought about a change in the overall situation of
mathematics education. These include the move to universal elementary
education in developing countries, the move to universal secondary education
in industrialised countries (where there have also been growing demands for
mathematical competence in an increasingly technologically and scientifically
oriented world) and from the experience gained with worldwide curriculum
developments such as the new mathematics movement. The tacit assumption
that what can be gained from mathematics can be gained equally in every
culture, independently of the character of the school institution, and of the
individual dispositions and social situations of the learner, turned out to be
invalid. New and urgent questions have been raised. Probably the most
important ones are:
- What kind of mathematics curriculum is adequate to the needs of the
  majority?
- What modifications to the curriculum or alternative curricula are needed
  for special groups of learners?
- How should these curricula be structured?
- How could they be implemented?

A lot of work has already been done all over the world in attempts to
answer these questions or to contribute to special aspects of the problem, in
particular:
- ICME IV yielded several presentations of results of research concerning
  universal basic education, the relationship between mathematics and its
  applications, the relationship between mathematics and language, between
  women and mathematics, and the problems of teaching mathematics to
  special groups of students whose needs and whose situations do not fit into
  the general framework of traditional mathematics education.
- The Second International Mathematics Study of the International
  Association for the Evaluation of Education Achievement (IEA) directed
  considerably more attention than the first study to the similarities and
  differences in the mathematics curriculum in different countries, and the
  different conditions which determine the overall outcome in mathematical

achievement. The IEA collected data on both the selectivity of mathematics and the differences between countries in the way they produce yield levels of mathematics qualifications. Although final reports on the Second International Mathematics Study are not available at the time of writing, preliminary analyses of the data have already produced useful results.

• In several countries national studies have been concerned with the evaluation of the mathematics education system. An important recent example is the Report of the Committee of Enquiry into the Teaching of Mathematics in Schools in England and Wales (commonly known as the Cockcroft Report, 1982).

• Last, but not least, there are many detailed studies, projects and proposals from different countries dealing with special problems such as:
  • teaching the disadvantaged;
  • teaching the talented;
  • teaching mathematics to non-mathematicians;
  • teaching mathematics in the context of real life situations;
  • teaching mathematics under atypical conditions, and so on.

The papers presented at ICME 5 addressed a variety of topics related to the theme, "Mathematics for All". Taken as a whole, they contribute to a better understanding of the problems of teaching mathematics successfully, not only to very able students, but teaching worthwhile mathematics successfully to all in a range of diverse cultures and circumstances.

Summary of Papers Presented to the Theme Group

The first group of papers dealt with general aspects of the theme, Mathematics for All. Jean–Claude Martin (France), in his paper analysed the special selectivity of mathematics education as a result of symbolism and mathematical language. The teaching of mathematics seems to have been designed to produce future mathematicians despite the fact that only a very small percentage of students reach tertiary level. This general character of mathematics education causes avoidable, system-related failures in the learning of mathematics, and often results in a strong aversion to mathematics. Martin proposed a general reorientation of mathematics education aiming at a mathematics which is a useful tool for the majority of students. The teaching of mathematics as a means of solving multidisciplinary problems by using modelling methods should restore student interest, show mathematics as being useful, enrich students' knowledge of related subjects and so better enable them to memorise mathematical formulas and methods, encourage logical reasoning and allow more students access to a higher level of mathematics.

In his paper, Bienvenido Nebres (Phillipines) discussed the same problem of the lack of fit between the goals of mathematics education and the needs of the majority in the special circumstances of developing countries. He offered a conceptual framework for discussing the specific cultural dimensions of the problem in these countries by using the distinction between vertical and horizontal relationships, that is, the relationships between corresponding institutions in different societies and the relationships between social or cultural institutions within the same society.

Historically, the establishment and growth of social and cultural institutions in developing countries is that they have been guided more by vertical relationships, that is, an adaptation of a similar type of institution to the mother colonial country, rather than by horizontal relationships. The result is a special lack of fit between mathematics education and the needs of the majority of the people. There is a tremendous need for researchers in mathematics education in developing countries to look at the actual life of urban workers, rural farmers and merchants and to identify the mathematics in daily life that is needed and used by people. It is then necessary to compare this needed mathematics with what is provided in the curriculum and to search for a better fit between the two. A cultural shift must be brought about in these countries. Mathematics educators, together with other educators and other leaders of society, should address the need for the social and cultural institutions to be better integrated with one another and to develop together in a more organic manner than in the past.

In a joint paper, Peter Damerow (FRG) and Ian Westbury (USA), examined the problem of designing a mathematics curriculum which genuinely meets the diverse needs of all students in a country. They argued that, by continuing to ignore the needs of all except a small minority of students, the curricula developed within the new mathematics movement proved to be no more satisfactory than their predecessors. Traditionally, mathematics curricula were developed for an elite group of students who were expected to specialise in the subject, and to study mathematics subsequently at higher levels in a tertiary institution. As education has become increasingly universal, however, students of lesser ability, and with more modest vocational aspirations and daily life requirements, have entered and remained longer in the school system in greater numbers. A major problem results when these students are exposed to a curriculum designed for potential specialists. This same type of traditional curriculum has frequently been transferred to developing and third world countries where, because of different cultural and social conditions, its inappropriateness for general mathematics education has only been compounded. So-called reforms, such as new mathematics, did little to resolve the major problems in that they merely attempted to replace one specialist curriculum with another.

The question addressed by Damerow and Westbury was how to cater both for the elite and the wider group of students for whom mathematics should be grounded in real world problem solving and daily life applications. One suggestion is that the majority would achieve mathematical 'literacy' through the use of mathematics in other subjects such as science and economics, while school mathematics would remain essentially and deliberately for specialists. This is effectively to retain the status quo. An alternative proposal insists that mathematics be kept as a fundamental part of the school curriculum, but ways of teaching it effectively to the majority must be found. The majority of students will be users of mathematics. Damerow and Westbury concluded that a mathematics program which is truly for all must seek to overcome the subordination of elementary mathematics to higher mathematics, to overcome its preliminary, preparatory character, and to overcome its irrelevance to real life situations.

The findings of the Second International Mathematics Study (SIMS) were used by Howard Russell (Canada) to argue that mathematics is already taught to all pupils at the elementary level in many countries. At the senior secondary level, however, the prevailing pattern in most countries is for mathematics to be taught only to an elite. At the lower level, the SIMS data suggest that promotion by age, rather than by performance, does not violate the concept of mathematics for all. The SIMS data also appear to provide support for the Cockcroft hypothesis that the pace of mathematics education must be slowed if sufficient students are to be retained in mathematics courses at the higher levels for it to be accurately labeled 'mathematics for all'. Alternatively, the content of the curriculum could be trimmed down as suggested by Damerow. Russell proposed a market-oriented rationale to construct such a core of material, particularly to meet the needs of the middle level students who will be required to use mathematics in their chosen work in the market place.

Afzal Ahmed, (UK) who was a member of the Cockcroft Committee, is now the director of a project for low attaining pupils in secondary school mathematics. He discussed the implications of the Cockcroft Report for the major issues addressed by the theme group. He pointed out that a suitable mathematics curriculum for the majority assumes greater importance as societies in the world become more technological and sophisticated. But at the same time, the evidence of failure at learning and applying mathematics by a large proportion of the population is also growing. The Cockcroft Report proposes a foundation list of mathematical topics that should form part of the mathematics syllabus for all pupils. In his discussion Ahmed focussed particularly on the classroom conditions which facilitate and inhibit the mastery of these fundamental topics.

In addressing the issue of universal mathematics education, Achmad Arifin (Indonesia) indicated that the level of community participation in the delivery of universal mathematics education should be raised through consideration of the many aspects of interaction within and between social and cultural institutions. He asked the question — which parts of mathematics can function as an aid to the development of an individual's intelligence and how should the parts chosen be presented? Any program to answer this question has to take into account three components of interaction. Firstly, depending on its quality, interaction through social structures can contribute to the improvement of people's abilities, especially by helping them appreciate mathematics. Secondly, a special form of social interaction, which he called positive interaction, can motivate mathematics learning and create opportunities to learn. Thirdly, school interaction itself can inspire, stimulate and direct learning activities. In developing countries, local mathematicians in particular are able to understand their cultural conditions, the needs, the challenges and the wishes of their developing nation. Taking into account the three components of interaction, they have the ability and the opportunity to spread and share their knowledge and to translate and utilise the development of mathematics in universal mathematics education for their nation.

In many countries, there is one mathematics syllabus for each year of the education system. Andrew Begg (NZ) questioned this practice and argued for the introduction of alternative mathematics programs to meet the varied needs of all students in a range of circumstances and with a range of individual aspirations. All such courses should contribute towards general educational aims such as the development of self-respect, concern for others, and the urge to inquire. Thus, mathematics courses should provide an opportunity to develop skills of communication, responsibility, criticism and cooperation.

Such an approach has implications for the way in which students are organised in mathematics classes, for the scheduling of mathematics classes, for the choice of teaching and learning methods, for the extent to which emphasis is placed on cooperation as against competition, for the use of group methods of teaching, and for the provision that should be made for students from diverse cultural groups. In this way, mathematics programs for all students should assist not only the achievement of mathematics objectives, but also the attainment of personal, vocational and humanistic aims in education. By matching mathematics programs to the needs of students, the development of self-esteem for every student becomes central in the mathematics curriculum.

The second group of papers was concerned with particular aspects of mathematics education related to industrialised countries.

Genichi Matsubara and Zennosake Kusumoto (Japan) traced the introduction of the teaching of western arithmetic to Japan in the late nineteenth century. At a time when universal elementary education was only just approaching reality in Japan, the government declared a policy of adopting western style arithmetic in order to enable the country to compete more successfully in a modern world. This move faced obstacles in its implementation because of the traditional use of the abacus and the widespread lack of familiarity with Hindu–Arabic notation. Further, in a developing national system of education, teachers were in short supply, and little attention could be given to teaching methods in the training courses. The paper emphasised the need to make such changes slowly and to take into account the situation of those closely involved with the changes if they are to be successful in modifying the curriculum for mathematics for all.

The extent to which the mathematics learnt at school is retained and used in later life was the subject of research reported in a paper by Takashi Izushi and Akira Yamashita (Japan). They reported on a 1955 study of people who had learnt their mathematics before the period in Japan in which mathematics teaching was focussed on daily life experience, and before compulsory education was extended to secondary schools. Although it was found that most people retained the mathematics skills and knowledge well, rather fewer claimed that this material was useful in their work. A second more limited study in 1982 confirmed these general findings in relation to geometry. It showed, broadly speaking, that younger people tended to use their school mathematics more directly while older people relied more on commonsense. The study covered a further aspect, the application of the attitudes of deductive thinking derived from the learning of geometry. The thinking and reasoning powers inculcated by this approach were not forgotten and were claimed to be useful in daily life, but not in work. Izushi and Yamashita concluded that the inclusion of an element of formal mathematical discipline in the curriculum is supported by Japanese society.

An attempt to create a modern course in advanced mathematics for those students who do not intend to proceed to university was reported by Ulla Kuerstein Jensen, (Denmark). The increase from about 5% in former years to about 40% in 1983 of an age cohort completing upper secondary education with at least some mathematics brought about an evolution toward a curriculum concentrating on useful mathematics and applications in daily life and mathematical modelling. This evolution led to the draft of a new curriculum which will be trialled under school conditions, beginning in the autumn of 1984. The origin of this development is based on new regulations for mathematics education for the upper secondary school introduced in 1961. These were influenced by ideas from the new mathematics and designed to serve the needs of the small proportion of students passing through upper secondary education at that time. But they soon had to be applied to the rapidly increasing number of students in the following years. Mathematics teaching, particularly for students in the language stream of the school system, was increasingly influenced by ideas and teaching materials for a further education program more closely related to usefulness for a broad part of the population than the usual upper secondary mathematics courses. In 1981, this development was legitimated by new regulations and, by that time, even mathematics teaching in classes concentrating on mathematics and physics became increasingly influenced by the tendency to put greater emphasis on applications. This has led ultimately to the draft of a new unified curriculum which is now going to be put into practice.

The central topic of a paper by Josette Adda (France) was an analysis of social selective functions of mathematics education. She reported statistical data showing the successive elimination of pupils from the 'normal way' at each decision stage of the school system until only 16% of the 17 year age cohort remain whereas all others have been put backward or relegated to special types of classes. These eliminations selectively hit socio-culturally disadvantaged families. Research studies have been undertaken to find out why mathematics teaching as it is practised today is not neutral but yields a high correlation between school failures in mathematics and the socio-cultural environment. They indicate the existence of parasitic sources of misunderstanding increasing the difficulties inherent in mathematics, such as embodiments of mathematics in pseudo-concrete situations which are difficult to understand for many pupils. On the other hand, it had been found that children failing at school are nevertheless able to perform authentic mathematical activities and to master logical operations on abstract objects.

Two papers were based on the work of the EQUALS program in the United States, an intervention program developed in response to a concern about the high dropout rate from mathematics courses, particularly in the case of women and minority students. The program aims to develop students' awareness of the importance of mathematics to their future work, to increase their confidence and competence in doing mathematics, and to encourage persistence in taking mathematics courses.

In the first of these papers, Sherry Fraser (USA), described the way in which the program has assisted teachers to become more aware of the problem and the likely consequences for individual students of cutting themselves off from a mathematics education. By working with teachers and providing them with learning materials and methods, with strategies for problem solving in a range of mathematical topics, together with the competence and confidence to use these, EQUALS has facilitated and encouraged a transfer of concern to the classroom and attracted and retained greater numbers of underrepresented students in mathematics classes. Since 1977, 10 000 educators have participated in the program.

Although the main focus of activity in the EQUALS program has been on working with teachers and administrators, needs expressed by these educators for materials to involve parents in their children's mathematics education led to the establishment of Family Math. Virginia Thompson (USA) described how this project has developed a curriculum for short courses where parents and their children can meet weekly to learn mathematical activities they can do at home together. This work reinforces and complements the school mathematics program. Although the activities are appreciated by all students, a major focus has been to ensure that underrepresented students, primarily females and minorities, are helped to increase their enjoyment of mathematics. The project serves to reinforce the aims of the EQUALS program.

The move over the past ten years or so towards applicable, real world and daily life mathematics in the Netherlands, inspired by the work of Freudenthal, was described by Jan de Lange, (Netherlands). Textbooks have been published for primary and lower secondary schools which reflect this view of mathematics, and research shows that the reaction of teachers and students has been very favourable. De Lange illustrated the vital role played by applications and modelling in a newly-introduced curriculum for pre-university students. Many teachers apparently view the applications-oriented approach to mathematics very differently from the traditional mathematics content. The ultimate outcome, de Lange suggested, may be that science and general subjects will absorb the daily life use of mathematics, and consequently this type of mathematics might disappear from the mathematics curriculum. That is, the ultimate for all students, as far as mathematics is concerned, could, in reality, become no mathematics as such.

Roland Stowasser (FRG) proposed the use of examples from the history of mathematics to overcome certain difficulties arising from courses based on a single closed system. Such courses increase mathematical complexity but do not equally increase its applicability to open problems. He stated that mathematics for all does not necessarily have to be directly useful, but it has to meet two criteria: on the one hand, the mathematical ideas have to be simple and, on the other hand, they have to be powerful. He illustrated these criteria through an historical example. Regionamtus formulated the problem to find the point from which a walking person sees a given length high up above him (for example the minute hand of a clock if the person walks in the same plane as the face of the clock) subtending the largest possible angle. The solution with ruler and compasses in the framework of Euclidean geometry is somewhat tricky. But, according to Stowasser, the teaching of elementary geometry should not be restricted to Greek tricks. For problem solving, he advocated free use of possible tools and the solution of the problem is very simple if trial and error methods are allowed. Thus the solution of an historical problem can represent the simple but powerful idea of approximation.

Allan Podbelsek (USA) listed a number of characteristics of a mathematics program suitable for all students and covering not only content, knowledge and skills, but also attitudes towards and beliefs about mathematics and the process skills involved in its use. For such a program, mathematics must be seen to be a unified, integrated subject, rather than a set of individual, isolated topics. He found that the Comprehensive School Mathematics Program (CSMP) developed over several years in the United States for elementary (K-6) level classes met these criteria successfully in almost every respect. Practical problems involved in the introduction of such a program as CSMP to a school included the provision of adequate teacher training for those concerned, meeting the cost of materials, securing the support of parents and the local community, and ensuring that administrative staff were aware of the goals of the program.

Those translating mathematical, scientific or technical material should have a basic knowledge of mathematics to do their job satisfactorily, yet, because of their language background, they are not likely to have studied mathematics to any great extent at school. This is the experience which led Manfred Klika (FRG), to a consideration of the nature and adequacy of present school mathematics programs. Conventional school programs, he claimed, do not prepare students to comprehend and make sense of mathematical ideas and terminology. The solution is to construct the mathematics curriculum around fundamental ideas. Two perspectives on this notion were offered: major mathematising models (such as mathematical concepts, principles, techniques) and field-specified strategies suitable for problem solving in mathematics (for example, approximate methods, simulation, transformation strategies).

A curriculum based on such fundamental ideas would result in more meaningful learning, and thus a more positive attitude to the subject. A course based on this approach has been established within the program for training specialist translators for work in technical fields.

The major concerns of the contributions reported above were problems of designing a mathematics curriculum which is adequate to the needs and the cognitive background of the majority in industrialised countries. The organising committee for the theme group was convinced that it is even more necessary to discuss the corresponding problems in developing countries. But it was much more difficult to get substantial contributions in this domain. To stress the importance of placing more emphasis on the development of mathematics education in developing countries, the work of the theme group terminated with a panel discussion. The discussion concentrated on the relation between micro-systems of mathematics education such as curricula, textbooks and teacher training and macro-systems such as economy, culture, language and general educational systems. Particularly in the developing countries, macro-systems often determine what kind of developments are possible on the level of micro-systems. The common conviction of the participants was that, in spite of the fact that it is often impossible to get a substantial improvement in mathematics education without fundamental changes in the macro-systems of education, micro-changes are possible. Indeed they are a necessary condition for making people realise what has to be done to get a better fit between mathematics education and the needs of the majority. This statement highlights the importance of the papers submitted to the theme group dealing with special aspects of mathematics education in developing countries.

Three reports were given by David Carraher, Terezinha Carraher and Analucia Schliemann (Brazil) about research undertaken in their country.

David Carraher reported on the results of a study investigating the uses of mathematics by young, schooled street-vendors who belong to social classes characteristically failing in grade school, often because of problems in mathematics, but who often use mathematics in their jobs in the informal sector of the economy. In this study, the quality of mathematical performance was compared in the natural setting of performing calculations in the market place with that in a formal setting similar to the situation in a classroom.

Similar or formally identical problems appeared to be mastered significantly better in the natural setting. The reasons for such discrepancies were discussed and it appears that the results of the analysis strongly suggest that errors which the street-vendors made in a formal setting do not reflect a lack of understanding of arithmetical operations. Rather they reflect a failing of the educational system which is out of touch with the cognitive background of its clientele. There seems to be a gulf between the intuitive understanding which the vendors display in the natural setting and the understanding which educators try to impart or develop.

Terezinha Carraher reported on the results of a second research project. Problems involving proportionality were presented to 300 pupils in order to find out whether a child already understands proportions if the routines being taught at school were correctly followed. The results indicated characteristic types of difficulties appearing in certain problems, some of which can be related to cognitive development. It is suggested that teachers' awareness of such difficulties may help to improve their teaching of the subject. If mathematics is to be useful to everyone, mathematics teachers must consider carefully all issues related to the transfer of knowledge acquired in the classroom to other problem solving situations.

The third paper, presented by Schliemann, highlighted the discontinuity between formal school methods of problem solving in mathematics and the informal methods used in daily life. This research study contrasted the approaches to a practical problem of quantity estimation and associated calculation taken by a group of experienced professional carpenters without extensive schooling, and a group of carpentry apprentices attending a formal school system and with at least four years of mathematics study. The results showed that apprentices approached the task as a school assignment, that their strategies were frequently meaningless and their answers absurd. On the other hand, the professional carpenters took it as a practical assignment and sought a feasible, realistic solution. Very few computational mistakes were made by either group but the apprentices appeared unable to use their formal knowledge to solve a practical problem. Schliemann concluded that problem solving should be taught in practical contexts if it is to have transferability to daily life situations out of school.

Pam Harris (Australia) identified a problem that occurs in the remote Aboriginal communities of Australia where teachers often get the feeling that mathematics is not relevant. Several reasons for these perceptions can be identified. Teachers often receive negative attitudes from other people so that they go to an aboriginal community expecting that their pupils will not be able to do mathematics. They also observe a lack of reinforcement of mathematics in the pupils' home life. For the most part, teaching materials are culturally and linguistically biased. Teachers feel discouraged because of the difficulties of teaching mathematics under these conditions. Nevertheless, Harris stressed the importance of mathematics because aboriginal children need to develop understanding of the 'second culture' of their country. They need mathematics in their everyday life, in employment, and in the conduct of community affairs. But to be successful, mathematics teaching in aboriginal communities has to allow for and support local curriculum development. Individual schools and language groups should make their own decisions on the use of the children's own language, the inclusion of indigenous mathematical ideas, priorities of topics, and sequencing the topics to be taught.

Kathryn Crawford (Australia) described a course which forms part of the Anagu Teacher Education Program, an accredited teacher training course intended for traditionally oriented aboriginal people currently residing in the Anagu communities who wish to take on greater teaching responsibilities in South Australian Anagu schools. The most important difference about this course and many others is that it will be carried out on site by a lecturer residing within the communities. From the beginning, development of the curriculum has been a cooperative venture between lecturers and educators on the one hand, and community leaders and prospective students on the other. The first group of students began the course in August 1984. The course design pays particular attention to the problems that arise where different cultures employ different conceptual schemes.

Thus, temporal sequences and quantitative measurement are dominant themes in industrialised western cultures but largely irrelevant in traditional aboriginal cultures. To overcome such difficulties, the focus of the problem is redirected from the 'failings' of aboriginals and aboriginal culture to the inappropriateness of many teaching practices for children from traditionally oriented communities. The course is based on a model designed to maximise the possibility of interaction between the world view expressed by Anagu culture and that of Anglo-European culture as evidenced in school mathematics. This is achieved by placing an emphasis on student expertise and contribution in providing information about Anagu world views as a necessary part of the course. For the first time, it seems that it is possible to develop procedures for negotiating meaning between the two cultures through this community based teacher training course.

Conclusions

The presentations summarised above are important contributions to the great program of teaching mathematics successfully not only to a minority of selected students, but teaching it successfully to all. But in spite of all these efforts, it has to be admitted that the answer to the question, 'What kind of mathematics curriculum is adequate to the needs of the majority?', is still an essentially open one. The great variety of issues connected with this problem which were raised in the presented papers makes it clear that there will be no simple answer. Thus, the most important results of the work of this theme group at ICME 5 may be that the problem was, for the first time, a central topic of an international congress on mathematics education, and that, as the contributions made abundantly clear, this problem will be one of the main problems of mathematics education in the following decade.

In summary, there are at least three very different dimensions to the problem which contribute to and affect the complex difficulties of teaching mathematics effectively to the majority:
- the influence of social and cultural conditions;
- the influence of the organisational structure of the school system;
- the influence of classroom practice and classroom interaction.

Cultural Selectivity

One of the major underlying causes of the above problem is that mathematics education in the traditional sense had its origins in a specific western european cultural tradition. The canonical curriculum of 'traditional mathematics' was created in the 19th century as a study for an elite group. It was created under the condition of an existing system of universal basic education which included the teaching of elementary computational skills and the ability to use these skills in daily life situations. There is a clear distinction between the aims and objectives of this basic education and the curriculum of traditional school mathematics aimed at formal education and not primarily directed at usefulness and relevance for application and practice. This special character of the canonical mathematics school curriculum is still essentially the same today in many countries.

The transfer of the european mathematics curriculum to developing countries was closely associated with the establishment of schools for the elite by colonial administrations. Under these circumstances, it seemed natural to simply copy european patterns. It is quite another problem to build a system of mass education in the third world and embed mathematics education in both the school situation and the specific social and cultural contexts of that world.

The papers summarised here clearly indicate some of the problems. Curricula exist which encourage students to develop antipathies towards mathematics; this is commonly the case in Europe. Further, such curricula have sometimes been transferred to countries where the social context lacks the culturally based consensus that is found in Europe, namely, that abstract mathematical activity is good in itself and must therefore be supported, even if it seems on the surface to be useless. It has been proposed that a sharp distinction should be made between applicable arithmetic in basic education and pure mathematics in secondary education on the one hand, and the integration of mathematics into basic technical education on the other. This argument raises the question of the relationship between mathematics and culture which may be the first problem to address when the idea of mathematics for all is raised as a basis for a program of action.

Selectivity of the School System

While the particular curricular patterns of different societies vary, the subject is still constructed in most places so that few of the students who begin the study of mathematics continue taking the subject in the last years of secondary schooling. The separation into groups of students who are tagged as mathematically able and not able is endemic. Curricula are constructed from above, starting with senior levels, and adjusted downwards. The heart of mathematics teaching is, moreover, widely seen as being centred on this curriculum for the able, and this pattern is closely related to the cultural contexts indicated above. However, even for industrialised societies, we must consider the problem of conceiving a mathematics which is appropriate for those who will not have contact with pure mathematics after their school days. Until now, we have made most of our students sit at a table without serving them dinner. Most attempts to face the problem of a basic curriculum reduce the traditional curriculum by watering down every mathematical idea and every possible difficulty so that it is possible to teach the remaining skeleton to the majority. There is only a limited appeal to usefulness as an argument or a rationale for curriculum building to avoid the pitfalls of this situation.

Students who will not have to deal with an explicit area of pure mathematics in their adult lives but will face instead only the exploitation of the developed products of mathematical thinking (for example, program packages) will only be enabled by mathematics instruction if they can translate the mathematical knowledge they have acquired into real life situations which are only implicitly structured mathematically. Very little explicit mathematics is required in such situations and it is possible to survive most without any substantial mathematical attainments whatsoever.

Is the only alternative to offer mathematics to a few as a subject of early specialisation and reject it as a substantial part of the core curriculum of general education? This approach would deny the significance of mathematics. To draw this kind of conclusion, we would be seen to be looking backwards in order to determine educational aims for the future. The ongoing relevance of mathematics suggests that a program of mathematics for all implies the need for a higher level of attainment than has been typically produced under the conditions of traditional school mathematics. This is especially true for mathematics education at the level of general education. In other words, we might claim that mathematics for all has to be considered as a program to overcome the subordination of elementary mathematics to higher mathematics, to overcome its preliminary character, and to overcome its irrelevance to life's situations.

Selectivity in Classroom Interaction
Some of the papers presented in this theme group support recent research studies which have suggested that it is very likely that the structure of classroom interaction creates ability differences among students which increase during the years of schooling. In searching for causes of increasing differences in mathematical aptitude, perhaps the simplest explanation rests on the assumption that such differences are due to predispositions to mathematical thinking, with the further implication that nothing can be done really to change the situation. But this explanation is too simple to be the whole truth. The understanding of elementary mathematics in the first years of primary school is based on preconditions such as the acquisition of notions of conservation of quantity which are, in their turn, embedded in exploratory activity outside the school. The genesis of general mathematical abilities is still little understood. The possibility that extra-school experience with mathematical or pre-mathematical ideas influences school learning cannot be excluded. Furthermore, papers presented to the theme group strongly suggest that the differences between intended mathematical understanding and the understanding which is embedded in normal classroom work is vast. We cannot exclude the possibility that classroom interaction in fact produces growing differences in mathematical aptitude and achievement by a system of positive feedback mechanisms which increase high achievement and further decrease low achievement.

It is clear that to talk of mathematics for all entails an intention to change general attitudes towards mathematics as a subject, to eliminate divisions between those who are motivated towards mathematics and those who are not, and to diminish variance in the achievement outcomes of mathematics teaching. This, in its turn, involves us in an analysis of social contexts, curricula and teaching, as it is these forces together which create a web of pressures which, in turn, create situations where mathematics becomes one of the subjects in the secondary school in which selection of students into aptitude and ability groups is an omnipresent reality almost from the time of entry.

THEME GROUP 2: THE PROFESSIONAL LIFE OF TEACHERS

Organisers: Thomas Cooney (USA), Fred Goffree (Netherlands), Beth Southwell (Australia)

This Theme Group was organised into five subthemes, each emphasising some aspect of a teacher's professional life. Convenors were appointed for each subtheme, with groups meeting for each of the four full working days of the Congress. For each subtheme, presentations of prepared papers provided the basis for discussions among the practitioners, teacher educators, mathematicians and researchers who participated in the Group. A brief overview of the activities and discussions for each subtheme follows.

Using Research

Convenors: Don Dessart (USA) and Tom Romberg (USA)

Papers presented by some twenty-six participants were used as a basis for much of the subsequent discussions.

Koichi Abe (Japan), Joan Akers (USA), Gerhard Becker (FRG), Jacques Bergeron (Canada), Dudley Blane (Australia), George Bright (USA), Philip Clarkson (Papua New Guinea), Arthur Clegg (UK), Claude Comiti (France), Tom Cooney (USA), Erik De Corte (Belgium), Charleen DeRidder (for Marilyn Suydam) (USA), Donald Dessart (USA), Don Firth (Australia), Gloria Gilmer (USA), Yoshihiko Hashimoto (Japan), Rina Hershkowitz (Israel), Murad Jurdak (Lebanon), Carolyn Kieran (Canada), Eio Nagasaki (Japan), Doug Owens (Canada), John Stewart (Australia), Malcolm Swan (UK), Tom Romberg (USA), David Wheeler (Canada), Doug Williams (Australia).

The papers addressed a variety of topics including the utilisation and impact of research on classroom practice and in the preparation of materials, the relationship between teachers and researchers, and problems related to disseminating research findings.

The following four common themes were central to the subsequent discussions:

• the usefulness of research for the teacher;
• the nature of research summaries;
• the utilisation of research summaries;
• the role of teachers in research.

Usefulness of Research

An informal survey of teachers in Australia, Scotland and the United States suggested that teachers generally possess a positive attitude towards research. Yet there seemed to be little evidence that research actually influenced classroom practice. Several factors may contribute to this lack of influence. Firstly, teaching is generally viewed as a skill in which an emphasis on reflection is minimised. Secondly, there is the question of whether research is useful in light of the fact that a considerable amount of the research consists of doctoral dissertations. Thirdly, it was felt that a strong relationship between research and teaching has not been adequately developed. Consistent with the view that research in the United States does not reach the practitioner was the finding that most teachers do not read research articles, although they are more likely to read succinct summaries of research. On the other hand, according to one survey, 91% of the Australian teachers surveyed indicated that some research results had been absorbed into their teaching practice.

Most participants voiced the opinion that research has produced conclusions that can guide teachers' decision-making processes. For example, it was felt that the accumulated findings on how children learn rational numbers and how they construct knowledge about addition and subtraction were useful for classroom practice. As well, evidence gained from large assessment projects that reveal inadequate problem solving skills is useful, at least in the sense of 'identifying and indicating a perceived importance that in turn supports greater emphasis on problem solving in practice'. The instruments used in the classroom processes component of the Second International Mathematics Study could be used as a means of promoting alternate interpretations in such content areas as fractions, ratio, proportion and percent, algebra, geometry and measurement.

It was also recognised that not all research can be evaluated solely on utilitarian grounds. The commonly used quantitative models require replications and accumulated findings in order to yield reliable generalisations; both conditions demand the passage of time and coordination among investigators. This is in sharp contrast to the field of mathematics in which the proof of a theorem decisively settles an issue.

Some participants emphasised the importance of generalisations which are formed in a 'natural way' from case studies or more qualitative research methodologies. It was emphasised that findings from this type of research can be beneficial in helping teachers envisage different ways of interacting with students, diagnosing learning difficulties, or reflecting on their own teaching styles and behaviour. It was also pointed out that teachers should be encouraged to be eclectic in their utilisation of research findings, picking and choosing among results when the contexts and findings of the research are harmonious with their specific contexts for teaching.

The Nature of Research Summaries

Various viewpoints about the nature of what should be released were expressed. Some felt that research findings should be made available as soon as possible to ensure recency of information. In this case, results from single studies could be provided, leaving the evaluation and interpretation of the study to the individual teacher. Others felt that only results that have stood the test of review, critique and sometimes replication should be published.

A particular case for consideration is the effectiveness of using games as a means of teaching mathematics. Questions were raised about the responsibility of the researcher to convey the 'ethos' of the research in order to promote proper interpretations of the effectiveness of games. The situational context of using games may be very important in determining the advisability of their use. This is particularly so in light of the inherent difficulty of incorporating games into the development of textual materials.

Utilisation of Research Summaries

Discussions on the utilisation of research summaries focussed on the need for written materials as well as the need for designing delivery systems for conveying research implications to teachers. Various utilisations of summaries were identified and discussed. The 'lowest' level consists of 'broadcasting' relevant findings. The means by which such ideas are internalised or used by teachers is basically left to chance. Metaphorically, it is like broadcasting or transmitting knowledge over the air waves; some teachers do not find the broadcast interesting and hence 'tune out'. Others listen but may not reflect upon the information in light of their own teaching, while still others utilise the information as a means for reflecting on and revising their own teaching. Presentations of research findings at professional meetings exemplify this lower level of dissemination.

A second level of utilisation involves communicating the contexts in which the research was conducted. It is hoped that teachers will see a certain ecological validity between the contexts in which the research was conducted and their own teaching situations in considering the research when making instructional decisions. The highest level of utilisation involves helping teachers construct a framework in which to integrate the research findings into their own conceptual schemes. This latter type of utilisation necessitates considerable interaction between researcher and teacher.

The Role of Teachers in Research

Defining the role of the teacher in the research enterprise is a complex question involving pragmatic concerns and philosophical issues that bring into question the very nature of research itself. Analytic methodologies borrowed from the hard sciences suggest that a certain distance be maintained between researcher and teacher as objectivity and the classical issues of validity and reliability are central to the acceptability of the research. Generality is viewed as a basic objective; biases are minimised through sample selection procedures. On the other hand, humanistic methodologies strive for understanding a teacher's idiosyncratic meanings with respect to their conceptions of mathematics and teaching. Here the teacher's role is more akin to that of a collaborator as the researcher deliberately minimises the 'distance' between himself/herself and the teacher. Action research exemplifies one type of humanistic methodology in which the teacher and researcher work in concert for a common goal.

It was emphasised that the desirability of having teachers participate in research activities may extend beyond the question of what constitutes research and relate to the professional development of the teacher by virtue of engaging in such activity. That is, not only can the teacher contribute to the creation of grounded theory but the teacher can mature and elevate his/her own expectations and professional aspirations by participating in a reflective research process.

There is a felt need for communicating with administrators and policymakers regarding the nature of research and the various roles teachers play. The cost of funding release time for teachers to participate in research projects and the anticipated benefits derived from such activity must be clearly communicated without overselling the benefits and then risking disillusionment and the withdrawal of support. The view was expressed that support for research can be obtained when the advantages of research are identified.

It was clear from the discussions that the linkage between research and its dissemination and utilisation needs considerably more attention and study than could be provided in these four sessions. Not only is the continuing process of research viewed as necessary for the enhancement of teachers' professional life but also it is necessary to focus attention on how research, both its processes and products, can increase the vitality of our profession and create a means of helping teachers utilise what we come to know about the teaching and learning of mathematics.

Inservice Teacher Education

Convenor: Bill Fitzgerald (USA)

Papers for this subtheme were presented by the following:
> Edith Biggs (UK), Ping-Tung Chang (USA), Peggy House (USA), Ramesh Kapadia (UK), Calvin Long (USA), Peter Reynolds (UK).

The sessions began with the papers which addressed a number of topics, including the following:
- What mathematical content should be included in courses for practising teachers?
- What pedagogical content should be included in courses for practising teachers?
- What types of inservice training programs seem to be most effective?

Subsequent discussions focussed on these and other issues as suggested in the following overview.

Inservice teacher education has become a very important part of the professional life of teachers, although the means by which inservice training is conceived and actualised varies considerably from country to country. Since the 1950s extensive inservice programs have occurred with mixed results. It is generally acknowledged that the 'new math' movement in the 1960s did not extensively influence classroom practice despite large inservice programs. Criticisms of inservice projects over the past decade or two suggest that many projects are too fragmented and neglect the important aspect of follow up activities once teachers complete a particular segment of training. This second criticism was raised in the context in which inservice work was separate from actual teaching situations.

In some developing countries there is a tendency to develop inservice programs that are national in scope and that have a unified set of objectives, as, for example, in Indonesia and Taiwan. One strategy for reaching many teachers in Taiwan is to use qualified teachers to form so-called 'seed teams' with the goal in mind that over a few years as many as 15 000 teachers could receive training. Follow up activities with supervisors are viewed as a very important feature of the program. The importance of supervision and follow up activities has been recognised for some time.

Edith Biggs (UK) in the 1960s organised inservice programs to inspire teachers to work in small groups with children and to create 'local' curriculum materials that would capture the imagination of the children as they worked in a cooperative manner. She developed the notion of 'support visits' to further assist teachers as they attempted to reorganise their teaching. The support visits were considered essential to the success of the inservice programs.

Reasons for Inservice Programs

The question of why inservice programs are important was addressed from several different perspectives. Included among the various motives were the following:

- to provide teachers with the time, means and support for developing their professional competencies;
- to increase teachers' knowledge of pedagogical approaches and their perceptions of what constitutes mathematics;
- to involve teachers in curriculum development including some aspects of quality control regarding curriculum materials development;
- to improve school resource utilisation;
- to enhance leadership development.

The view was expressed that teachers are interested in inservice programs because they are looking for solutions to the many problems they face. Hence, the effectiveness of inservice programs is closely related to the extent to which teachers can relate theory to practice.

A Developmental Approach to Inservice Education

When thinking about the professional growth of teachers, it is helpful to think of the teacher as a 'developing professional'. In this sense inservice should not be viewed as a single occurrence but rather as a continuous process in which teachers are helped to realise their full potential. Peggy House (USA) has used Bloom's taxonomy of educational objectives as a global framework to develop means of assessing teachers' professional growth. The concept of development includes the notion of teachers moving from an acquisition of 'theoretical' knowledge about teaching mathematics to the point where knowledge is internalised and eventually used as a basis for reflection about their own teaching.

The highest level of development would include a state in which the teacher's knowledge could be used in a generative way to help solve problems encountered in classroom practice. This type of development is viewed as an ongoing process throughout a teacher's professional life.

Problems Related to Beginning Teachers

It was recognised that beginning teachers have special problems in becoming professional teachers. At least three acute problems exist. First, they lack experience in coping with the daily administrative and logistical demands placed upon them. Second, they frequently have considerable difficulty in transforming their 'theoretical' mathematical and pedagogical knowledge into appropriate classroom activities. Third, there is generally a lack of an organised support structure from peers or mentors when they struggle to cope with the first two problems. As a result, there is often a feeling of isolation which sometimes turns into disillusionment about teaching to such an extent that they leave the teaching profession.

Qualifications of Teachers and Inservice Programs

In some countries, for example, the UK and the USA, inservice courses lead to additional qualifications for teaching mathematics. The diplomas received on completion of carefully defined programs raise the status and sometimes salaries of the teachers. The inservice programs which lead to such diplomas and more extensive qualifications seem to be quite successful as reported by a number of participants.

The notion of inservice is broadened in some countries to include training for non-mathematics teachers who wish to become adequate teachers of mathematics. This is particularly significant where there is a shortage of mathematics teachers as is presently the case in the United States and Australia. A basic issue associated with the retraining process is the question of what constitutes adequate training in both mathematics and pedagogy for teaching mathematics.

Computers and Inservice Education

Convenor: Dick Shumway (USA)

The sessions were marked by lively discussions on ways teachers can use the computer with students and the problems of inservice development of such capabilities. Papers were presented by Dick Shumway (USA), Alfinio Flores (Mexico), and Art White (USA). The specific issues addressed were the nature of mathematics, remaking the mathematics curriculum, and problems of human resources development. The basic assumptions underlying these sessions were that:
- computer use is desirable;
- big changes are likely;
- there are lots of people to train;
- it may be an ideal opportunity for teachers to learn more about mathematics and the learning of mathematics.

In a brief overview, Shumway discussed mathematics, learning mathematics, computer models of mathematics, and learning computer mathematics. There appears to be a consistent congruence between mathematics and good mathematics learning, and computer models and the learning about and constructing of computer models. Some of the best thinking about the nature of mathematics and the learning of mathematics is dramatically reinforced and stimulated by computers and the use of computers to do mathematics.

Short programs of less than 10 lines in BASIC and designed for computers priced less than $100 in the USA were used to illustrate mathematics to increase the power of computational programs; drill and practice programs for estimation; the division algorithm for 'infinite' decimals; approximating area and defining the definite integral; simulation programs to explore the meaning of 'random'; graphing arbitrary dilations of functions such as $\sin(x)$, $x\sin(1/x)$, and $\sin(\exp(x))$); and parametric representation and graphing of functions such as conic sections. The examples were used to illustrate an emphasis on doing mathematics and the elegant, important mathematics discovered and used in writing short programs with inexpensive machines. It became clear that 'the' nature of the mathematics curriculum has to be reconsidered in the perspective of doing mathematics with computers.

In discussing the revision of the curriculum, Flores noted that, with a computer, topics such as variables and probability can be introduced earlier. For topics such as graphing and statistics, computers can take the burden from calculation. Exploration and the discovery nature of learning are more readily accessible and possible with a computer. He also noted the interesting properties of computer arithmetic that can be explored, together with the issues of algorithmic thinking, such as efficient and inefficient algorithms and powerful algorithms. Further discussion is necessary on what topics can be deleted from the curriculum.

The computer offers a rich context for problem solving investigations and provides a sharp contrast to teaching mathematics using only the textbook. To illustrate how the computer can be used to promote both the learning of basic content and problem solving skills, a few small computer programs were presented and discussed. The programs had been analysed by young children as they were asked to predict the programs' outcomes. Two such programs written in BASIC were the following:

```
(1)   10 FOR N = 1 TO 20          (2)   10 FOR N = 1 TO 20
      20 PRINT N + N                    20 PRINT 2*N
      30 NEXT N                         30 NEXT N
```

The programs were designed to promote conceptual development in addition and multiplication respectively. This approach is in marked contrast to the traditional textbook-oriented approach.

This contrast is further exemplified by considering slightly more complicated programs such as:

```
10 FOR N = 1 TO 100
20 LET X = 2*N - 1
30 LET S = S + X
40 PRINT S
50 NEXT N
```

The use of this program enabled students to discover the relationship between square numbers and the sum of consecutive odd numbers. What was argued, however, is that such a program does not necessarily relate to a child's conception of adding consecutive odd numbers. The use of visual representation was suggested to help insure an intuitive orientation to the problem.

Similarly, students can study programs in which the simulation of throwing

dice is represented.
```
10 FOR N = 1 TO 20
20 LET R = INT (6*RND 1)
30 PRINT R;" ";
40 NEXT N
```
Children associate a certain 'magic' with throwing dice as they experience games where dice are used. What is important then, is not just the presentation and analysis of such a program, but rather presenting the program in a context that relates its outcome to the children's experiences with dice. Nevertheless, there was testimony to the fact that students can profit from an analysis of such programs without encountering extensive learning problems.

The above examples illustrate how curriculum can be designed to capture both the use of computers and a problem solving orientation to instruction. Other issues considered include the language of computer programs; the relationship between simulations and reality; numerical versus geometrical insights; and environmental considerations for using the computer.

The last two sessions were devoted to problems of inservice education and school practices associated with implementation of computer uses of the types proposed. White shared his experiences with these issues while working in Trinidad. There were numerous comparisons and contrasts made between experiences of participants in their own countries.

At the concluding session, the following recommendations were made:
- Teachers of mathematics at all grade levels should incorporate short (3-10 line) computer programs into their teaching for learners both to write and to modify.
- Curriculum developers and publishers should incorporate computer programs related to the mathematical concepts of each topic into their materials.
- Small, inexpensive microcomputers should be used for effective teaching and learning of significant mathematical concepts and processes.
- All teachers of mathematics and/or teacher trainers should have a computer for their own professional use at home and/or in the mathematics classroom. This includes teachers of mathematics from primary to adult groups.
- Courses for training teachers should be model experiences of the use of computers in learning and teaching mathematics.
- The computer should be used in teaching mathematics so that it provides a learner controlled environment in which the learners are engaged in generating their own concept structures and problem solutions.
- The educational system should provide support to facilitate mathematics teacher use of the computer in the classroom. This support should include a computer resource person and inservice programs for teachers to increase and improve computer use for learning mathematics.

Becoming an Effective Teacher

Convenors: Mary Lindquist (USA) and Geoffrey Wain (UK)

The discussions for this subtheme focussed on the following aspects of effective teaching:

- knowledge of the learner;
- knowledge of mathematics;
- methods of evaluation;
- development of teaching strategies;
- self-awareness.

Papers were presented by the following individuals:

John del Regato (USA), Janet Duffin (UK), Mary Gilfeather (USA), Bengt Johannson (Sweden), David Moncur (UK), Helen Pengelly (Australia), Bob Roberts (Australia), Gentaro Sasaki (Japan), Alan Starritt (UK), Paul Trafton (USA), Ron Welsh (Australia).

To enable a teacher to experience and develop expertise in each of the aspects mentioned above, access to a wide range of support services and resources is essential. For instance, the process of self-awareness only occurs if a teacher is equipped with a fairly sophisticated range of analytical frameworks that permit self-maintenance. The viability of self-monitoring and associated professional development by a teacher acting alone is questioned even though one aim of teacher education is often stated as preparation for a teacher to act professionally.

There are many support structures with which a teacher might interact in order to become more effective. The most appropriate form for each individual teacher will depend upon the circumstances relating to the school, the geographical area and the requirements of the employing bodies, as well as other factors. A teacher needs to be embedded in a social structure designed in some way to bring about increased effectiveness.

The planning, implementation and evaluation of effective teaching were the central themes considered in the discussions. It was noted that there is a growing body of research in these areas but more research is needed, particularly classroom-oriented research.

Planning for Effective Teaching

Some aspects of planning for effective teaching include yearly planning, building personal resources, using textbooks wisely in unit planning, and decision making for daily planning. It is important to plan classroom work using a wide variety of methods and to emphasise means by which classrooms become more child centred and less teacher centred.

Research in Sweden involving a very detailed analysis of classroom interactions revealed that most students were actively engaged for only very short periods of time and that teachers, using questioning techniques, were able to 'pilot' pupils through material for which they were inadequately prepared. These findings have implications for planning instruction.

Implementing Effective Teaching Techniques

It seems possible to implement improved forms of teaching by structuring the curriculum so that it stresses a wide range of process skills. These include interpreting information, collecting data, selecting a strategy, carrying out practical work and communicating various means of solving problems. This is the approach taken by the Munn and Dunning Development Program in Scotland where the whole process was centered around problems set in a real world context, as those contexts provided meanings for the mathematics studied. Results included greater acceptance of calculators, more awareness of the need for a variety of classroom organisations, better presentation of school-produced materials, more practical work, and a different view of mathematics. It is not yet clear, however, how lasting the changes will be.

A new series of textbooks has been designed for use in schools in Papua New Guinea with the hope that the texts will improve teacher effectiveness in the primary schools. Teachers are being involved in detailed inservice activities before using the new books. An important aspect of using the texts is that the teachers are involved in an evaluative program to improve the materials and to increase teacher professionalism.

Evaluating for Effective Teaching

A new initiative in a local education authority in England has laid down a professional structure for the first year of teaching. Beginning teachers are provided a clearly defined set of support procedures which include observations of experienced teachers in several schools. A clear indication of the criteria that would be used by a visiting advisor in assessing the effectiveness of the teacher is also provided. The criteria include a knowledge of a wide range of teaching approaches, a knowledge of resources and support systems, and an awareness of the needs of children. This new initiative is designed to embed the new teacher in a firm framework of professionalism to ensure a maximum level of teaching effectiveness.

In one Australian state the extent to which new teachers are helped by their teacher training experiences is being investigated. Items to be considered include:

* understanding of basic mathematical concepts;
* knowledge and ability to plan a range of appropriate learning activities;
* knowledge of curriculum materials;
* the development of positive attitudes toward mathematics;
* ability to organise mathematics teaching and learning situations;
* ability to evaluate student learning;
* knowledge of audio/visual aids;
* ability to account for individual differences.

While results of this investigation are not yet available, it was interesting to note that in the reports from both the United Kingdom and Australia, emphasis was on creating structures that place specific requirements on teachers with a view to maintaining a minimum level of teacher competence.

Social Contexts and Philosophies that Affect Teachers

Convenors: Max Stephens (Australia) and Perry Lanier (USA)

Discussions during the four sessions were based on papers presented by the following individuals:

Brian Donovan (Australia), Ms. Koeswachjoeni (Indonesia), Stephen Lerman (UK), Marilyn Nickson (UK), Neil Pateman (Australia), Max Stephens (Australia), Chich Thornton (Australia), Piet Verstappen (Netherlands), Stephen Willoughby (USA)

The basic assumption that guided the presentations and discussions was that the planning and practice of mathematics teaching takes place in an institutional context shaped by educational and political traditions embedded in a particular society and influenced by philosophical and political beliefs about the purposes of schooling.

Among the many factors that affect mathematics education at various levels, the following were noted:
- what is considered the appropriate work for teachers and for students;
- how schools are organised;
- how a mathematics curriculum is developed, communicated to teachers, implemented by teachers, and learned by students;
- how research is approached;
- how change and reform in mathematics teaching and learning are conceived.

Indeed, thinking about the social contexts and underlying philosophies adds an often forgotten but fundamental dimension to all of the subthemes of the professional life of teachers. An analysis of innovative curriculum projects over the past few decades, as well as various attempts to change teachers' thinking and attitudes about mathematics teaching, suggest that social context plays an important but often subtle role in shaping the development and utilisation of curriculum materials. As Christine Keitel emphasised, 'A mathematics education not shaped by these influences does not exist, and a mathematics education ... stripped of these influences cannot be studied except on an abstract level'.

Sometimes pressing societal needs overshadow the impact of other factors that affect the teaching of mathematics. A large inservice training program in Indonesia illustrates this circumstance. The fundamental aim of the program, stated by the Government, was to achieve equal opportunity for students at all educational levels. The notions of investigative learning and teacher/student discussions (as opposed to teacher dominated lectures) were emphasised. To promote such teaching methods, an extensive and intensive inservice training program was used, including 'onservice' visits to teachers' classrooms.

Perceived impediments to the program's success included the low level of initial training of many participants, the practice of many teachers to work a second shift, large class sizes, as well as teachers' own beliefs about mathematics and mathematics teaching. Experience suggests that greater attention will need to be given to addressing the problem of supporting teachers in resolving potential conflicts between their perceptions of teaching and the project's goal of promoting more inquiry modes of instruction.

One of the problems in assessing the impact of contextual factors on the teaching of mathematics is the implicit nature of associated social and philosophical considerations. For example, where calculators are commonplace in homes, elementary mathematics teachers are still reluctant to permit their use in schools for fear students will fail to learn the basics. This resistance may have its roots in sometimes personal, sometimes common 'theories' of learning. Teachers' implicit notions about the nature of mathematics and how mathematics should be taught and learned, notions which potentially mitigate against the use of calculators, touch upon such basic epistemological issues as what constitutes mathematics, and even the most fundamental question of what constitutes knowledge.

These epistemological questions about knowledge emerge from interactions between subjects (human beings) and objects (reality). The view was expressed that knowledge is not only objective in nature but that it also has to be shared as a condition for the creation of new knowledge. Thus knowledge has both a subjective and a public aspect. A distinction which some find useful is between objective knowledge and personal knowledge.

One's view of knowledge has a direct impact on how one conceives the task of teaching mathematics. For example, a formalist view of mathematics (founded in logical positivism), leads to a view of mathematics in which systems of logical, coherent propositions based upon the language of well-defined symbols is emphasised. Such a view neglects the practical day-to-day, real life meanings often attributed to mathematics and which children intuitively bring to the classroom. Formal definitions, operations and rules are presented in a context which emphasises the intuitive notions that students possess. As a consequence, the mathematical environment often involves sterile meanings devoid of real life applications. The resultant teaching and learning activities become exercises in drill and practice in which the manipulation of symbols constitutes the primary focus.

An emphasis on problem solving and more investigative activities can be based on a quite different conception of knowledge. Central to this view is *activity*, in which individual and social aspects are considered. In this case, mathematics education is based on a 'growth and change' conception in which the view is taken that mathematics is more a human activity than a mere body of logically connected propositions. From this perspective the teaching of mathematics emphasises interaction and social processes since meanings are a function of shared communication among individuals.

What relevance does such an analysis have for the teacher? Perhaps it is enough for teachers to simply be aware of the implications of various epistemological influences and potential relationships between foundational considerations and classroom practice. Consideration of such relationships may reveal conflicts teachers feel but may not explicitly address. For example, consider the conflict expressed by many elementary teachers who feel a discrepancy exists between the curriculum to be taught and the development of their students. This perceived discrepancy dissolves to the extent that mathematics is viewed as something internal to students and more intuitive mathematical ideas are promoted.

But the issue is even more complicated as individuals within a society have quite divergent views about what and how mathematics should be taught. Such diversity has been illustrated by a study of a measurement project in an Australian primary school. An innovative program emphasised an activity-orientation to teaching. Funds were allocated to the school to increase educational opportunities in an underprivileged neighbourhood. However, while middle class parents supported an enriched, activity-oriented curriculum, working class parents stressed quite a different view, namely, the acceptance of authority and the acquisition of basic skills. In addition, the teachers interpreted the curriculum in a 'traditional' way. That is, they held the view that teaching low achievers meant teaching in short, isolated periods, giving many directions, and expecting children 'to do their best'. In short, the distribution of mathematical knowledge in school conferred unequal benefits upon children and these differential benefits were related to institutional arrangements within the school as well as to wider social-cultural conditions.

In conclusion, the view was expressed that attempts to improve mathematics education must take into account the social, political and epistemological contexts in which mathematics education occurs, including teachers' conceptions and the societal expectations of both parents and students regarding schooling outcomes. Finally, the means of addressing such concerns should be within a 'change and growth' framework which models a desired view of mathematics. The following two considerations are basic for effecting change in both pre- and inservice teachers:

(1) strategies for teachers' acquisition of knowledge and skills to reflect upon and examine their own teaching, its underlying assumptions and the constraints which impede change; and

(2) strategies to support teachers, especially groups of teachers as they attempt to enact constructive changes as a result of (1).

THEME GROUP 3: THE ROLE OF TECHNOLOGY

Organisers: Rosemary Fraser (UK), Hartwig Meissner (FRG), Tony Ralston (USA), David Roseveare (UK), Jury Mohyla (Australia)

Films, video, calculators, computers, TV, OHPs, slides, tape are becoming available resources to a growing proportion of the teaching profession, with newer developments such as video disc also within sight. This theme was concerned with how best to understand the use of these media with their wide ranging educational materials and to begin to identify or, in some cases, report on effective roles for them. There was also an urgent need to consider their effect on the current mathematics curriculum itself and their effect on our patterns of learning and teaching. Each of these diverse media has its own spectrum of possibilities and questions; accordingly it was decided to offer 17 working groups each considering different topics but grouped under 6 major headings:
- Challenges to the Curriculum;
- Algorithms and Programming;
- Television, Video and Film;
- Classroom Dynamics;
- Teacher Education;
- Miscellaneous.

The summaries of the reports of the groups form the main part of this section of the Proceedings, following a brief account of the Congress presentation, 'Technology in Action'. The full working group reports, together with a range of contributed papers will be available in a separate publication. The reports reflect the need for diverse but systematic explorations of a host of possibilities.

Inevitably this Theme Group made very considerable demands on the local organisers for the provision of equipment. The Chief Organisers would like to place on record their thanks to all those who provided and all those who helped locate, transport and install equipment on our behalf.

Congress Presentation: 'Technology in Action'

Convenor: Jury Mohyla (Australia)

The areas of interest of the various working groups were indicated. The emphasis of the session was to provide brief but dynamic illustrations of technology in action. Audience participation was often demanded and generously given. The presider set the scene in historical context referring to 'Techne', the term coined by the Ancient Greeks to define the complexity and diversity of human activity that today is embodied in the term 'Technology'.

Hartwig Meissner (FRG) used an ordinary liquid crystal display hand calculator (with part of its back replaced by transparent plastic) on an overhead projector to enable the audience (a whole class) to view the display. With a video camera linked to a large video projection screen, and an untreated calculator, the display and/or keyboard could similarly be seen. To demonstrate the potential hidden in calculators, the audience had to work on two primary school estimation problems where the calculator display was used implicitly to get hints for a next-best guess. Analysing the guesses afterwards, the audience experienced having used, unconsciously, intuitive concepts of direct and indirect proportionality, of percentages, of monotonic change, convergence and limits.

Ipke Wachsmuth (FRG) used an Apple microcomputer with disk drive to demonstrate a Prolog program TERRI, originating from cognitive research but with potential for teacher education. The program models responses, supposedly from an uncertain pupil, not only to 'straight' mathematical questions such as 'Which is greater, 1/4 or 1/3?' but also to questions asking *why* s/he (the pupil) gave a particular answer, for example, '1/4 is greater than 1/3 (sic) because they have the same number on top and 4 is greater than 3'. The pupil responses were not necessarily mathematically consistent but were modified in the light of what wordings were used or what questions had already been asked by the user.

David Stoutemyer (USA), using a 16mm silent film record of the real- time output of a line-printer, demonstrated software which performed symbolic manipulation, including the factorisation of polynomials and the algebraic evaluation of indefinite integrals.

Dale Burnett (Canada) discussed a range of programming languages that are generally available. He commented on their various advantages and disadvantages and how they might enhance the teaching and learning of mathematics.

Rosemary Fraser (UK), using a BBC microcomputer with a disc drive, illustrated the way in which a single microcomputer can be used as a teacher's assistant in the classroom. Two programs, PIRATES and EUREKA, were used to illustrate the power of this dynamic assistant. Discussions on the role of the teacher, the students and the micro followed the simulation of a classroom where the conference audience agreed to participate as the 'pupils'. Children's work that had resulted as follow up activities to EUREKA was shown and it illustrated clear role-shifting with children taking over computer and teacher roles, resulting naturally in problem solving and open-ended activities. The microcomputer was seen to take over or assist in the managing, explaining and tasksetting roles, releasing the teacher to spend more time counselling and working alongside the children.

David Roseveare (UK) showed excerpts from school television, each of which provided an audiovisual experience otherwise unavailable in the classroom:
- the electronic manipulation of real-life images (*Landscape of Geometry*, OECA, Canada) — to point up the difference between real life (complicated) and geometry (simple).
- the large-scale studio demonstration (*Mathscore*, BBC, London) — also raising the question of stop-start *use* of video and the production of video intended specifically for such use.

- dramatisation (*Maths Counts*, BBC, London) — also raising the question of whether 'sugaring the pill', while motivating, can also be divisive within a school.
- documentary (*Maths at Work*, BBC, London) — also see Action Group 4.
- animation (*Advanced Level Studies: Statistics*, BBC, London) — including computer-originated animation (BBC Micro) videorecorded in real time, using inferior graphics.

Finally, an excerpt from *Micros in Schools* (BBC Open University, UK) for teachers and parents, raised the question of editorialisation in making film/ video documentaries about learning activities.

Technology Working Group Reports

Challenges to the Curriculum

Calculators for Developing Countries and for Developed Countries
In the first two sessions, the group work centred on national reports. Some countries are in the stage of developing materials and curricula. In Italy calculator activities are produced by the RICME Project (1976/80), the Genova Project (1982), and the GRISIMM Project (1982). Although calculators are marketed extensively by Japan, they still remain outside of the school curriculum. The traditional use of 9×9 recitation and the Japanese abacus (soroban) remain educational goals, though there are also experimental classes with calculators, especially in probability and statistics. In Lesotho there is no institutionalised use of calculators at the school level, but the production of instructional materials will start within the next few years. One of the largest calculator projects started in 1976 in Sweden. It is known as the ARK-Project. It aimed to provide an analysis of the consequences of the introduction of the pocket calculator, including their use as a numerical aid, their influence on current methodology, and their impact on contents of courses. In Puebla Mexico a mathematics laboratory including calculators has been established in a primary school to provide a manipulative-based mathematics curriculum to facilitate the progression from concrete to abstract understanding of mathematics. In Papua New Guinea a numeracy project was set up specially to help adults develop computational ability and to bridge a gap in basic number skills. In South Africa several research studies and projects on calculators have been completed since 1976. In the new centralised core curriculum the calculator is now officially included as a computational tool and an instructional aid. In England many investigations were started to explore the potential of the calculator. The Durham Schools Council project, for example, produced excellent materials 'Calculators Count'. In West Germany two major calculator studies explored the impact of calculators on number sense (Lange), and on arithmetic skills (Wynands). But mainly calculators in Germany are used only as computing aids.

The national reports were complemented by survey reports on calculator activities and research results in Europe (from Lucilla Cannizzaro) and in North America (from Donna Berlin). Summarising, we can state :
1. There has been much less research activity concerning calculators since ICME IV (Berkeley) than there was between ICME 3 (Karlsruhe) and ICME IV.

2. Research investigations since 1980 have not produced dramatic new insights or results, they mainly confirm the findings of previous investigations.
3. The overwhelming majority (> 95%) of investigations indicate that the use of calculators has no negative effects on mathematics achievement in terms of the traditional curriculum and traditional tests.
4. There are many countries where more than 80% of children have calculators available, yet the school curricula ignore the existence of calculators or calculator use in mathematics education is not allowed in many grades.

In the third session group members presented non- traditional calculator activities to demonstrate the educational power hidden in calculators (guess and test, mental arithmetic training and exploration of mathematical concepts). An overview was given of how traditional curricula might change.

In the last session the group discussed future demands. Primarily there is no need for new teacher or student materials or for research upon the effects of the calculator use in mathematics education. The experts have already done their job. Our deep concern is that a large majority of the educational society still ignores the onmipotent existence of calculators. What will be the role of paper and pencil algorithms? What will be the value of teaching fractions? *The objectives must be redefined! Mathematicians, educators, administrators, text book writers must all be alerted to their responsibilities.*

Technology and the Secondary School Mathematics Curriculum
Computers and other electronic information technology are becoming standard problem-solving tools in business, industry, government, and personal affairs. Thus secondary schools are being challenged to rethink the content objectives and instructional approaches of all curricula — especially in mathematics. One group, with representatives from 10 countries, examined three dimensions of potential technology-induced change in secondary school mathematics curricula:

1. Possibilities for improved teaching of traditional mathematical topics.
2. Opportunities for new selection and organisation of mathematical topics — new priorities among traditional concepts and skills, new topics from computer-related discrete mathematics, and access to important mathematical ideas that previously have been too difficult to teach effectively.
3. Problems of teacher education and curriculum change presented by the technology challenge. On each issue the group members sought a survey of outstanding work from around the world, analysis of the strengths and weaknesses in that work, and recommendations for needed research and development. The extent and style of classroom use of computers varies widely, even among countries with similar educational systems and similar access to technology.

Furthermore, the emergence of low-cost and high-powered computing technology is such a recent phenomenon that the state-of-the-art in its use for education is changing almost daily.

Improved Teaching of Traditional Mathematics
Developments in computer-enhanced teaching of traditional mathematical skills and concepts fall into several major strategy categories.

First, many teachers believe that students who write computer programs to guide mathematical calculations will, in the programming process, acquire unprecedented insight into the structure of relevant concepts and skills. A second form of computer-enhanced instruction uses pre-programmed packages to provide students with tools for numerical or graphic exploration of mathematical ideas. The interactive capabilities of computers has encouraged development of a third strategy to enhance mathematics teaching, with vivid simulation and game formats embodying problem solving situations that lead to discovery of new concepts and skills. Despite the exciting promises in the above instructional innovations, made possible by computers, the most common form of computer use to enhance mathematics instruction is some form of drill and practice of traditional skills.

Many of these programs do present such practice in a new style, and the most sophisticated are beginning to provide insightful diagnosis of student learning difficulties. However, few could provide and manage mathematics education as teacher substitutes. With few exceptions, computer-based instructional aids is still a 'cottage industry'.

In considering actions that would be most useful in helping schools exploit the potential of computers to improve instruction in traditional content areas, the group made the following recommendations:

1. Software packages should emphasise things that teachers cannot normally do without computer assistance.
2. Each country or state should have libraries of carefully evaluated software and teaching ideas that make use of that software. Communication networks for dissemination of that information are equally important.
3. One very useful component of such a software library would be a collection of mathematical sub-routines which teachers could draw on to construct their own personalised classroom software.
4. To overcome the substantial financial risks involved in production of high quality educational software, it seems advisable for governments to provide seed money to meritorious development ideas.
5. Government agencies can also provide an important service by supporting, or at least coordinating, research to study the broad question of effectiveness of computer aids in teaching and the value of specific strategies in particular.
6. Efforts should be made to minimise the problems of machine incompatibility, either by thoughtful design of the software itself or by standardisation of hardware requirements.
7. The introduction of computers to mathematics education brings an opportunity, but also a risk of losing very capable teachers to the computer studies curriculum.

New Mathematical Content and Priorities
The near universal access to calculators and computers that is emerging in many countries offers a broad range of opportunities to reorganise and redirect the priorities of traditional courses. Some old favourite topics will decline in importance, other topics now make a legitimate claim for space in the curriculum, while still others have gained new stature because of their importance to the new discipline of computer science. In assessing the validity of proposed curricular changes, it is important to keep three questions in

mind:
- What are the requirements of mathematical literacy in a highly technological society?
- What is the essential mathematical basis of technology, engineering, science, and the computer information sciences?
- What mathematical content and instructional approaches are essential to develop student creativity and problem solving abilities?

Several representative ideas were discussed, including realistic problem solving, emphasis on understanding, the algorithmic point of view, graphic display methods, new approaches in geometry, statistics, and discrete mathematics.

Teacher Education

To meet the challenge of computing in mathematics education, the crucial change agent will be classroom teachers. At the present time, they may be classified according to their level of interest and knowledge as computer literate, computer illiterate, or computer conscious. While a variety of practices exist for educating teachers in the integrated use of computers, each method has some shortcomings. It is important that effective inservice materials and strategies be developed and shared so that teachers can confidently incorporate computers fully into their mathematics programs. In summary, the school response for computer use in secondary mathematics has only begun to be realised, in even the most advanced and active countries. There are a few impressive prototypes with potential, but a great need remains for development of effective software, alternative curriculum materials, and inservice education of teachers.

Symbolic Mathematical Systems and Their Effect on the Curriculum

The second revolution in scientific computing is here. Although we expect this revolution will have far-reaching effects throughout science, the focus of this report is its impact on mathematics education. Widely available technology (in particular, many microcomputers) can support computer algebra systems which will do to algebraic manipulation what the hand-held calculator did to arithmetic. Capabilities of these systems are symbolic as well as numeric and include, for example, expanding polynomials and collecting like terms, symbolic differentiation and integration, arbitrary precision arithmetic, matrix algebra, symbolic solution of systems of algebraic and differential equations, and operations on abstract mathematical structures.

This raises immediate and long-term philosophical and practical concerns for mathematical education. How can this new power be utilised, for whom, and at what points in the curriculum? What effect will the automation of symbolic as well as numeric calculation have on mathematical thought processes?

We can distinguish three broad areas in which computer algebra will affect mathematics education:

(a) Courses *about* computer algebra. Since the techniques of computer algebra are based on much traditional mathematics as well as significant recent research, courses on (as distinct from courses using) computer algebra at the graduate or advanced undergraduate level will be important in their own right. A focus on computer algebra may also provide a new framework for an exploratory approach to the teaching of much mathematics in the present tertiary curriculum.

(b) Courses on *how to use* computer algebra. Utilisation of computer algebra as a tool will have major impact on the practice of mathematics in science and technology. Instruction on how and when to use it, its capabilities, and its limitations, will become an important part of mathematics service courses, and will also play a role in the teaching of mathematics specialists.

(c) Pedagogical uses of computer algebra. The most profound effects will be felt in the secondary and undergraduate mathematical curricula. In contrast to (a) and (b), the details of how computer algebra will be used in these contexts are not yet clear. What *is* clear, however, is that beyond providing routine training in symbolic techniques, computer algebra can dramatically enhance and enrich mathematical instruction and learning — for example:

- as a vehicle for mathematical discovery;
- by extending the range of examples which can be studied;
- as a programming environment ideally suited to mathematics;
- by emphasising the inter-relationships between different mathematical representations;
- as an aid to preparation and checking of instructional examples;
- by promoting a hierarchical approach to the development of concepts and algorithms.

Implementers of computer algebra systems need and invite suggestions on how to tailor such systems to meet educational requirements of varying student populations.

A more precise idea of the capabilities of the systems and some ways in which they are already being used in education may be gained from references, but the best method of gaining a feel for their power and likely impact is to try one, such as mu-MATH (available on microcomputers so the most readily accessible), MACSYMA, MAPLE, REDUCE, and others.

Algorithms and Programming

Programming Languages for Mathematics Education

In examining the relationship between programming languages and mathematics education, it is important to consider the relative emphasis which should be placed on computer science and on mathematics, even though there are strong areas of overlap between the two disciplines. From the perspective of the mathematics educator it seems clear that computer languages should be seen essentially as tools rather than as objects of study in their own right. A characteristic of the technology which is especially important in this context is the opportunity it offers for the dynamic visual representation of mathematical concepts.

The impact of computer technology on mathematics education is complex and defies simplistic description or recommendations. In the recent past, at least in the so-called developed world, there has been very rapid growth in public accessibility to computers, mainly because of the advent of the microcomputer. This growth, which shows little sign of abating, has been paralleled by that of an active software industry which has produced numerous versions of 'classical' languages such as BASIC, Logo, Pascal, Pilot, Fortran and APL. More recently, other languages, such as Prolog and

Smalltalk, have been the subject of much discussion. As has been true historically with other educational innovations, there would seem to be considerable difficulty associated with the transfer of this technology to the developing world.

Two principal roles that a programming language could play in mathematics education can be identified: (1) as an arena for student-directed exploration, and (2) as a medium for the development of teacher or curriculum support materials. Languages such as APL and Logo appear to be particularly well suited for 'doing' mathematics, permitting the student to engage in self-initiated and open-ended exploration. APL was originally developed as an alternative to conventional mathematical notation and has been recommended by some for more common use, even if one does not have access to a computer. Logo was constructed with young learners very much in mind and is particularly well-suited for the elementary grades although it has, in its full implementations, the capability for use at much higher levels. Examples are beginning to appear which illustrate a different pedagogical approach where one begins with a rich problem area or microworld and uses this as an entry point to relevant mathematical topics.

Algorithmic Approaches to Mathematics Education

1. General issues in the algorithmic approach:
- the advent of the computer forces us to reassess and reinterpret the idea of the 'algorithm', not only in mathematics but also right across the curriculum.
- an algorithmic approach goes well beyond the definition of ALGORITHM — namely, a finite sequence of instructions which, when carried out, will solve in a finite number of steps all problems in the category for which the algorithm is designed.
- an algorithmic approach is limited to finite systems. Infinite systems do not lend themselves to algorithmic treatment unless first approximated by finite systems.
- some algorithms, such as those in elementary school for addition, subtraction, multiplication and division, do have to be learned even though they may be 'rote'. These are the mathematical equivalent of 'walking', and are a necessary preliminary to any mathematical development.
- as teachers, we have not always clarified the concepts in our own minds nor have we communicated the concepts to students appropriate to their level.
- a good algorithm does not just get an answer but enhances understanding of a concept. Developing one requires a strong understanding of the process.
- emphasis on algorithms is frequently misunderstood and arguments such as 'why bother when calculators have them built in already?', and 'students get answers without understanding what they are doing', still persist.

2. The language of algorithms in the teaching of mathematics:
- the characteristics of algorithms should reflect the age group, ability level and, pedagogical purpose being served. There is a HOW?, WHEN?, WHERE?, WHY? with every use.
- an algorithmic approach requires good symbolism. Even the use of 'pidgin' BASIC was seen as a good vehicle for the synthetic division algorithm.

3. Efficiency of an algorithm:

- what is the importance of efficiency? (such as, long division vs. synthetic division in division of polynomials).
- non-efficiency may be didactically useful.
- an algorithm to help learning is not necessarily the same as an algorithm for expediency.
- for a very weak child, any algorithm it can use is a good one, whether understood or not.
4. Problem solving with and without algorithms:
- there is a need to distinguish between what has to be learned by heart and what has to be arrived at by logical processes. (It is an observed fact that children can multiply faster than they can add — is this due to drilling in tables?)
- algorithms must not eliminate the need for students to think.
- some algorithms are better 'programmed into' children, such as formula for the area of a rectangle (primary), solutions of quadratic equations (secondary), in order to free them for more advanced work.
- other problems have to be solved without the aid of an algorithm, such as 'word' problems, which require creative thought.
- a student who learned and understood an algorithm will 'recover' it much more quickly after a period of non-use than one who only used it without understanding earlier.
- not knowing an algorithm versus knowing that it is not possible to find an algorithm should be expected only of sophisticated students.
- in revision and enrichment, solution by several different algorithms may be useful but not in the early stages of dealing with a problem.
- there is a place for students to create their own algorithms.
5. Known experiments in the development of algorithmic thinking:
West Germany is in the process of introducing courses in the theory of algorithms. Blagovest Sendov (Bulgaria) outlined an experiment in his country in which great stress is placed on skills in symbolic expression and algorithmic thinking at an early primary age.

Technology and Cognitive Development
The group focussed on three major issues:
1. The use of technology to further the cognitive development of learners of mathematics;
2. The use of technology to achieve a better understanding of the processes involved in children's development of cognitive structures relevant to mathematics;
3. Possible disadvantages connected with the use of technology.
   It appeared that 'technology' was almost exclusively thought of as (micro-) computers. 'Cognitive development' was taken to be a long-term process of change in children's mathematical intellectual behaviour. A questionnaire completed by the group members served as a background of this report. Contributions reflected interest and concern with students of all ages from early elementary through tertiary level.
   1. Programming a computer may stimulate thinking experiences that are beneficial to the learning of mathematics, especially problem solving. Children can come to observe and become reflective about their own thinking and thus use it more effectively (technology as a 'thinking aid'). The dialogue style of

interaction with some more advanced systems might even enhance that feature, as was reported by several participants. Other points emphasised possible uses of computers for explorations and simulations. A challenge for technology is to develop methods that can support imaginative thinking about mathematical situations and relationships. Examples showed how the graphic capabilities of microcomputers can be used to achieve dynamic visualisations preparatory to the development of important concepts.

2. In discussing ways that technology can be used to foster understanding of the processes involved in children's cognitive development with respect to mathematics, special emphasis was placed on computer systems evolving from artificial intelligence research and technology that attempt to model students' knowledge of mathematics and behaviour in mathematical situations. It was realised that such models can help to obtain a more precise notion of 'cognitive structure' which would provide a basis for a more profound understanding of the process of building-up such structures, that is, cognitive development.

3. Among the possible disadvantages of using technology, loss of basic skills and deprivation from natural environments were suggested. The fact that the computer might be overused, misused by inexperienced teachers, used for inappropriate topics, or to repeat things which can be done in other ways, was thought of as undesirable. Approaches with reduced ways of interaction or individual exploration and CAI programs that rely more on rote than on reason seem counterproductive to cognitive development.

In summary, the group was unanimous in the feeling that students may develop a different perspective on mathematics as a consequence of working with the computer. It is not yet clear what cognitive structures are being developed by exposure to computing. Examples have demonstrated that technology can be very useful, depending on how it is used. While it should not replace any other medias, the computer may certainly complement them. Being in the early stages of exploring ways of using technology, we need to search for a balanced use of computers.

Television, Video and Film

Film and Video in Mathematics Education

Film and Video about Mathematics Education

The majority of participants in these two groups opted for Film and Video in Mathematics Education. The Film and Video about Mathematics Education group started with just two members, but grew in size daily. Members of both groups attended general viewing sessions and film/video-related Project Presentations — Emmer (Italy), Hadar (Israel), Koumi and Jaworski (UK). The two groups combined for the final session.

A fundamental problem facing those who wish to discuss film and video (film/video) material is that it needs to be viewed, in real time, in advance of the discussion itself. When a film/video lasts for a total of only 5 minutes, the viewing can reasonably be incorporated into the time allocated for discussion. But many educational f/vs are of 20–25 minutes' duration. To show only short excerpts will give some impression of style and content but it cannot adequately give a feeling of the overall *structure*, nor of whether it has achieved its stated aim. This is particularly true when the film/video forms an integral part of a larger teaching scheme, as with distance education.

In attempting to overcome this problem, we encouraged participants to view film/video material at ICME 5 *outside* the Theme Group sessions. This was not easily achieved in practice and we have to acknowledge that our expectations exceeded our achievement. Participants have since asked for more predigestion of material by the group organisers and it may be that we should have narrowed considerably the range of material under discussion.

On the positive side, this was the first ICME to have included specialist working groups devoted exclusively to film and video. (1984 also saw the First International Congress on Mathematical Movies, held at the University of Turin.) An encouragingly large number of participants were either already involved in film/video production or were shortly about to start work in this field.

Koumi (UK) drew up a suggested checklist of some 35 points to be borne in mind when devising an educational TV program (not necessarily mathematical), grouped under 9 main headings: Hooking, Signposting, Scripting, Coherence, Sympathy with viewer, Active (mindful) learning, Attuning (making the viewer more receptive), Range of viewer ability/preference, Ending. Examples (good and bad) were taken from BBC Open University programs to illustrate these points.

One participant called for more discussion of material known (by its producer) to be *poor*. One such example was bravely brought to the group, Film and Video about Mathematics Education, and subjected to scrutiny: the (amateur) producer has since made a tape/slide presentation of the same material (for teacher education), claimed to be more effective but lacking the portability of videotape.

Another poor example was shown in the group, Film and Video in Mathematics Education. It consisted of a single shot of a teacher writing on a blackboard. The teacher/producer stated that it had been made simply to obviate the need for repeating the same lecture every year. Most participants felt that, nevertheless, more thought should have gone into the presentation, for example, by doing all the blackboard writing in advance of the videorecording, and revealing it progressively, with consequent gains in legibility and in flexibility of pacing.

Because utilisation procedures are different from one country to another, discussion tended to focus on production details within each film/video rather than the place of film/video within a particular educational structure. Not all countries have national or even regional curricula, not all production companies are exclusively concerned with educational material, some teachers rely heavily on broadcast material, others have access to good video distribution systems, and so on.

The possibility of setting up an international Newsletter and maybe an exchange scheme whereby producers of mathematical film/video (and their advisors) can view each other's material between one ICME and the next is being explored. One discussion which might be continued between ICMEs is the clear distinction which emerged between tertiary level material and that for schools (primary and secondary). The examples of the latter which were shown, from Israel and the United Kingdom, concentrated much more on motivating students than did the tertiary level material from the Open University (UK) that is, the OU material was much denser cognitively. Participants felt that one reason for this was that the material for schools was intended as a classroom resource, to be introduced and followed up by the teacher; in contrast, OU programs are intended for distance learning by individual adult students without a teacher. This rationale for the distinction in styles may or may not be valid; perhaps school film/video should increase its cognitive element while tertiary film/video should increase its motivational element.

A guest at the final session was Robert Davis (USA) who presented a 20 year old film recording of a real-time videotaped lesson conducted by himself. It is a tribute to the production procedure that this class-observation material is as valid today as in the 1960s; the fact that we all became involved in the teaching procedure suggests that the video recording had done its job well, despite our efforts to examine it in its own right. The medium is definitely *not* the message, but it can be an extremely effective messenger.

Classroom Dynamics

Feedback to the Learner
The group spent some time viewing software, both to explore the varying types of feedback provided and to provide a common software base to support discussion. Throughout the sessions, attention was always strongly focussed on the use of technology to enhance mathematical education rather than on the mere exploration of computers. The results of our deliberations can best be summarised by a definition, some dimensions for classification, and a diagram of the dynamics of a feedback system.
Definition: *Feedback* is a response, from the computer or teacher or other learners, to the learner's interaction with them.
Classification: The following dimensions for classification and the examples given are designed to highlight aspects of the interactions between the computer and the students which will need to be considered when the focus is on possible styles of feedback to the learner. There is no attempt to make value judgements; the aim is to give an indication of the spectrum of possible feedback. Although we use the term 'classification', the field of computers in mathematics education is in its infancy and fast evolving. Developments in hardware and software, and our understanding of their implications for teaching and learning, will continue to grow and change.
1. Psychological effects of feedback. (behavioural — cognitive; attitudinal — feedback being competitive and providing satisfaction)
2. Mode of presentation of feedback/sensory modality. (sight — text, graphics, colour, special effects, animation; sound — music, speech, special effects; touch — special peripherals)
3. Instructional mode — feedback variation: drill-and-practice, tutorial, enquiry, simulation, text, game, computer-managed instruction, problem solving, word processing, laboratory, guidance, demonstration)
4. Interactions ('A/B' here indicates 'A *to* B') between different parties. (computer/learner, computer/teacher, teacher/learner, learner/teacher, teacher/computer/learner, learner/learner)
5. Degree of help of feedback. (limited/extensive, tutorial, diagnostic/

    remedial)
6. Predictability of the feedback. (unexpected/predictable, anticipatory/non-anticipatory)
7. Frequency of the feedback. (spaced [rarely receiving], massed [frequently receiving])
8. Machine capability or limitations with regard to feedback.
9. Correctness of responses. (correct, non-committed, incorrect)
10. Type of user considered or not considered by feedback offered. (special needs of the learner)

    It was felt that the desirable features of feedback would include graphics, sound, colour, animation, humour, variety, clarity, engaging positiveness, simple responses and the promotion of other activities. Undesirable features would include lengthy texts, mathematical symbols, complex keyboard response and the need to move to other screen displays.

    It should be noted that, as with all curriculum material, the educational aims have top priority. For practical reasons, some desirable features will frequently be excluded and some undesirable features included. However, it was felt that due attention should be paid to this rather idealistic categorisation.

Feedback Systems — one approach

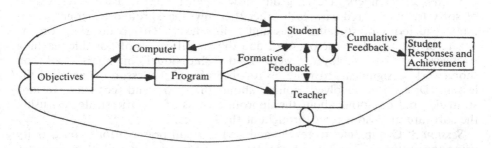

This diagram highlights the difference between formative (or immediate) feedback occuring during the actual use of the computer and the cumulative feedback of the total student involvement with the teaching package. Additionally, the outer loops emphasise the use of feedback in the development process — indeed, the continual process of development — undergone by teaching material.

    Conclusion: As indicated above, the group considered mathematical education rather than computing. Indeed, the general feeling was that, as with more traditional materials, the majority of the feedback would take place between the student and teacher, and student with fellow students. However, our analysis should provide useful guidelines for software designers, publishers and users.

Roles of the Microcomputer in the Mathematics Classroom
Reviewing computer programs for their educational value presents a serious problem as it is not the program alone but possible uses of the material that should be explored. In this group a very practical approach was taken. Several

computer programs, PIRATES, EUREKA, SNOOK (ITMA Longmans UK), LES LIGNES DROITE (Marie–Louise and Serge Hocquenheim, France) and GREENGLOBS (Sharon Dugdale, USA) were explored by small subgroups. The variety of ideas and actual lesson presentations that were worked out will be presented elsewhere. To share the group's teaching ideas it would be necessary to describe the programs in order to put the reader in context. Space prevents this here. However, members of the group found the organisation of the working sessions a useful model on which to base IN-SERVICE training that aims to introduce teachers to the idea of using the microcomputer as a 'teaching assistant'. It was thus decided to report here the way in which the group worked and to refer you to the fuller publication to read the details of the teaching ideas.

What actually happened — four sessions

Session 1: The group members subdivided into six sub-groups, each at a different computer station and looked at a piece of software to investigate its potential for use at various grades and ability levels and to share this investigation with each of two other groups. After 30 minutes, the sub-groups moved round to new stations and began to select specific software to be used as the tool and springboard for developing a lesson. At the end of Session 1 participants signed up for particular groups on the basis of the software they wished to work with, and met to plan action for Session 2.

Session 2: Members of each group now began more extensive investigation of software and shared with each other their own background experiences and interest in terms of students' ages and ability levels. Group decisions on the nature of the activity to be planned, and the identification of possible teaching strategies to be employed, were made. Individual members took on 'homework' assignments to write up rough drafts of the lesson or pieces of the lesson. Discussion involved points about the pros and cons of various strategies and questions about the appropriateness of specific strategies, using the software in various roles throughout the lessons.

Session 3: Groups met to review and revise lesson presentations and activity using the software. The biggest problem was to incorporate all ideas — these varied according to age and ability range of students considered. Each group agreed on the major portion of the lesson to be presented to the whole working group in the time allotted to them. Participants were encouraged to submit their own written ideas as well.

Session 4: Presentation of the 'lessons' involved role-playing by the participants who became involved in the lessons with great enthusiasm. Thus all participants had an intuitive feel of the work of each small group. This allowed the maximum exposure for each person to many of the possible roles played by the micro in the mathematics classroom. Many of the participants had not been aware of such a dynamic approach and expressed enthusiasm and declared an intent to work in this manner with groups at home.

The computer programs most certainly acted as a catalyst to the discussion of teaching and learning, promoting a good exchange of ideas and information on different national approaches.

Managing Learning

Our brief was to discuss the different learning environments which made use of a microcomputer and to consider how these would effect students' learning. This was clearly a very broad topic where a number of different factors contributed to the total environment.

The software itself can be thought of as an environment. This was most evident with systems like LOGO and with adventure games, but the idea extends naturally to smaller 'microworlds' where the software consists of simulations, games or tutorial materials.

Hardware also directly affects the learning environment. Computer networks offer different possibilities from stand-alone machines. The type of hardware purchased by a school, and its physical location, have a strong influence on the way computers are used to promote learning — and indeed, if they are used at all for this purpose.

However, there can be little doubt that the most important factor in the learning environment is still the teacher. Teachers' attitudes, perceptions, experience and knowledge of computer-assisted learning (using the term in its broadest sense) still control what really happens in a classroom, even if they may sometimes choose to delegate some of their responsibilities to the machine.

In order to establish some common points of reference, some time was spent looking at three pieces of software: DARTS and GREENGLOBS and 'L' — a mathe-magical adventure (Association of Teachers of Mathematics, UK). A number of short video recordings of teachers and children working with computers, which illustrated some of the contrasting styles of classroom organisation were also viewed.

In summarising the discussion, it must be noted that there were differences on a number of fundamental issues. The relative merits of having a single learner at the keyboard, a small group around the keyboard, and the whole class use of microcomputers where the teacher is present all the time were argued. The relative merits of author languages, where teachers largely design their own materials, and of pre-packaged software, where teachers have less direct control over content produced divided opinion. However, it was clear that there is potential in all these approaches.

There was agreement on a number of important points. The existence of computers must change what is taught in mathematics, and the use of computers in the classroom must change how it is taught. This pressure from new technology was seen as a positive force in mathematics education. The immediacy of response of computers, the way they often free teachers from low-level cognitive tasks, and many other computer talents, combine to make the computer a teaching aid of formidable potential. Its motivational power seems considerable and is especially striking with children who have behaviour disorders. However, there was some caution in discussing the long-term motivational value of the computer: while generally optimistic the group were well aware they were discussing forces they did not entirely understand.

Optimism was tempered by an awareness of a great many practical obstacles which impede the successful adoption of computer-assisted learning in mathematics. Shortages of hardware and software frustrate teachers' awareness of the possibilities, and attitudes to any innovation change very slowly — it might take a whole generation before the full benefits reach the classroom.

Teacher Education

Computers and Inservice Education: Big Problems or Big Solutions
This group reported under Theme Group 2, Professional Life of Teachers.

Computers and Pre-Service Teacher Education
Great similarities exist between the different countries: within the next five years, every school plans to be equipped with computers. Often there are more computers than teachers able to use them. Often there are more teachers willing to use computers than computers available. Some special computer science courses for teachers and students have been developed. Many uncoordinated developments have occurred. Adequate international dissemination and reflection is required.

The group noted that we are training teachers who will teach for 30 or 40 years. We cannot give them a definitive body of knowledge but we have to enable them to adapt and evolve. At the present state of development, there is a strong link between pre-service and in-service training. In-service training is, in fact, 'pre-service' because of the constant evolution of informatics and of pedagogy. Computer science is more than a technical tool, it is also a tool for thinking, reasoning, and problem solving, and it can lead to different thinking styles. Questions of means are as important as ends. Computer science is embedded in the curricula, it is not a complement to it. Student teachers must be taught, not only computer science, but also pedagogical aspects of its use; not only programming, but also algorithmic thinking. There is a need to re-centre the role of algorithms in mathematics, to stress the main algorithmic concepts and methods for problem solving, and the different ways informatics appear (data banks, work processors and expert systems).

Computer science changes the philosophical foundations of mathematics. It affects what mathematics is to be taught, the development of symbolic systems, doing more experimental mathematics, new skills, observation, visualisation, simulation and numerical verification. Computer science changes pedagogical styles. The computer will alter the pupil-teacher relationship.

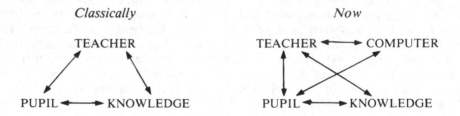

The working group expects considerable turbulence in the development of mathematical and computer science education due to the increase of information technology. National and worldwide meetings will be helpful in identifying true directions. The working group is also aware of deep philosophical questions which information technology raises in our society. Teachers have to be alert and ensure that students learn how to use information technology as a powerful tool and how to control it completely.

Miscellaneous

How to Start with Computers?
The microcomputer appears to offer the richest possibilities for mathematics. One should realise, however, that the microcomputer is only the third stage in the development of computers in education: central processing, computers in schools, computers in classrooms, individual student computers. It should also be realised that the microcomputer is not the cheapest solution and that its size makes it difficult to have more than a few in an ordinary, general purpose classroom. But if it seems reasonable and feasible to introduce computers into mathematics, then it should be done as soon as possible.

The group was adamant that students' introduction to computers should not be in mathematics. It was felt that art, music, word processing and social studies were much better places for this to occur. It was also felt that studying programming and language syntax was relevant in mathematics only when this would lead to the investigation by the student of non-trivial mathematical problems. Introduction of microcomputers may lead to changes in classroom organisation, for example, individual student timetables or work schedules. One valuable class project is the production of the class' own computer users manual.

There will usually be only one microcomputer in a classroom to begin with. High resolution graphics and disk storage were deemed to be the only initial hardware considerations for mathematics. It was noted with approval that there was now available software designed for these conditions. The software observed by this group was ITMA (UK), RIME (Australia), and Maths for the Real World (Australia). Each of these was only a part of a curriculum package, either teacher guidance or student text, and was designed to present students with concept building challenges rather than with ready made answers. In selecting software, it should be borne in mind that its purpose is to help students learn mathematics and that it should be usable in more than one way, allow the user to remain in control, and have the potential to reveal mathematical richness. Quite often, simply written programs or seemingly unattractive or irrelevant material, for example, simple simulations, drill and practice or games have proven to be beneficial in teaching students or motivating them to learn. It was felt that producers should develop licensing systems to allow schools to make copies for use in single machines or in networks.

Teachers about to begin with computers in any field will need technical and educational knowledge. They may begin with a few activities and build upon these. For the future, however, it is essential that teacher training institutions ensure that their students are able to use computer materials in their teaching and that they are able to appraise such materials and their promoters' motives positively and critically.

Technology for Education of the Handicapped
This group did not attract participants and was therefore cancelled.

Technology as an Aid to the Teaching of Problem Solving
This group is reported under Theme Group 7, Problem Solving.

References

Buchberger, B., Collins, G.E., and Loos, R. (eds), (1983) *Computer Algebra: Symbolic and Algebraic Computation*, Second Edition. Springer- Verlag, New York.

Stoutemeyer, D.R., (1979) 'Computer Symbolic Mathematics and Education: A Radical Proposal', *ACM SIGSAM Bulletin 13*, pp. 8–24.

Yun, D.Y.Y. and Stoutemeyer, D.R., (1979) 'Symbolic Mathematical Computation', *Encyclopedia of Computer Science and Technology* (ed. Belzer, J., Holzman, A.G., Kent, A.), Vol. 15, pp. 235–310. Marcel Dekker, New York.

THEME GROUP 4: THEORY, RESEARCH AND PRACTICE IN MATHEMATICAL EDUCATION

Organisers: Alan Bell (UK), Jeremy Kilpatrick (USA), Brian Low (Australia)

Theory, research, and practice belong together, but the separation of professional roles of teacher and researcher produces a tendency to separate development. The discussion focussed on those ways in which research activity, whether performed by professional researchers or not, interacts with teachers' approaches and practices.

Nine groups were formed to allow the 300 theme participants to move into groups of 15 to 20 for discussion. The groups covered areas in which there was sufficient research material to allow useful interaction between classroom teachers and researchers. Each group contained one or more leaders who organised presentations of research for the small group discussions. The reports from these groups follow.

Primary School Mathematics

Conveners: Tom Carpenter (USA), Gerard Vergnaud (France)

A great amount of research activity in recent years has focussed on aspects of primary school mathematics, especially on the nature of the counting process, the recognition and renaming of large numbers, and the solution of verbal arithmetic problems, especially those requiring addition or subtraction of small numbers.

The methodology for much of this work has consisted of:
- a theoretical analysis of the conceptual fields;
- an analysis of the procedures that children use to solve problems within each domain;
- the construction of models of the underlying cognitive structures which might explain the children's procedures.

Counting and Numeration. The Piagetian emphasis on the invariance of numerosity over changes of configuration has, until recently, tended to obscure the fact that *counting* is a complex, yet fundamental acquisition. Its developments into 'counting on', without recounting an initial set, and into the grouping of numbers, eventually by tens, leading to understanding of the numeration system, all need careful nurturing. The learning and recall of subtraction facts is treated by many pupils (36% of 12 year olds in one study) by counting; the thorough development of addition facts before subtraction followed by emphasis on the inverse relationship between these can improve this situation.

The learning of all these ideas is inhibited by too early symbolisation and recording. Physical materials to help this learning (as developed both in the west and in Japan) remain useful at the stage of learning written algorithms, but the steps of the physical and written procedures need to be closely related in the children's activity.

Word Problems in Addition and Subtraction. These have been classified into four main categories — Change, Combine, Compare and Equalise, and into 20 subcategories. The ability to solve these problems develops alongside the work described above. Children's methods move from direct modelling of the problem structure with counters to the use of increasingly efficient counting methods and of known number facts. In one study, 70% of errors were in choice of operation, rather than in computation. Improvement in solving these problems may come from identifying the solution methods associated with the different problem structures, and constructing 'didactical situations' in which the manipulation of these and other didactic variables (such as size of the numbers) can provoke progress to higher level methods. In order to do this, it is essential for the teacher to understand the children's normal approaches, as revealed by the research.

The ability to solve logical design problems may be facilitated similarly by presenting progressively more difficult problems in a concrete medium (mechanical flip flops).

Tertiary Level Mathematics

Convener: Shlomo Vinner (Israel)

Much of the theory and research in mathematics education at the tertiary level lies within the general framework of error analysis and identification of students' misconceptions. It is reported that very often students apply rules and procedures that are not suitable to the given problems, but that have been selected because there are some superficial similarities with problems to which these rules and procedures can be applied.

With respect to mathematical concepts, there is a gap between the mathematical definitions and the students' concept images. Students' primary intuitions of certain mathematical notions do not change when they first meet the concepts in tertiary level mathematics, and these intuitions may conflict with the mathematical definitions. The conflict is probably a result of the short time students have in which to learn and adjust to the new concepts. A considerable amount of research is available, particularly on concepts of limit, real numbers and the calculus.

How might the results of this research be used to develop different approaches to teaching some of the concepts involved in tertiary level mathematics? In general, it was argued that for the majority of students the concepts should be introduced in a way which is sensitive to the students' intuitions and existing schemas. This can be achieved in many cases by avoiding unnecessary rigour, formalities, and over-subtle mathematical notions. In other cases, special attention should be given to recognising and resolving students' conflicts.

A particular example of one approach is an introduction to group theory, based on having students attempt many concrete activities and examples. The formal definitions are introduced much later, after students have had sufficient experience with the group structure and have been able to modify and develop their intuitive schemas. While some teachers doubt that the method can be used with large groups found in mathematics courses which have a service function, others do not see this as a major problem.

These are just a few examples of the many aspects of the learning of tertiary level mathematics which need investigation.

Fractions — Rational Numbers

Conveners: Tom Kieren (Canada), Merlyn Behr (USA)

Results of research. Surveys in several countries suggest that, in general, current programs of instruction are not yielding a high, or even a reasonable level of performance with rational numbers. Various changes in textbooks appear not to have led to improved performance. Radically different approaches to fractions do, however, seem to produce different types of understanding.

Children's understanding of whole numbers and the schemas they have developed for working with whole numbers have overpowering effects on their work with common and decimal fractions. Given a problem in which, say, 4 is to be divided by 12, pupils may make any of a number of transformations of the problem to allow a whole number answer. In the short term, instruction can counteract this tendency, but in the long term children are likely to revert to older schemas. Some researchers see the dominance by whole number schemas to be a problem of language; the language of instruction must build on naturally occurring language or children will not continue to use it.

The formation and development of the concept of unit fraction needs to include consideration of both discrete and continuous situations. Not all situations are equally easy for children to understand, and they do not automatically build number ideas from them. When they are given only a single quantitative model of fraction, their understanding is hampered. Tasks involving fractions should include an emphasis on the number idea and should make use of the associated (and, at first, informal) language.

Some research on operations with fractions suggests that, although the multiplication algorithm is easily taught and learned, it can lead to poorer learning of the addition algorithm. Performance may be improved if the addition algorithm is taught first. Such suggestions, however, depend on how instruction relates to a better understanding of how children build up rational number ideas. Children may have 'missing links' in their understanding. Links needing attention are those between symbols and their meanings, between natural knowledge structures and formal knowledge structures, and between syntactical structures and mathematical-logical structures.

Implications for Practice. Teachers should be aware of pupils' viewpoints concerning fractions. The teacher should be like a consulting architect — knowing the ground of a pupil's knowledge and helping the pupil to build a knowledge structure on it. To achieve this, teachers need to follow paper-and-pencil testing with class discussions and individual interviews. The assessments should tap pupils' intuitive knowledge of fractions in the real world, allow for variations in language, and provide mathematical and contextual diversity.

During instruction, teachers should provide intuitive knowledge-building experiences with great attention to language use, mental imagery, and mathematical variety. Various games in which pupils estimate the relative sizes of fractions or order them on a number line can be used.

Algebra

Conveners: Lesley Booth (UK), Nicholas Herscovics (Canada)

It seems that children continue to have difficulty in understanding algebra. Some of this difficulty seems to be due to problems in understanding algebraic concepts and algebraic symbols; some to the fact that children often do not seem to 'reason algebraically'; and perhaps most of all to the fact that children often do not understand what algebra is all about and what the purpose might be in studying it. Teachers and researchers alike have been concerned with trying to identify the reasons for these difficulties and in attempting to develop teaching approaches that will help children overcome them.

Research into children's problems in learning algebra has suggested various possible sources of these difficulties. For example, attention has been drawn to the various ways in which children may interpret letters in algebra and to the meanings they may give to symbols, such as the equals sign or conjoined terms such as xy or $4x$. Children's construction of conventions, such as those for the use of brackets and the order of operations, may similarly lead to problems in algebra, as can their conceptions of a *term* or *side of an equation*. Attention has also been drawn to children's difficulties with such algebraic thinking processes as generalisation, reversibility and flexibility. In addition, children's difficulties in algebra stem in part from their conceptions in arithmetic and from their use of informal procedures in that domain. Work on children's equation-solving procedures has similarly illustrated the use by children of informal or intuitive procedures. The following aspects are embodied in the teaching approaches of work done in Canada, the United States, England, Israel, Australia and South Africa:

- Providing situations and experiences that create a need for algebraic expression; for example, problems involving number generalisation (such as that the sum of three consecutive numbers is always a multiple of 3).

- Making a similar study of the various laws governing arithmetic notation (such as $a - (b - c) = a - b + c$).

- Drawing direct parallels between valid ways of transforming numerical expressions ('arithmetic identities') and operations on corresponding expressions in which a number is 'hidden' by a letter.

- Exposing and discussing the distinction between algebraic expression regarded as (a) an elaborate 'name' for a number; and (b) an instruction to perform certain operations on a number yet to be chosen.

- Easing the transition from a numerical generalisation to its symbolic algebraic expression by the use of verbal statements and other placeholders ({ }).

Probability and Statistics

Conveners: Hans–Joachim Bentz (FRG), Manfred Borovcnik (Austria)

Descriptive statistics — methods of representing data graphically — occur to some degree in both elementary and secondary courses. The basic concepts of probability are generally taught in the secondary phases though there have been experiments with their introduction at earlier ages. Sampling theory and the study of frequency distributions have been reserved for the specialist levels of post-16-year-old or even tertiary level courses.

Research on understanding of these ideas has not been extensive but has expanded in recent years. Misconceptions in probability are not as easily detected as in other areas of mathematics because of the special nature of stochastic intuition. Research shows also that there are quite different approaches to the basic concepts of probability, such as a definition in terms of long-run frequency, a subjective approach, or an 'a priori' approach. However, much of the research has some difficulty with the wording of questions and in the experimental contexts used for the questions.

Important ideas that arise from the research include the value attached to children gaining background experience for the understanding of stochastic concepts and the place of notions of randomness in the school curriculum.

A number of curriculum projects have been developed for children in the 11 to 16 years age range. A major aim of these projects is for the children to gain experience in stochastic situations and an important characteristic is their attempt to reduce dependence on algebra, arithmetic and combinatorics as much as possible. Topics that are seen to be suitable for the lower secondary school syllabus are the misuse of statistics, sampling (representative rather than random), and the distinction between inference and deduction.

The use of teasers, paradoxes and fallacies in stochastics is a valuable teaching device. Such problems may be used to clarify ambiguous stochastic situations, to introduce basic concepts, and to assist in the formulation and interpretation of results.

Many problems still remain in the teaching of probability and statistics. The use of continuous distribution functions, the concept of dependency, and the relation of physical and stochastical dependence are some of the problems needing further research. Although the concept of 'fairness' appears to be well-developed in children aged 11 to 16 years, one remaining problem is how to link probability and statistics concepts for these children.

Geometry, Spatial Awareness, and Visualisation

Convener: Michael Mitchelmore (Jamaica)

Traditional geometry teaching has dealt mainly with the properties of plane and solid figures as an organised deductive system. But there is an expanding awareness of the need to move more slowly towards deductive activity by first experimenting more informally with objects and figures themselves. More radically, there is interest in developing pupils' ability to visualise objects and spatial situations, large and small, and to manipulate them mentally.

Spatial awareness. Tasks such as giving directions to a stranger, sorting plane shapes and solids and fitting them together to create desired figures or patterns, making up objects from drawings and vice versa, all help to develop spatial awareness, particularly when they involve first predicting or imagining what is required, then checking it by construction. Informal experience has shown that pupils can perform these tasks with enjoyment and show improvements in these abilities from the age of 4 upwards. The Japanese curriculum also includes the use of straightedge and compass at age 8.

Some recent research in Papua New Guinea distinguishes the ability for 'interpreting figural information' from 'visual processing' and shows that it is relatively easy to improve pupils' weaknesses in the former, but that the latter is a deeper-lying ability more dependent on cultural influences.

Some attempts have been made to construct full curricula for developing spatial awareness, but these have not so far been fully tried or evaluated. Computer-based materials, including 2-D and 3-D displays are also being developed.

Stages in Geometry and Proof. The van Hiele levels can be seen as an elaboration of the British Stages A, B and C; they also have links with Collis's SOLO taxonomy, as well as with Piaget's general developmental model. Level 1 is that of global recognition and naming of shapes; in Level 2 the properties of figures are analysed, and in Level 3 figures are organised according to their common properties and definitions made. Level 4 concerns the role of postulates, theorems and proofs, and in Level 5 the pupil becomes aware of how different axiom systems determine different abstract geometries.

Research so far shows that tests constructed to determine pupils' levels also predict performance in relevant aspects of the geometry course. It also confirms that most pupils (and student teachers) do not progress beyond Level 2, though it should be possible with suitable instruction to bring more of them to Level 3.

3-D Geometry. A useful approach is to consider the relation between form and function in common objects (for example, a bucket); map reading, giving directions and 3-D drawing should also be taught.

Transformation Geometry. These ideas are, from the mathematical viewpoint, more general, intergrative and applicable, and psychologically more intuitive and accessible than traditional geometry. The spontaneous development of concepts of symmetry can be observed from children's paintings. Building transformation geometry into a rigorous deductive system presents difficulties, but a course following the 'stages' approach would leave such matters for older and abler pupils. A current problem is the lack of knowledge of these topics on the part of teachers.

Large-Scale Surveys of Mathematical Attainment

Convener: Kath Hart (UK)

Large-scale surveys have developed in recent years, both in extent and in the design of items and analysis of results, so as to provide more useful diagnostic information. Used in this way, they can offer new insights into pupils' mathematical reasoning and provide more realistic appraisals of attainment on which to base:

- suggestions for classroom practice;
- directions of curriculum change;
- teacher training programs;
- future research.

Concepts in Secondary Mathematics and Science was a research project in England that sought to formulate levels of understanding (hierarchies) in 10 mathematical topics currently taught in secondary school. This was done by testing 300 children with word problems in a one-to-one interview setting, followed by the use of the same problems in a paper-and-pencil test format with 10,000 children. Of particular significance in the results was the indication of certain erroneous responses being given by large numbers of children from different schools and of different ages. Such evidence led to a deeper investigation (Strategies and Errors in Secondary Mathematics) into the causes of the errors and possible remedies that could be suggested to teachers.

The Assessment of Performance Unit (APU), also based in England, is a government funded project charged with monitoring mathematical attainment in the schools. Over five years a large sample of 11 year olds and an older group of 15 year olds have been tested on concepts, skills and attitudes to mathematics. A unique aspect of the APU has been the introduction of the testing of practical work in which the child is asked to do some mathematical problems that require the manipulation of materials or data.

The International Study of Education Achievement (IEA) is carried out on many children in a variety of countries. The use made of the results by the participant countries will obviously vary. One country at least, Thailand, has tried to use the results to evaluate the effectiveness of changes in the curriculum. Large-scale surveys of numeracy have been completed also in Australia and New Zealand, some of them related to aspects of the IEA studies in mathematics.

An important result from large-scale surveys is that some mathematical concepts are much more difficult for children than is commonly believed. Although the survey is a very suitable method for detecting such widespread errors across countries, it cannot be used to infer children's forms of understanding. Detailed interviews and case studies are needed to follow up a survey.

There are some difficulties in the dissemination of results from surveys. It is often difficult for teachers to abstract important information for their teaching and curriculum decision making from the wealth of data available from a survey. Many teachers cannot see the practical relevance of the results for their teaching. Also, some of the data may be politically and socially sensitive, and so it is not disseminated by the government bodies responsible for the surveys.

However, large-scale surveys can provide rich data bases for teaching and curriculum decisions, and the newer developments in oral and practical testing have exciting possibilities for extending understanding of children's mathematics achievement.

Problem Solving, Proof and Process Aspects of Mathematics

Conveners: Ed Silver (USA), Claire Dupuis (France)

Mathematical competence includes, as well as the possession of a range of relevant concepts and skills, the capacity to approach situations in a mathematical way; that is, to deploy strategies for problem solving in the broader sense, including investigating situations, representing situations, formulating problems, proving and systematising. Research in this area includes studies of pupils' proof activity, investigations of pupils' problem solving processes, and experiments in the teaching of problem solving and proof.

What specific examples can be given of theory and research that are relevant to practice? In what ways is the research relevant?

Research in the area can be useful for educational practice because of its underlying theoretical constructs and the problem tasks used in the research. Both of these enable transfer to the classroom. Two particular important areas of research are metacognition and pupils' perceptions of problem similarity. The research on metacognition has suggested the importance of giving instructional attention to pupils' beliefs about mathematics and problem solving and also to their ability to monitor and evaluate their progress in a solution episode. The research on perceptions of problem similarity has suggested attention to helping pupils *organise* their problem solving experience in ways that are useful and appropriate. Moreover, this research suggests many tasks that could be used in the classroom.

What specific examples can be given of promising practices in the teaching of problem solving, and what is the connection between these examples and existing research and theory?

A great many distinct activities are conjoined under the general heading, 'the teaching of problem solving'. Some of these activities involve teaching *for* problem solving, in which the particular problems are the focus of instruction because they represent an important class of problem to be mastered in the curriculum. Other activities involve teaching *with* problem solving, in which problems are used as convenient vehicles for reinforcing concepts or skills that are central to the curriculum, or, less frequently, problems are used to introduce or motivate important curricular topics. A third class of problem solving activities involves teaching *about* problem solving, in which the problems provide a convenient vehicle for acquainting pupils with certain aspects of the nature of mathematical problem solving.

There appears to be a great deal of interest in this topic in many countries, including Australia, Canada, the United Kingdom, Jamaica and the United States. A large number of curricular materials have been produced in these countries, and a list of references to some of these materials is provided in the supplementary document. However, not only is the connection between these materials and existing theory and research slight, but also the use of these materials by teachers is sporadic. One hopeful sign is that research and researchers appear to be having some influence on recent innovations in assessing problem solving on state or national examinations. To the extent that problem solving is represented in these examinations as an important goal, practitioners and curriculum developers will certainly reflect that emphasis in educational practice.

Dissemination of theory and research to teachers is a difficult task. In order to be effective, communication of research must affect teachers' knowledge of problem solving, their beliefs about themselves and about problem solving, and their attitudes toward problem solving. Only when these have been adequately affected can one hope to see a change in teachers' actions in teaching problem solving.

In many countries there is a growing interest in research on the part of practitioners. The time may be ripe for building permanent bridges between researchers and practitioners in the area of problem solving.

Classroom Studies, Teaching Styles, Attitudes

Conveners: Douglas Grouws (USA), Gilah Leder (Australia)

Research under this heading concerns teacher-pupil interactions, social and organisational aspects of the classroom, attitudes to mathematics and questions relating to language use and development.

In the management of teacher-pupil interaction, the 'wait-time' between posing a question by the teacher and calling on a pupil to answer is a significant variable; times less than 1 second are usual, but times of 3 to 5 seconds provoke higher level cognitive activity. Another result is that in dividing the available class time between teacher explanation and discussion ('development time') and pupil practice, giving rather more than half the time to 'development' is associated with greater learning. Regarding class organisation, there is interest in group work in which small groups of pupils are assigned different tasks; the faster pupils either taking each topic farther than the slower ones, or else continuing at a faster pace through the entire course. Whether or not this leads to greater pupil progress depends on the nature of the tasks assigned and of the teacher-pupil interactions. In some cases, class teaching is more strongly associated with higher level cognitive demands and group work with a preponderance of 'managerial' teacher-pupil contact. The research on teacher-pupil interaction suggests that teachers need to vary their method of presentation much more than commonly occurs. The advantages of small group interactions should be explored further.

One of the few attempts to *change* teachers' styles and methods is the Missouri Mathematics Program. Teachers' effectiveness (measured by the progress of their 9 year old pupils) was improved substantially when they attempted to follow a plan in which they expounded clearly, were task focussed, non-evaluative and relaxed yet brisk-paced.

Language. Mathematical progress is related to linguistic power, except when the mathematical tasks are mechanical. Some teaching developments are aimed at finding and using effective language-provoking situations; games, calculator activities, familiar and realistic situations are helpful. Learning mathematics in an unfamiliar language presents special difficulties. A personal case study illustrates this.

Attitudes and Self-Concept. Research among Thai pupils shows a desire for more explanation of lesson objectives, and of poorly understood points and difficult exercises. Some Japanese university and college students perceived mathematics as difficult, important, of average interest, but not very much liked.

Case studies of individual pupils can provide unique insights into the influence of social and academic factors on their achievement and self-concept, increasing the teacher's awareness of how school and mathematics appear to the pupil. The primary/secondary school transition is a critical time

in this respect. Sex differences in mathematical learning are related to the effect of socialisation experiences. A survey of articles in the popular press revealed three common themes relevant to a female 'fear of success' factor:
- females need to work harder than males to achieve the same goal;
- females have to balance success and interpersonal relations;
- success somehow happened; it was not expected or sought after.

Teaching Mathematical Strategies. Comparative teaching experiments in which classes of 10 year olds were taught problem solving (a) with direct teaching of named strategies along with the solving of relevant problems, and (b) by experience of solving problems, with subsequent reflection on the successful methods, showed improvements in problem solving performance with no great difference between the groups except for somewhat greater flexibility related to the second treatment. In another experiment, the teaching of modelling to 16 and 17 year olds was successful, the more so if mastery criteria were applied.

Researchers generally work within a particular characterisation of effective teaching, and their conceptualisations of a good teacher continue to be multidimensional. They are searching for techniques that will allow the rich classroom data source to be quantified realistically and accurately. A major problem is to strike a balance between extensive and intensive monitoring of the behaviours of interest.

References

Bell, A.W., Costello, J., and Kuchemann, D.E. (1983). *A review of research in mathematical education. Part A. Research on learning and teaching.* Windsor, UK: NFER–Nelson.

Driscoll, M. (1982). *Research within reach: Secondary school mathematics.* Reston. National Council of Teachers of Mathematics.

Driscoll, M. (1983) *Research within reach: Elementary school mathematics.* Reston. National Council of Teachers of Mathematics.

Ginsburg, H. (Ed.), (1983) *Development of Mathematical Thinking.* New York. Academic Press.

Hart, K. (Ed.) (1981). *Children's understanding of mathematics: 11–16.* London: Murray.

Jones, G. et al, (1984). *Research in Mathematics Education in Australia,* Brisbane, MERGA.

Lesh, R. and Landau, M. (Eds.) (1983). *The acquisition of mathematics concepts and processes.* New York: Academic Press.

Silver, E.A. (Ed.) (in press). *Teaching and learning mathematical problem solving: Multiple research perspectives.* Philadelphia: Franklin Institute Press.

THEME GROUP 5:
CURRICULUM DEVELOPMENT

Organisers: Geoffrey Howson (UK) and John Malone (Australia)

1. Introduction

Much activity has taken place in the past twenty or so years under the heading of 'curriculum development'. Yet, despite enormous effort and a great expenditure of money, very few changes of a fundamental nature have taken place. Changes in content have been accepted or rejected in a manner reminiscent of body organ transplants. At times waves of enthusiasm for curricular changes in teaching practice have swept through classrooms. All too often though, as the waves have receded, calm has returned and with it a return to well-tried orthodoxy.

Such curricular changes — whether permanent and successful or not — are great consumers of resources. They use economic, human and intellectual resources, and, moreover, resources of confidence in the public at large. None of these resources is unlimited. Accordingly, deliberate curriculum development becomes a matter of the wise expenditure of limited, even scarce, resources. If this is to be successfully accomplished, it is essential that we obtain a greater understanding of the pragmatics of the processes of curriculum change. Only then can we hope effectively to inject intentionality into curriculum development.

Yet the need for planned curriculum development is more apparent now than ever it has been. The effect of the computer and micros on what might be done in school is enormous, yet even more important have been the consequences for society in terms of new employment patterns, new expectations and changed motivation. The world is entering a second industrial revolution which is proceeding far faster than its predecessor. New and great demands will be placed upon schools: demands that will only be met by accelerated curriculum development.

In this group, therefore, attention was focussed on the processes of curriculum development and on the varied roles which those within the school system are called upon to play.

2. Curriculum Development in an Historic Perspective

It is helpful when considering the present position in curriculum development to realise that curriculum development has a long history which can serve to illuminate the processes of change. Institutionalised mathematics education can be traced back some 5000 years (Høyrup, 1980). Originally, the mathematics communicated would appear to have been concerned with simple utilitarian topics.

By 1800 BC the scribal profession became more independent as a social body, the school gained more autonomy, and 'pure mathematics' arose in that problem solving now embraced whimsical problems as well as realistic applications. The division between mathematics as a science and as a teaching subject developed further during the great age of Greece when mathematics teaching became associated more with a love of wisdom than with practical affairs — a bias which persisted within certain sections of education. It was, of course, a feature of mathematics education until the present century, that in all civilisations, Greek, Roman, Islamic, Indian, Chinese, ..., it was the prerogative of 'the few'. For example, the municipal schools which began to emerge in western Europe from the 13th century gave almost no mathematical instruction. The economic and political changes in 15th and 16th century Europe influenced mathematics education in two ways: the rediscovery of the work of Greek scientists enriched and improved mathematics education at the universities, and, beginning in Italy, mathematicians became more and more specialists rather than polymaths. Common public schools were established in many countries but in these mathematics had no, or only a marginal role, their curricula being dominated by the classical languages. There were, however, new social needs to be met. Arithmetic was needed by the developing mercantile and trading classes and training in practical skills and computational techniques was supplied by mathematical practitioners who established their own schools and provided textbooks written not in latin but in the vernacular.

Social and economic changes, then, resulted in an enlargement of the mathematics curriculum, in a rethinking of the position of mathematics in existing institutions, and in the creation of new scholastic institutions to meet specific needs.

This pattern is one that can be observed on other occasions. At the end of the 18th century the philosophy of the enlightenment led to the establishment of secular public school systems in France and Prussia. At the same time Pestalozzi initiated important developments in the methods of mathematics teaching, emphasising 'child-centred' and 'natural' learning as propounded by Rousseau. New military academies were founded in several countries, each emphasising the teaching of mathematics, and the establishment of the Ecole Polytechnique had wide-ranging effects, including a recasting of the concept of the organisation of mathematical knowledge. As the Industrial Revolution gathered momentum, further initiatives appeared. In England the inability of existing schools to respond to societal needs led to the establishment of a new type of school in which mathematics played a larger role. In Prussia, Humboldt formulated the concept of *Bildung*, a fusion of knowledge and process. Gradually, mathematics education became more widely available and was presented in a new context within technical schools.

By 1900 school systems in Europe, North America, Japan and elsewhere were adapting rapidly to new needs, and great changes were taking place within mathematics education. In 1908, ICMI was established. (Countries associated with it included, for example, Argentina, Australia, Chile, China, Egypt, Mexico, Peru and Turkey.)

After two world wars, many countries again made extraordinary efforts to adapt school systems and, in particular, mathematics and science curricula to new social, political, cultural, scientific and technological contexts. Many social groups and forces joined the reform movement which produced a great variety of goals, approaches, and contributions from widely differing scientific, theoretical, and pedagogical positions. In the western world attempts were made to introduce 'mathematics for all'. The results were less significant than intended, yet, nevertheless, the movement created and stimulated the notion that curriculum development is a continuing necessity. Changing contexts demand new responses from educational systems, and it is the lesson of history that if existing institutions cannot adapt quickly enough, then alternatives will arise which are not bound by tradition and can more readily meet the perceived needs of society.

The study of curriculum development as an historical process is in its infancy. Preliminary attempts to derive lessons from the movements of the 1960s and 1970s have been made but there is still much to be done. On-going research within curriculum development is difficult to organise (although more could be attempted in the way of study of pragmatic development as it occurs). It can be said then in a very real sense that history is the laboratory of the researcher into curriculum development.

3. The Position Now: Changes from 1976

One feature of the ICME held at Karlsruhe in 1976 was the presentation of a report on 'Trends in Curriculum Development'. It is of interest to consider the changes which have taken place in the past eight years and to identify possible deficiencies in the arguments presented there.

First it is important to view 1976 in an historical perspective. It can be seen as a watershed in curriculum development. Since then there have been few, new large-scale projects. Moreover, there has been considerable retrenchment in many countries as a result of the economic crisis: in Holland the IOWO has disappeared as has the Schools Council in the UK; in France the IREMs have been engaged in a perpetual battle for survival, whilst in most countries centres for mathematics education have been starved of funds. Changes have been demanded of schools, but the financial resources available to facilitate change have been severely limited.

Already by 1976 grave defects in the RDD (Research, Development, Dissemination) model of curriculum development were evident. Teachers were seen to have a much more important role to play than that first technologically-based model allowed. Local, small-scale projects began to appear, yet even at that time doubts were raised about the ability of these to effectively carry the full burden of development. Some of the problems which developed will be considered below in the section on the role of the adviser. The drawbacks of such schemes are mainly their cost (in terms of the provision of specialist support), the slow rate at which such locally-based schemes permeate the system, and indeed the load which they place on the already over-worked teacher.

The RDD model saw the teacher as the *implementer* of the developer's plans. The local model saw the teacher as *curriculum developer*. A more realistic model is probably that in which the teacher acts as *moderator* or *interpreter* of a centrally designed curriculum. This will be considered in Section 5.

The pupil was recognised in various ways in 1976. It was realised that curricula had to be designed not only for the academic elite, but for all ability ranges and for the socially and physically handicapped. The problem of 'mixed-ability' classes was also raised. These matters are further considered in Section 7 below. Here, however, we note a factor that was largely omitted from consideration in 1976: the role which pupils play in curriculum development, a role which has been thrown sharply into relief by the social changes that have taken place in the last eight years. In many countries the economic crisis has brought with it greatly increased rates of unemployment amongst the young, the result of which has been to produce new responses so far as motivation is concerned. Mathematics has frequently been seen by students as the key to a good job; in many countries attainment in mathematics is used as a sifting device. This has traditionally supplied motivation to students — even if they have not derived satisfaction or pleasure from their mathematics lessons, they have still seen them as serving personal ends. Such motivation has been a powerful aid for (and one often misused by) mathematics teachers. Now a pass in mathematics is no longer a guarantee of employment. Indeed, there are growing signs of disenchantment with schooling amongst adolescents in many countries. There is a greater need to consider problems of motivation, and, correspondingly, to design curricula with changed needs in mind. It was reported that in some countries new types of examinations were being created which would give all pupils the opportunity to gain a 'prized' certificate. However, concern was expressed not only over the likely temporary nature of such motivation (for the value of a certificate will soon come to be judged by its marketability), but also by the way in which such examinations place even more shackles on teachers — shackles which might be being imposed with the best of intentions, but which will not be shed easily.

Yet pupils affect the curriculum in another manner. In each classroom the teacher and the pupils come to an implicit agreement concerning expectations and the work to be accomplished. Unawareness of this fact can lead the teacher to set, and to have imposed upon him/her, goals which are too low. In this sense pupils are powerful, but neglected agents of curriculum reform. The way in which so many have eagerly grasped opportunities to develop and extend interests in computing is clear evidence of this.

Another 'trend' remarked upon in 1976 was that 'curriculum development must become a gradual cumulative process rather than a frantic pendulum-swinging exercise'. Elsewhere, it was proposed that change should be modelled by analytic continuation (a gradual extending of boundaries) rather than by catastrophe theory (non-continuous jumps). Matters have changed! In effect, the emergence of the micro and the enormous expansion of cheap computing facilities — and of graphic possibilities — in less than a decade has been little less than a commercial, industrial and social 'catastrophe'. Gradual cumulative change at the rate at which we have known and come to accept it is unlikely to meet society's needs. History would point to one conclusion — the establishment of new types of educational institutions, possibly independently run and divorced from the state system, which do meet perceived desires. In a world   in   which   divisions   between   rich   and   poor,   employed and unemployable, are growing wider, there is a risk of new educational/social divisions being created if curriculum developers cannot solve the (seemingly insurmountable) problems they are set.

In 1976 it was asserted that it was then being recognised that, 'There is a need to consider more closely the processes of innovation and to base curriculum development upon better elaborated theories of teaching and learning and upon improved theories of curriculum development' (p.159). Since then there has been much valuable research done on teaching and learning, but, as was stated earlier, research on curriculum development and theory-building within that area has been neglected. We now neglect it at our peril.

4. Models for Curriculum Development

Several models for curriculum development have been presented in recent years — many in the general context of educational theory. Here it is necessary to ask if and in what ways curriculum development in mathematics differs from that in the mother tongue, a foreign language, and a science subject, and whether these differences are sufficient to necessitate special models for mathematics. It was argued that this is the case; for example, the role in mathematics teaching of textual material and of exercises (tasks and activities), the emphasis placed upon attainment in mathematics by employers and the public at large, its standing as a 'hard' ('well- defined', 'objectively-evaluated' rather than merely 'difficult') subject, the fact that it is taught to all and taught by many who are not well-qualified and who often do not teach it as their principal subject, all serve to set mathematics apart.

Certainly, it is now becoming much clearer to identify those forces which serve to foster curriculum change — advances in mathematical and educational thought; concern over poor performance; new technology; social, economic and political changes; market forces (ranging from individual authors to publishers and manufacturers of computers). An accompanying range of instigators can also be described as can the media of curricular change — platform reports (for example, the Cockcroft Report in England and Wales, the Dunning Committee Report in Scotland, the Lichnerowicz Commission in France, ...) which lead to the preparation of new materials, and the redesigning of examination systems. The barriers which must be overcome if changes are to be successfully effected can be described at a reasonable level of complexity — resistance can come because of differences in values, attempts to retain power, psychological reasons, and, of course, sheer unreadiness in terms of knowledge and financial and material resources. Change will only come successfully if accompanied by increased awareness and knowledge on the part of the teachers, the provision of needed resources, and the recasting of evaluative and assessment procedures so that they facilitate rather than act against change.

Such statements as those above would seem nowadays to have assumed the status of truisms, yet they are still frequently ignored. However, whilst revealing what the barriers are and how the forces of change might assemble in order to approach them, they tell us little about how the barriers will be broken down. Here history reveals a number of insights and possibilities. One is that the most effective solution is when the barriers are taken down by the defenders. That is, rather than marshal yet more arguments externally in favour of change, it can be more rewarding to build up the strength and confidence of those from whom change is demanded. Often a different strategem has been employed. Teachers have been convinced of the worthlessness of what they are doing, existing procedures have been derided and innovation offered as a panacea. The results of such strategies have frequently been disastrous. Teachers have embraced changes but have lacked the knowledge and understanding to make them successful. Recent changes as described in Mozambique appear to have been far more successful because, after a political and social upheaval, administrators, educators and teachers were united in a desire to supplant the old system by something new. In general, though, even when the need for change is universally agreed, there is no consensus on what the solution should be. Here the centralised system possesses advantages, for it is possible (if not always the custom) for an acceptable compromise to be mediated and then submitted to teachers. In decentralised systems, however, developers are apt to devote scarce resources to arguments with each other, leaving the schoolteacher even more confused.

Finally, it would seem important to study in greater depth the incentives and influences which will encourage teachers actively to engage in change and to try to alleviate that most obvious and practical barrier to true involvement — lack of time.

5. The Teacher's Role

The mathematics curriculum we teach, the freedom for innovation we possess and the materials we use are fashioned by our education system's philosophy. For instance, the teachers' role in curriculum development varies from country to country mainly because of national differences in the responsibilities assigned teachers and of the different expectations of them. Curriculum developers who believe that teachers are an essential component in the process tend to produce source materials which rely substantially on correct teacher interpretation and development; innovators who are wary of teacher input tend to produce materials that reduce substantially the teachers' role, and developers from the countries remaining which believe that teachers should not be involved at all tend to produce a curriculum that offers the teacher few innovative opportunities but expects a high level of teaching competence. Happily the latter situation is uncommon, due, no doubt, to the growing awareness of the interplay among such factors as the aims of education, the nature of knowledge, the role of the teacher in the classroom and the nature of the learner. Further, it has become clear within the last ten years that this changing concept of curriculum needs to be supported by procedures that oblige persons at all levels in the educational hierarchy to participate in different ways in curriculum activity. Evidence exists that the program compiled by the curriculum developer (the intended curriculum), the program taught by the teacher (the implemented curriculum), the program viewed by an observer (the perceived curriculum), and the program absorbed by the pupil (the achieved curriculum) can vary widely and significantly.

Considerable differences can occur between the educational goal perceptions of administrators, teachers, pupils and parents, and this dissonance in perception is most likely to be expressed in an ineffective school curriculum.

Because the students in our schools possess an ever-widening range and variety of abilities, aptitudes, interests and personal-social adjustments, curriculum development now requires that the classroom teacher be able to vary the content and methods of a subject to the abilities and interests of each class.

Teachers are best able to achieve this goal if, first, they are part of the overall curriculum development enterprise referred to above, and secondly if they are allowed the freedom to play the role of moderator or interpreter of the curriculum generated through the efforts of the enterprise in which they participated. The moderation involves, among other things, constantly seeking to match methods of instruction with subject matter. Consequently, the teacher should not only be supplied with curriculum materials and other resources, should not only be acquainted with educational research, but should also participate in curriculum development at least at the level of action research.

In summary then, curriculum development should reflect the unity and harmony of theory and practice. It should be flexible enough to allow a teacher's educational theory to function as an analytical tool with which the teacher can judge the adequacy of his moves and his pupil's, and also to be in harmony with the practice as set down by the curriculum developer. It should provide opportunities for teachers in a school (individually or as a group) to turn theoretical insights into actual classroom practices so that the curriculum actually conducted in the classroom can be derived from the teachers' personal definitions of learning.

The process of curriculum development should also be sensitive to the capacity of the school system to assimilate and use research findings and to create innovative practices; it should attempt to raise this capacity by using a curriculum development methodology in which classroom teachers and their pupils, curriculum workers, teacher educators and the wider community participate, with equal worth, in curriculum development.

Further, the teacher who, through a variety of activities, contributes to the curriculum enterprise should be made aware that his/her contribution is of equal worth to that of other teachers, curriculum developers, pupils, administrators and parents, so that he/she becomes motivated to work on educational issues and problems of pressing concern and eager to play a role in raising the capacity of a school system to assimilate and create innovative practices within that system.

6. The Adviser's Role in Curriculum Development

In Section 3, reference was made to the RDD model which saw the teacher as the instrument of the curriculum developer's plan. The essence of this model was that the curriculum package was handed down for the teacher to implement. A period of teacher inservice (if funds permitted) was the system's usual concession to the implementation strategy. The user's lack of involvement in the development of the materials, and the inadequate training provided in their use, almost certainly guaranteed that the intent of the materials would be misinterpreted or at least distorted.

The last ten years has witnessed a swing away from this model to either the local (school-based) scheme, referred to in Section 3, or to the curriculum development enterprise model explained in Section 5. It is in the first of these two models that the mathematics adviser can exercise the greatest individual influence, for here school personnel with whom the adviser works are able to take on the key responsibility for the provision of the program. This approach has produced one very fortunate by-product in many of the countries in which it has taken hold: it has increased dramatically the number of practitioners who are able to talk about, develop and competently implement worthwhile curriculum reforms in their own classroom. The hitherto near-total reliance on external experts has been diminished as groups of teachers within schools and from amongst schools have assumed the responsibility for solving their own curriculum problems. Within this scenario, and in spite of the lopsided adviser/teacher ratio (1:700 was quoted in some places), the adviser can initiate change in several different ways.

Perhaps the major and most visible means of adviser influence, evidenced in the Australian state of Victoria, derives from Lewin's early pioneering work on action research during the 1940s. This has been translated in a number of different ways — some systematic, some ad hoc — but each possessing the distinctive feature that those affected by planned change take the prime responsibility for deciding on the course of action likely to produce improvement, and for evaluating that change. The role of the adviser in this scenario is clearly one of facilitator and critical friend.

A more direct approach on the part of advisers, and one which has proved successful world-wide, is represented in their effort to assist in the establishment of the 'lighthouse' or 'magnet' school. Here the adviser locates and brings together an enthusiastic group of teachers within a single school. This group is encouraged to set new standards on the assumption that one such innovator group will influence other groups in nearby schools. The effect of this influence on a large educational system tends to result in the development of a 'conscience' within teachers, resulting in an individual commitment to improvement rather than mass change.

Other contributions by the adviser include:
• the fostering of local networks of teachers who can share experiences and of schools which can share developmental tasks;
• the provision of a focus through which national and international literature and materials reach the teaching body.

Advisers will continue to explore alternative strategies to assist teachers in the process of curriculum development and implementation. To date, the most successful strategies have been those which allow teachers to review and reflect on their own practices, and to make changes which they themselves can monitor and evaluate.

7. The Differentiation of Curricula

Students learn mathematics at vastly different and non-uniform rates. As a result there is a wide divergence in the attainment of children in any age cohort. This problem was traditionally solved by only promoting those students who had indicated that they had attained a certain degree of proficiency appropriate to their present grade: it is a method still used in some countries, for example, Hungary and the USSR. These countries, therefore, offer essentially the same mathematics curriculum to *all* students. Thus, in the USSR all students will meet axiomatic geometry and the differential and integral calculus. In England and Wales the Cockcroft Committee (DES, 1982) recommended that mathematics curricula should be designed in three wide bands — for the low-attainer, the average and the high-attainer — and that the interests of the first named should determine the first steps in the process.

Practices with regard to differentiation, then, differ widely. They were the subject of discussion by one of the working groups. Here, approaches in Tanzania, Greece and the United Kingdom were presented in some detail and others were mentioned in discussion. The differences in practice were considerable. Differentiation can, of course, take place between schools, within schools or within classrooms. Often this last type of differentiation makes use of 'individualised learning'. This can be offered in a variety of forms, from the computer-based PLATO (USA) to booklets which are to form the basis for individual or small group activity. The SMP 11–16 Project offers such booklets for the first two years of the course — assuming that differentiation will take place within a single classroom. In later years, however, it assumes that classes will be streamed, that is, differentiation will take place within the school and individual classes will, as far as possible, be homogeneous. A second United Kingdom project, SMILE, adopts a looser form. In this, the teacher has greater responsibilities for ordering and determining the pupil's work, always bearing in mind:

- how well previous activities have been carried out;
- the teacher's view of the pupil, based on classroom interaction;
- the teacher's judgement about the type of activity which would be beneficial.

The variation in the role played by the teacher using 'individualised' methods can, therefore, be considerable. This statement also applied to the pupil, for 'individualised' methods now are very different from those encountered in the early days of programmed learning. Teacher-pupil interaction within the SMP and SMILE schemes is not confined to administrative exchanges.

The differences to be found in practice clearly reflect different theoretical standpoints adopted by curriculum designers. Group discussion, however, centred on practice rather than the underlying theories and assumptions concerning the exact role which teachers should play. Yet theory is always tempered by practice and it was accepted that differentiation often takes place as a result of teachers deliberately omitting particular syllabus items — sometimes from a feeling of insecurity or because they do not esteem those special topics highly.

The extent to which differentiation can take place, and how open that differentiation can be, differs considerably according to the political and social context. It is, then, a difficult subject to discuss internationally, yet, nevertheless, one of considerable importance.

References

DES, (1982) *Mathematics Counts* (Report of the Committee of Inquiry), HMSO, London.

Howson, A.G., (1979) 'Curriculum Development' in *New Trends in Mathematics Teaching, Vol.IV*, UNESCO, Paris.

Howson, A.G., Keitel, C. and Kilpatrick, J., (1981) *Curriculum Development in Mathematics,* Cambridge University Press, UK.

Høyrup, J., (1980) *The Influences of Institutional Mathematics Teaching on the Development and Organisation of Mathematical Thought in the Pre-Modern Period,* IDM Bielefeld, FRG.

Group Organisers for the Congress sessions were:
Jean Dhombres (France) (unable to be present at ICME 5), Geoffrey Howson (UK), John Malone (Australia)

Discussion Group Leaders:
Barry Fenby (Australia), Geoffrey Howson (UK), Nigel Langdon (UK), John Malone (Australia), Bill Newton (Australia), Roland Stowasser (FRG)

This survey paper was based on written contributions by Rowan Barnsley, Peter Brinkworth, Desmond Broomes, Peter Cribb, Jean Dhombres, Barry Fenby, John Gaffney, Patricia Hess, Christine Keitel, Nigel Langdon, John Ling, Curtis McKnight, Steve Murray, Bill Newton, James Power, Tom Romberg, Alan Starritt, Roland Stowasser, Karen Usiskin. We are grateful to all of these and to those who made contributions to the various discussions.

THEME GROUP 6:
APPLICATIONS AND MODELLING

Organisers: Richard Lesh (USA), Mogens Niss (Denmark),
David Lee (Australia)

Introduction

The work in Theme Group 6 had three main objectives:
(1) To identify and illuminate the current state of and trends in:
  • applications and modelling in the mathematical curricula at different
    educational levels in different countries;
  • applications and modelling as a subject for research, development and
    debate by mathematics educators and educationalists.
(2) To exchange among theme group attendees experiences of applications and
    modelling and views on issues related to applications and modelling.
(3) To generate, if possible, a set of recommendations or policy statements
    addressing priority issues for applications and modelling in the curricula at
    different educational levels.

The organisation of Theme Group 6 was designed to encourage active
participation during working sessions, taking advantage of different
backgrounds and experiences. To this end, during the working sessions,
participants in the Theme Group worked in subgroups according to the matrix
structure described below:

Topic Areas \ Ed. Levels	E. Elementary and middle school	F. Secondary and high school	G. College and university	H. Teacher education; adult ed.
A. Purpose, role and perspective	cell groups			
B. Curriculum issues				
C. Psychological research; testing and measurement				
D. Resources, incl. technology				

Each topic area (a–d) and each educational level (e–h) was led by a
Convenor. The eight sub-groups and their Convenors were:

(a) Purpose, role and perspectives of applications and modelling. *Jan de Lange* (Netherlands)

(b) Curriculum issues associated with applications and modelling. *J.N.Kapur* (India)

(c) Psychological research and testing and measurement. *Terezinha Carraher* (Brazil)

(d) Resources for applications and modelling including technology. *Keith Hamann* (Australia)

(e) Elementary and Middle School (Grades K — 7). *Bernadette Perham* (USA)

(f) Secondary and High School (Grades Circa 7 -12). *Gabrielle Kaiser* (FRG)

(g) College and University. *Avi Bajpai* (UK)

(h) Teacher Education — Adult Education. *Chris Harman* (Australia)

At the first session each of the nearly 200 Theme Group 6 participants chose to work in one of the four educational level groups (f–h). For each group the session started with a twenty minute presentation by the subgroup leader. The presentations, supported by one to three page handouts, gave a background briefing and outlined issues which working groups might address in the remainder of the session.

After the opening presentation, each subgroup was split into working groups with an average size of 15. The discussions of the working groups were guided and recorded by pre-selected chairpersons and recorders.

The second session was similar to the first except that the subgroups were organised according to topic areas rather than educational levels. Again, the sessions began with a brief agenda-setting presentation by the subgroup leader.

In the third session, attendees joined small working groups focussing attention on specific topics at a specified educational level of interest. That is, the groupings corresponded to the cells of the 'topics by level' matrix. Cells which did not attract a sufficient number of attendees were merged with other cells.

The fourth session began with four parallel thirty minute presentations of cases on applications and modelling projects from which participants were invited to choose one. The presentations were as follows:
'Mathematics in Industry: from the Workplace to the Classroom' — the CAM-project (Careers And Mathematics). *Pat Costello, Peter Jones and Brian Phillips* (Australia)
'Mathematics at Work' — an audio tape/slide report of a project on mathematics in education and industry sponsored by the Sheffield Regional Centre for Science and Technology. *Connie Knox* (UK)
'The Savings Transfer Effect of Teaching Mathematics', based on an empirical study of the transferability of experiences in mathematical modelling to the learning of physics. *Dinesh Srivastava* (Australia)
'Mathematical modelling' — a presentation of excerpts from a series of Open University video productions on mathematical modelling activities and complex open-ended real-life situations. *John Jaworski* (UK)

In the remaining hour of the Theme Group sessions, working groups were reconvened as educational level subgroups to generate recommendations. Those recommendations are summarised in the concluding section of this report.

As well as the four Theme Group sessions, two additional summary sessions on applications and modelling were held during the general Congress. The first summary session consisted of two state-of-the-art lectures:
'Applications and Modelling as Part of the Mathematics Curriculum.' *Mogens Niss* (Denmark)
'Applied Problem Solving from the Point of View of Psychological Research'. *Richard Lesh* (USA)

The aim of the second session on the Wednesday was to summarise as far as possible the outcomes of the Theme Group's work. The summary was prepared by David Lee (Australia) with additional contributions by sub-group leaders, Bernadette Perham (USA) and Gabrielle Kaiser (FRG).

In the belief that the principal characteristics and the important issues of applications and modelling in mathematics curricula are much the same through all educational levels, the following exposition is organised according to the topic area categories (a) to (d) listed in the first section. Sections 2 to 5 are intended to refer in general to all the educational levels (e) to (h).

Purpose, Role and Perspectives of Applications and Modelling

This topic area deals with issues which are external to any given course.

The reasons which underly the inclusion of applications and modelling in mathematics curricula seem to be of four different, but not mutually exclusive, kinds.

P1. The ultimate reason for teaching mathematics to students, at all educational levels, is that mathematics is useful in practical and scientific enterprises in society. But experience shows us that the potential of mathematics is difficult to realise for persons who have not had access to a broad set of experiences applying mathematics and building and analysing mathematical models. Therefore, to foster in students the ability to utilise mathematics in complex situations, it is necessary to include applications and model building activities in mathematics curricula.

P2. A similar, but not identical line of reasoning reaches the same conclusion on slightly different premises. It is a fact that mathematics is being used extensively and increasingly throughout society; sometimes with very good reason, sometimes with less justification. The role of mathematics is crucial to the shape of our societies and influences everybody. Therefore, it is important that all students be provided with prerequisites for understanding, assessing and handling the use of mathematics in problem situations outside mathematics itself.

P3. Incorporating applications and modelling in the curriculum substantially assists the acquisition and understanding of mathematical ideas, concepts, methods and theories, and provides illustrations and interpretations of them. Furthermore, applications and modelling offer practice in the exercise of mathematical techniques. One mathematical idea can model a large variety of applications.

P4. In recent years it has been established that only a minority of students are attracted and motivated by issues and activities in pure mathematics. For a majority of students, mathematics lacks visible relevance to their present and future lives. To convince these students that mathematical activities are worthwhile and to motivate their studies, applications and modelling components should be included in mathematics curricula. This also may help to make mathematics more open and friendly. In addition to this, there is in some countries a similar need to convince governments, authorities, and colleagues from other subjects, that mathematics can be taught in a relevant way for socially important purposes.

Of these four reasons the first three imply a positive interest in applications and modelling for their own sake or for their capacity to improve mathematics teaching, whereas the fourth reason is rather defensive, serving to protect or strengthen the position of mathematics. It may, of course, transpire that in a given curriculum context several of these reasons apply simultaneously. In the pursuit of one or more of the four purposes, applications and modelling may serve different specific goals:

G1. To help students activate mathematics in unstructured, complex open-ended problem situations where problems must be identified, models built and decisions made.

G2. To help students apply mathematics in problem situations which are already mathematically structured, that is, where a mathematical model has already been chosen and built prior to the application of mathematical methods and results.

G3. To enable students to analyse, characterise and assess established or new applications of mathematics and mathematical models from practical, epistemological and mathematical points of view.

G4. To help students acquire knowledge of, and insight into, the characteristics and properties of a number of standard mathematical applications and models within different fields.

The precise selection and emphasis given to these goals is dependent upon the corresponding reasons for including applications and modelling in the curriculum.

To stimulate Theme Group discussion, the following issues had been identified and listed in the pre-Congress material:
- For what categories of students should applications and modelling be part of their mathematics education, and why?
- What purpose should applications and modelling serve, what role should it have?
- What aspects of applying mathematics and building models should be emphasised?
- What sorts of applications and modelling are appropriate for:
  — different ability groups?
  — different countries?
- What are the barriers preventing applications and modelling courses

occupying reasonable positions in curricula?

Participants were in overall agreement on the necessity and desirability of including applications and modelling activities in mathematics curricula at all levels. However, a general need for clarification of the meanings of the terms 'application and modelling' and their inter-relations was felt. An application illustrates the utility of mathematics probably in an existing model; modelling refers to the process of formulating a mathematical structure thought to correspond to a real situation. As to the purposes of applications and modelling in curricula P1 to P4, all were considered important. Most participants thought that the value of applications and modelling was greater than merely providing motivation (P4).

Many argued that P1 should be given the first priority. All specific goals G1 to G4 gained some support and it was generally agreed that goals should not be limited to providing knowledge of standard applications and models (G4). A majority of group participants were in favour of attributing particular importance to the first goal, that of dealing with unstructured problem situations.

In the working sessions, the barriers to having applications and modelling occupy a reasonable position in curricula were touched upon again and again. Several barriers were identified — the ones considered most important being:

Teachers — First of all, there are those teachers who are against applications and modelling activities because such activities inevitably reduce the mathematics content in a limited syllabus and substitute it with material which, in their view, is of secondary or even inferior quality. While teachers of this kind abound all over the world, not very many of them seem to have attended Theme Group 6. For those teachers who do want to include applications and modelling in their teaching there are several difficulties. Although collections of case studies containing useful and stimulating material exist, this material provides only partial assistance in the generation of complex, open-ended problem situations where neither the problems nor the methods to tackle them are clearly defined from the outset. This makes teachers feel that they have to transgress the limits of their competence and enter unknown territory. Applications and modelling is not a systematic science nor a trade with established procedures and rules, but rather an art. Furthermore, applications and modelling activities are time-consuming, both with respect to the teacher's own time and with respect to the time allotted to mathematical activities in the total curriculum.

Colleagues — Colleagues in mathematics or in other subjects whose courses are affected by the inclusion of applications and modelling activities in the curriculum are likely to resist such changes. Depending on where the curriculum authority lies, such resistance might turn out to be an effective barrier. Pressure from other institutions receiving graduates from the one in question may exacerbate this problem.

Evaluation — Traditional tests and examinations, especially the written ones, represent a major barrier to applications and modelling, a barrier to which much attention was paid during working sessions. The problem has two dimensions. Firstly, if the content of examinations assessing courses which contain applications and modelling components addresses only the pure mathematical issues, the applications and modelling activities will tend to be squeezed out and occupy a marginal position. Secondly, the usual forms of examinations do not allow applications and modelling activities to be evaluated according to the actual processes and substance involved, the more so, the more complex and open-ended the problems being considered are. Since examination conditions and traditions are difficult to change, this is a considerable barrier.

Curriculum Issues

This topic area is concerned with issues which are internal to any given curriculum. Applications and modelling activities can be introduced in mathematics curricula through several different approaches. The discussion can be focussed by the following categorisation:

A1. The *two compartment approach*. In this approach the mathematics program (which may range from school terms or individual courses to a total education program) consists of two parts. In the first part a traditional pure mathematics course is given, while the second part deals with applications and modelling (within one or several areas) related to the mathematics developed during the first part. Applications and modelling activities may range from applying results and methods in ready-made models to entire modelling projects.

A2. The *multi-compartment* or *islands of applications approach* is related to the two-compartment approach but addresses smaller units than entire programs. Thus small units of new mathematical topics/concepts/results are followed by an applications and modelling sequence which is followed by a new mathematics unit and so on alternatively.

A3. The *mixing approach*. Applications and modelling activities and the formation of new mathematical concepts and theories are woven together in dynamic interaction. On the one hand new pieces of mathematics are motivated and illustrated by applications and modelling situations, and on the other hand applications and modelling problems are described and solved by new pieces of mathematics. It is a characteristic feature of this approach that the mathematical syllabus proper is given, at least in outline, from the beginning.

A4. The *integrated mathematics subject approach*. In contrast to the approach above, there is no fixed mathematical syllabus in the fourth approach. Instead, each sequence of mathematical activity is initiated by a situation for which the necessary mathematical support is sought and developed with teacher and literature assistance.

A5. The *interdisciplinary integrated approach* — hard version. This approach differs from the previous one in that the integration takes place within an interdisciplinary framework. Although it certainly will include mathematical activities, these are intended to be subservient to the needs of the problems considered.

Apart from the first one which is a total approach, these 'pure' approaches can be combined in numerous ways — for instance, by being assigned to different phases of the total program, thus creating a rich variety of different ways in which applications and modelling may be incorporated in curricula.

In addition to the question of basic approach, applications and modelling give rise to a large number of curriculum issues. The pre-Congress material identified the following ones:

* What kinds of applications and models should be considered in a given curriculum?
* How should applications and modelling activities interact with other mathematical activities?
* How could and should 'applied problem solving' contribute to the understanding of internal mathematical content?
* Which new mathematical topics are becoming important and which obsolete?
* What is the role of applications and modelling in subject matter areas outside mathematics?
* What organisational structures would facilitate the acquisition of a varied set of experiences in applying mathematics to complex problem solving situations and in building mathematical models?
* Which aspects of model building processes and applied problem solving need particular attention?
* Some problem situations are artificial and simple, but amenable to a satisfactory mathematical treatment while others are real and complex, and difficult to tackle. How should these two types of problem situations be incorporated in the curriculum and how should they be balanced?
* Which approach or combination of approaches (refer A1–A5) should be chosen for applications and modelling activities on different educational levels?

In the Theme Group sessions not all approaches listed above were discussed. This holds in particular for A5. Participants seemed to share the view that the first approach (A1) is insufficient and unsatisfactory as a means for pursuing the purposes and goals described in the previous section. The general opinion was that none of the 'pure' approaches are optimal, but that all approaches A2, A3 and A4, might be valuable and appropriate in different contexts and at different times, and that each of them might well be present at different phases within the same course.

It was agreed that there was a place for both the ʻartificial, simple applications situations and the realistic ones. The emphasis should be directed towards the latter type. In this connection, it was stressed on several occasions that the mathematical aspects of the problems should not be dealt with. It is crucial that students obtain experiences with the entire process of modelling and application from their first encounter with a fuzzy, unstructured situation for which no sharply defined problems have been identified to the final evaluation of the results obtained — and perhaps around the sequence again in several cycles.

Finally, there was general consensus on the desirability of inter-disciplinary sequences, involving not only situations from other disciplines but also professionals from them, for instance, scientists, engineers, economists, industrialists and planners. This would also help teachers to transgress the frontiers of their subject which is of central importance for the success of applications and modelling activities in the mathematics curriculum.

Psychological Research and Testing

The discussion was focussed around previously distributed questions including:
* What is known about the psychological processes involved in applied problem solving and model building?
* What is known about activities capable of fostering applied problem solving capacities?
* What research problems are in focus currently? What should be in focus in years to come?
* What should we test in connection with model building and applied problem solving, and how can we test it?

Priority issues identified were remarkably similar for each of the levels from elementary school through to university. Three distinct kinds of concerns were apparent:

(1) Some participants were interested primarily in 'teaching mathematics so as to be useful', that is, applications are used to teach topics already in the curriculum. Here, the most relevant psychological issues have to do with identifying ways in which basic mathematical ideas become more meaningful through applied problem solving experiences.

A large share of the research relevant to this issue derives, directly or indirectly, from the work of Piaget. For some topics such as early number concepts, rational number or proportional reasoning concepts, some probability or statistics concepts and elementary spatial/geometric concepts, significant advances have been made in recent years. In general, the conceptual areas which have been investigated most thoroughly are 'natural concepts' developed in 'natural environments', whereas participants in the applications and modelling group were most interested in ideas and conceptual amplifiers which may only develop in artificial, mathematically rich environments. Problem solving behaviour of students working in isolation was less interesting than situations in which a variety of realistic resources such as a computer may be available. Psychological research in such environments is at a relatively primitive state; more work in this area clearly is needed.

(2) Some participants were interested in introducing new 'applied mathematics' topics such as discrete mathematics, statistics, graph theory into the curriculum, at the expense of other 'obsolete' topics. Possible effects of these changes to topics remaining in the curriculum would need to be investigated.

Other psychological research related to the insertion of new topics into the curriculum comes from studies observing people in 'real world settings' to see what mathematical ideas they really use. At every educational level, and for nearly every kind of student's career aspirations, participants argued that there is a poor match between the mathematics students are taught and that which they are likely to use. Some of the most interesting psychological research related to this issue comes from studies investigating similarities and differences between experts and novices in various fields. For many students, 'school math' and 'used math' evolve as semi-autonomous systems. More research investigating the nature of 'used math' is needed for a variety of 'expert' populations.

(3) Some participants were interested in instruction aimed at providing students with experiences in modelling, quite apart from issues about the 'applied/traditional/pure' nature of the content involved. Here, the relevant psychological issues are that little has been done to investigate the processes, heuristics, strategies, skills and understandings required in modelling activities.

On the one hand, modelling seems to be a particular kind of problem solving activity. On the other hand, the processes needed for students to 'go from clearly specified givens to well defined goals' represent only a portion of those needed when the entire conceptualisation, that is, model of the problem situation, must be constructed and refined through several cycles. While the decisions students must make to 'solve' a real problem may be clear, the mathematically-derived information that can inform these decisions seldom is well specified, and the given information may exist only in a form that is not immediately usable. Relevant information may need to be generated rather than just being selected.

In modelling situations, mathematics is not simply used, it is created (or at least modified, reorganised, specialised or otherwise adapted to fit particular problem situations); solutions are not simply found, they must be constructed.

Research on mathematical modelling is virtually non-existent, although an overwhelming body of research is relevant. Furthermore, existing research is not necessarily a basis for unbridled optimism for modelling enthusiasts. It is known that even youngsters are able to create mathematical models (that is create mathematics) in some kinds of everyday settings. It is also known that many of the processes, skills and heuristics which have been emphasised in problem solving research and instruction do not generalise well when they are embedded in the larger domain of modelling. Many of Polya's heuristics, for example, have been shown to be either irrelevant or counter-productive in typical modelling situations.

One of the foremost reasons that ICME participants gave for being interested in applications was to increase student motivation. Applied problems are commonly viewed as a way of responding to the question, 'Why are we learning this?' Unfortunately, many teachers who are experienced in the use of applied problems have found that:

(1) Problems which 'turn on' some students can be equally powerful forces to 'turn off' others. Because applied problems are often difficult, repeated failure experiences can be damaging to student morale. Furthermore, a problem viewed as 'relevant' by one student (or teacher) may be viewed as highly 'irrelevant' by others. The issue of relevance begs the question, 'Relevance for whom, and for what purpose?' Mathematics for informed citizenship, or for intelligent consumers, may be viewed quite differently from mathematics for menial employment, or opportunities in technological professions, or higher education, or for entry into the natural sciences or the social sciences. The point here, insofar as these difficulties have to do with psychology, is that 'task variables' have not been investigated for applied problems in the way they have been for routine textbook word problems. Research is needed to identify a matrix useful for generating, varying or selecting problems specifically designed to achieve particular educational objectives. Lack of clarity concerning task variables and objectives forces teachers to rely on a 'shot gun' approach which works for some students but not for significant

numbers of others.

(2) The preceding difficulties are amplified because applied problems tend to be more time consuming than their 'cleaner' counterparts. Therefore, time devoted to applied problem solving often must be subtracted from instructional time devoted to regular instruction. An instructor can argue, 'What good is an idea if students never learn how to use it?' On the other hand, those to whom teachers are accountable (administrators, school boards, parents) can argue, 'First teach them underlying ideas, then, if any time remains, teach students to use them'.

Problems related to accountability are severe for many teachers who want to devote instructional time to applications and/or modelling. Discussions at ICME made it clear that, at every school level, we have not been sufficiently clear about what it is we expect students to gain from instructional time devoted to applications. Even in cases where some degree of clarity has been achieved, convincing ways to measure these desired instructional outcomes are seldom available. Perhaps it is true that many applied problem solving goals cannot be fragmented into tiny discrete pieces which are measurable using standardised tests, but discussions at ICME identified processes, skills and understandings which can be measured if specified more clearly. More systematic work is needed in this area.

Resources, Including Technology

The pre-Congress notes listed the following questions on resources and technology:

- What resources such as problem banks, computer hardware and software, and instructional materials are needed or would be useful for implementing applications and modelling work?
- What is the impact of technology on modelling and applied mathematics instruction?
- What functions should computers have in applications and modelling contexts (as a tool for making real problems solvable and for simulation of problem situations and problem environments)?
- How do computers influence the mathematical modelling aspects of applications?
  The group leader expanded the questions to include:
- What classroom environments are needed when applications, modelling and problem solving are the methods of teaching and learning?
- How do we support and develop teachers for this task of teaching applications and modelling?
  The Congress group discussions elicited some partial responses reported briefly below.

The resources needed for the successful teaching of applications and modelling may be categorised as follows:

R1. Teachers with sufficient background and confidence to guide modelling situations. There needs to be a recognition that teaching modelling and applications involves a different teaching paradigm; there are no longer single correct answers and no one individual will have the answers. Teachers must be given adequate encouragement and opportunities to prepare for such teaching experiences.

R2. Course materials and collections of case studies appropriate to the educational level of students. A distinction is sometimes made between structured modelling— developing experience by working on existing case studies — and projective modelling where the problem is novel, the approach creative and the outcome unknown in advance. Collected case studies are valuable for work in structured applications and modelling and as precursors to projective modelling. A number of collections of case studies are now published.

R3. Calculators and computers with appropriate software packages. A first requirement is for calculators and computers to assist in the analysis of real data arising in practical applications. Simulation packages provide a useful opportunity to explore some modelling situations in a classroom. Symbol manipulation software has important potential.

R4. Mechanisms for contact with industry and other problem sources. Whilst published case studies in books and journals are valuable, there is no substitute for experience with applications at their source. Some examples of enterprise and cooperation with local industries were recounted and recommended.

R5. Opportunities for cooperative teaching of mathematics with other subjects to provide modelling and applications situations and support. Other subjects may provide the motivation and source data for modelling experiences. The interdisciplinary teaching approach has already been briefly discussed in the curriculum design section.

R6. Films, videotapes and materials which evoke problem situations. These resources offer an important opportunity to demonstrate real applications in context. The Challenge of the Unknown series (J.C. Crimmins Co., New York) released at the Congress was cited as an exciting example.

Conclusions and Recommendations

Despite many differences in nationality, position in the educational system, teaching and research experiences, Theme Group 6 participants featured a remarkable homogeneity in their views on applications and modelling in the mathematics curriculum. From this it cannot be inferred, however, that mathematicians and mathematical educators in general share these views.

There is a considerable gap between those advocating the development and use of applications and modelling on the one hand, and the main stream of mathematics instruction on the other hand. But this gap is not entirely due to differences in opinions; much of it originates from genuine difficulties which would not be easy to overcome even if the homogeneity of the Group was representative. This is illustrated by the present report. It shows that, although several important issues were dealt with, many more were left untouched, including the clearing up of basic concepts. The field of applications and modelling in mathematics education gives rise to many complicated questions including curriculum development, development of instructional materials and resources and, very importantly, research, all of which deserve further attention and effort. This work should not only address the problem of reducing the gap between the pioneers in applications and the mainstream but also aim at improving the situation for those pioneers.

The following recommendations should be seen in the context of the preceding remarks:

Recommendations

A. Distinctions and inter-relations between the concepts 'applications', 'modelling', and 'problem solving' should be made clear. Furthermore, processes which are realisations of these concepts should be identified on all educational levels.

B. Mathematics education whether it is for school pupils and students, for mathematics minor and major students, for students of other disciplines, or for prospective mathematics teachers should encompass applications and modelling components and activities, of a constructive as well as analytical and critical nature.

C. Modelling and applications activities should permeate the entire mathematics curriculum from its very beginning, rather than be isolated islands in an otherwise pure mathematics program. Additionally, there should be longer sequences devoted to applications and modelling activities which allow students substantial involvement in tackling fuzzy, complex and extensive applications and modelling problems. Student generated problems should play an important part.

D. The above recommendation (C) should not be taken to imply that pure mathematics components are superfluous. On the contrary, the usefulness and use of mathematics in external problems rely on its properties as a discipline. In an applications and modelling context it is important to provide students with an understanding of the difference between reality, a model and mathematics. Hence, in any curriculum, attention should be paid to the balance between applications and modelling components and systematic mathematics studies.

E. The chief resource for applications and modelling in the mathematics curriculum is the teacher. For any mathematics teacher (including University teachers), applications and modelling should form part of pre- and in-service education. This education should develop applications and modelling awareness and skills with the teacher as an individual, as well as provide knowledge of and insight into psychological processes related to applications and modelling activities.

F. Many applications and modelling case studies suitable for various levels already exist in publication. There are basically two sorts of case collections and problem banks. One presents a number of applications and models within a given field/discipline of application, the other portrays a variety of applications and modelling cases involving nominated mathematical concepts or topics. Case collections of these and other types are most valuable resources for applications and modelling activities, and more should be published. They are, however, only helpful to some extent. A study of cases can in no way substitute for modelling open-ended problem situations.

G. Appropriate uses of hand-held calculators and, in particular, computers, should be explored.

H. Interaction and cooperation between mathematics instruction and the world outside, such as other disciplines, industry, commerce, administrative and planning agencies, organisations, etc, is needed as an important resource for applications and modelling teaching in particular, and mathematics education in general. Such interaction and cooperation should be stimulated and organisational structures should be sought.

I. The role of applications and modelling in the curriculum should not be hindered by current conditions and traditions of assessment and examinations. The evaluation procedures based on, and closely related to the processes, skills, and products of applications and modelling activities should be developed and adopted. It is recommended that colleges and universities, which by tradition have more freedom in curriculum matters than schools in most countries, take the lead in making a breakthrough in this respect. More generally, individual institutions and individual teachers should be given more freedom in design of curricula and in planning instruction and assessment than is usually the case today.

J. The theory of teaching mathematics applications and modelling on different levels should be established. Researchers should conduct studies to this end.

Acknowledgements
The above report, prepared by the Organisers, has benefitted considerably from the written and oral presentations of the subgroup leaders and the working session reports produced by the recorders.

References
Anderssen, R.S., de Hoog, F.R. (eds.) (1982) *The Application of Mathematics in Industry*, Wijhoff.
Andrews, J., McLone, R. (1976) *Mathematical Modelling*, Butterworth.
Beck, U., Biehler, R., Kaiser, G. (1983) 'Review of Applications in School Mathematics in the Federal Republic of Germany'. *Occasional Paper 34*, IDM, Bielefeld.
Bell, M.S. (1979) 'Teaching Mathematics as a tool for problem-solving', *Prospects*, IX(3), (UNESCO).
Bender, E. (1978) *An introduction to mathematical modelling*, Wiley.
Biehler, R. (1982) 'The role of applications of mathematics in the theory and practice of mathematics education in the Federal Republic of Germany, a report on literature, projects and ideas', *Int.J.Math.Educ.Sci.Technol*, 13(2).
van der Blij, F., Hilding, S., Weinzweig, A.I., 'A synthesis of national reports on changes in curricula', in Steiner, H.G., below.
Boyce, W. (ed.) (1981) *Case Studies in Mathematical Modelling*, Pitman.
Broomes, D. 'Goals of Mathematics for Rural Development', in Morris, R., et al. below.

Burghes, D., Wood, A. (1980) *Mathematical Models in the social, management and life sciences*, Ellis Horwood.

Burghes, D., Huntley, I., McDonald, J. (1982) *Applying Mathematics: A course in mathematical modelling*. Chichester, Ellis Horwood/Wiley.

Burkhardt, H. (1981) *The Real World and Mathematics*, Blackie, Glasgow.

Christiansen, B. (1975) *European Mathematics Education — The Past and Present*, Copenhagen Royal Danish School of Education Studies.

Crimmins, J.C. (1984) *The Challenge of the Unknown*, J.C. Crimmins Co., New York, (Film).

Costello, P., Jones, P., Phillips, B., *Mathematics and Manufacturing*, The Institution of Engineers, Australia.

Davis, P.J., Hersh, R. (1981) *The Mathematical Experience*, Birkhauser, Boston.

Dorfler, W., Fischer, R. (eds.) (1976) 'Anwendungsorientierte Mathematik in der Sekundarstufe II,' *Proceedings of the I. Kartner Symposium fur Didaktik der Mathematik*, Klagenfurt.

Dym, C., Ivey, E. (1980) *Principles of Mathematical Modelling*, Academic Press.

El Tom, M.E.L. (ed.) (1979) 'Developing Mathematics in Third World Countries', *Proceedings of the International Conference* held in Khartoum, March 6-9, 1978, North-Holland.

Fletcher, T.J. 'A Framework for the discussion of the place of applications' in *The Teaching of Mathematics* (report), in Steiner, H.G., below.

Fletcher, T.J. *Applications of Mathematics in English secondary schools*, in Steiner, H.G., below.

Ford, B., Hall, G.G. (1970) 'Model-building — An educational philosophy for applied mathematics', *Int.J.Math.Educ.Sci.Tech.*, 1(1).

Genzwun, F., Hodi, E., Lasztoczi, G., Urban, J., Varga, T., (1980) *Secondary school mathematics in Hungary*, in Morris, R., below.

Glaser, A. (1978) *The German Democratic Republic: Mathematics for a Polytechnic Society*, in Swetz, F., below.

Haberman, R. (1977) *Mathematical models*, Prentice-Hall.

Howson, A.G., McLone, R. (1983) *Maths at work*, Heinemann, London.

Jacoby, S., Kowalik, J. (1980) *Mathematical modelling with computers*, Prentice-Hall.

James, D.J.G., McDonald, J.J. (eds.) (1981) *Case studies in Mathematical modelling*, Stanley Thornes, Cheltenham.

Jurdak, M., Jacobsen, E. (1981) *The evolution of mathematics curricula in the Arab States*, in Morris, R., below.

Kaiser, G., Blum, W., Stober, W. (1982) *Dokumentation ausgewahlter Literatur zum anwendungsorientierten Mathematikunterricht*, Fachinformationsze rum Energie, Physik, Mathematik, Karlsruhe.

Kapur, J.N. (1984) *Mathematical modelling in High Schools*, UNESCO and Indian National Science Academy, New Delhi.

Larson, L. (1983) *Problem solving through problems*, Springer.

Lesh, R. (1983) *The learning and development of mathematical concepts and processes*, New York, Academic Press.

Lesh, R. (1983) 'Conceptual Analyses of Problem Solving Performance'. In E. Silver (Ed.), *Under Prepresented Theme in Research on Problem Solving*. Philadelphia, Franklin Institute Press.

Lesh, R., and Akerstrom, M. (1981) 'Applied problem solving: A priority focus for mathematics education research'. In P. Lester (ed.) *Mathematical Problem Solving*. Philadelphia, Franklin Institute Press.

Magnier, A. 'Changes in secondary school mathematical education in France over the last thirty years', in Steiner, H.G., below.

Mmari, G.R.V. (1978) 'The United Republic of Tanzania: Mathematics for Social Transformation', in Swetz, F., below.

Morris, R. (ed.) (1980) *Studies in mathematics education* 1. UNESCO, Paris.

Morris, R. (ed.) (1981) *Studies in mathematics education*, 2. UNESCO, Paris.

National Council of Teachers of Mathematics, (1980) *An Agenda for Action —Recommendations for School mathematics of the 1980s*, NCTM, Reston, Virginia.

Niss, M. (1977) 'The crisis in mathematics instruction and a new teacher education at grammar school level', *Int.J.Math.Educ.Sci.Technol.*, 8(3).

Osborne, M.R., Watts, R. (1977) *Simulation and Modelling*, Univ. of Queensland Press, Brisbane.

Open University, *Modelling by mathematics*, TM 281.

Pollak, H.O. (1979) 'The interaction between mathematics and the other school subjects', in *New Trends in Mathematics Teaching*, IV, UNESCO, Paris.

Polya, G. (1945) *How to solve it*, Princeton.

Saaty, T., Alexander, J. (1981) *Thinking with models (Mathematical models in the physical, biological and social sciences)*, Pergamon.

Shabanowitz, H. *Educational reform and mathematics in the Soviet Union*, in Steiner, H.G., below.

The Spode Group, *Solving real problems with mathematics*, Cranfield Press.

Steiner, H.G. (ed.) (undated) 'Comparative studies of mathematics curricula change and stability 1960–80', *Materialien und Studien*, Band 19, IDM, Bielefeld.

Swetz, F. (1978) 'The People's Republic of China: Mathematics for the Proletariat', in Swetz, F., below.

Swetz, F. (ed.) (1978) *Socialist Mathematics Education*, Burgundy Press, Southampton (USA).

UNESCO, (1972) 'Applications of Mathematics', Chapter VII of *New Trends in Mathematics Teaching*, III, Paris.

Wickelgren, W. (1974) *How to solve problems*, Greeman.

THEME GROUP 7: PROBLEM SOLVING

Organisers: Hugh Burkhardt (UK), Alan Schoenfeld (USA), Susie Groves (Australia) and Kaye Stacey (Australia)

Congress Activities

Under the umbrella of this Theme Group, several activities occurred. The Theme offered an Overview of Problem Solving, a series of more specialised Invited Talks, six Problem Solving Workshops, some Oral Communications and twenty Working Groups which met for the duration of the Congress. In addition, a variety of problems was published in the daily Congress newsletter and prizes for a light-hearted competition for 'teaching solutions' were provided by Objective Learning Materials and by Doubleday.

The central activities in the Problem Solving Theme were the meetings and reports of the Working Groups. Each group consisted of about fifteen people, who worked together for four sessions of one and a half hours on a topic of their choice. The following nine topics, aimed at aspects of the Theme, were offered:

PS1   Problem Solving: Getting Started
PS2   Problem Solving in the Curriculum
PS3   Problem Solving and the Real World
PS4   Teacher Education for Problem Solving
PS5   Research and Development in Problem Solving
PS6   The Detailed Analysis of Problem Solving Performance
PS7   Technology, Teaching Style and Problem Solving
PS8   Problem Solving Competitions
PS9   What is Problem Solving?

The mode of operation of the groups is described below in the section on Working Groups. The outcome of all these activities, which involved the more than 300 Congress participants who chose this Theme, are outlined here.

An Overview of Problem Solving

This overview was presented by Hugh Burkhardt (UK) and Alan Schoenfeld (USA) in two parts; the first on the processes of problem solving, and the second on teaching problem solving.

The Processes of Problem Solving

From occupying a single session at ICME IV, originally listed under 'unusual aspects of the curriculum', Problem Solving has become a major Theme at ICME 5. This reflects an almost universal perspective that drill-and-practice in the 'basics' of mathematics, together with the standard pedagogical practice of providing explanation, followed by worked examples, followed by practice exercises, are not sufficient. Students master the 'basics' but do poorly on problems that differ even slightly from the examples on which they have been trained. (See, for example, the Cockcroft Report and data from the 1982 National Assessment of Educational Progress, USA). There has been increased interest in the *processes* of mathematical problem solving, in addition to the mastery of mathematical subject matter and technique. We focus here on three aspects of mathematical thinking: mastery of general strategies or heuristics, control or 'executive' behaviour and belief systems.

Heuristic Strategies. In *How to Solve It* and his subsequent works on problem solving, George Polya described a variety of 'rules of thumb' for making progress on unfamiliar problems. Such *heuristics* include drawing a diagram, examining special cases, exploiting related problems, using analogy, and many others. There is mounting evidence that:

• competent problem solvers use such strategies, often unconsciously;
• even talented college students are generally unaware of such strategies and do not use them;
• students *can* master such problem solving techniques — but the degree of explicitness and attention required to elucidate the strategies and to give students adequate training should not be underestimated, though it almost always is.

Control or Executive Behaviour. The notion of reflecting upon our own behaviour, of including the monitoring and assessing of our progress while we are solving problems was the theme of Kilpatrick's plenary session ('Reflection and Recursion'), and the reader should seek further elaboration there. Competent problem solvers maintain, as they work, a distance from their work which allows them to evaluate it 'on line', and act accordingly; they control, to a degree, the way they work. In contrast, many students seem to view their minds as independent, autonomous entities that are in a sense beyond their 'control'. Much of students' failure to solve problems is not a result of lack of knowledge, but the result of bad 'management'. They go off on 'wild goose chases' and, by not curtailing them, deprive themselves of the opportunity to use the knowledge they do possess.

Belief Systems. Part of using mathematics or 'thinking mathematically' consists of approaching the world from a mathematical point of view. Students learn their points of view about mathematics largely from their classroom experiences with it, and some of these can cause difficulties. Among beliefs about the mathematics commonly held by students are:

• All mathematics problems can be solved in ten minutes, if they can be solved at all;
• Mathematics is passed on 'from above', by experts, so one must take a passive role in learning it;
• Formal mathematics has little to do with thinking or discovering.

Needless to say, such beliefs affect the way the students use, or fail to use, the mathematics they are taught.

While the preceding paragraphs have focussed on research results, it should not be assumed that such research is, or should be, divorced from practical applications. Suggestions from the research have been incorporated with success into curriculum materials and into teaching practice. Conversely, real classroom experience can and should serve as a well-spring of ideas for elaboration and exploration through research.

Teaching Problem Solving

Teaching problem solving is harder than teaching mathematical technique. It is harder mathematically, pedagogically and personally. The usual expository 'single-track' style, used by every mathematics teacher, will not do; its elements — explanation by the teacher, with illustrative examples, followed by many imitative exercises — put the pupil in a passive, imitative role. Pupils with difficulties are given help in the same way, by re-statement and demonstration. The whole point of problem solving is for the students to *tackle an unfamiliar task, finding their own way*; further teacher explanation destroys this task as a problem. In problem solving the teacher's task is inevitably 'multi-track', with pupils pursuing different lines of thought with mathematical consequences that may be unclear, and needing guidance that still leaves the problem in their hands.

What can be done to help teachers handle this situation in the classroom, particularly ordinary teachers? A great deal of teaching material that emphasises problem solving has been produced in recent years, particularly in Australia, the United States, and the United Kingdom. It generally consists of three elements:

• practice on problems;
• taught strategies for problem solving or heuristics;
• guided reflection by the student.

Most of the materials have simply been drafted by groups of teachers from their own experience; they aim to provide ideas and suggestions for using these new and *unfamiliar* teaching activities. Almost all offer far less support than the materials (textbooks and so on) that teachers use for the *familiar* parts of the curriculum. In a few cases, research based studies of what happens when they are used in the classroom have been a part of their development and evaluation. In only one case has this involved a sample representative of typical teachers.

The design and collection of problems that are effective for teaching problem solving is important. Each should be well matched to the student, providing some success for everyone involved and yet a challenge for the most able. Each should have different facets, linking various aspects of mathematics and its applications. Their technical demand should usually be low, needing only mathematical techniques that are well and long absorbed by the students; the main 'load' on students should be strategic, asking them to find effective lines of attack on an unfamiliar situation. The explicit teaching of strategies is effective; it also provides teachers with some place for their explanatory skills. Further, it is generally safe to assume that if a strategy or technique is not taught explicitly, students will not learn it. We have already mentioned the importance of students' reflecting on what they have done, what they are doing, and why; classroom methods for promoting this are less well developed. An emphasis on encouraging pupil explanation helps in this and many other regards.

'Teacher Lust' — the urge to explain, to tell the children more — is the greatest danger in teaching problem solving. Teachers play many roles in the classroom; they may be grouped as 'manager', 'task setter', and 'explainer' (which all mathematics teachers play regularly), and 'counsellor', 'fellow pupil' and 'resource', which are found only in that small proportion of classrooms where problem solving is a regular activity. 'Counselling' is discussing with children how they are tackling the problem. The key style shift in teaching problem solving is moving teachers from 'explaining' into 'counselling' roles and children into 'explaining' and 'managing', and 'task setting' roles. These role shifts are not easy and teachers need support in making them. Well designed materials carefully developed in the classroom, the use of video and of the microcomputer programmed as a 'teaching assistant' have all been shown to provide powerful, and, taken together, adequate support.

Other fundamental issues include real problem solving, different pupil abilities, and the role of inservice training. Real problem solving presents all these same challenges together with the extra factors that arise from the 'modelling' of the real world. In addition, real problem solving leads to discussions of children's own problems; which teachers may or may not welcome. Less able and gifted pupils have their special needs in problem solving, as in other aspects of mathematics. Inservice training has an obvious part to play in promoting problem solving, and may very well be crucial in generating a positive attitude to change. Very few school systems have sufficient facilities, and this situation should be improved; however, foreseeable resources for inservice training cannot provide the major support that teachers need to absorb this new activity into their teaching. The materials really need to be designed with the essential support built in.

The Working Groups

The Problem Solving Working Groups were experimental in format. While prototypes for such groups have been found successful elsewhere, the number was then smaller and their composition more homogeneous. This was a first attempt to implement such a working group format on a large scale at an International Congress. For that reason, the format of the Working Groups and the discussion of our experiences with them is given attention in section (b) below. Section (a) summarises the Reports of the Groups at ICME 5.

(a) What the Working Groups Produced

Each of the twenty Working Groups produced a report on its own chosen topic from PS1 to PS9. In these proceedings, there is space for only a brief outline of what emerged, emphasising the issues of current concern, and highlighting a few of the interesting and unusual points.

On reading the reports, it is clear that there are several different aspects of mathematics education for which the phrase, problem solving is used. The Theme Organisers focussed on *unfamiliarity* to the student as an essential feature. This specifically excluded, for example, what are called in many places 'word problems'. There is also a spectrum of purposes for activities labelled problem solving, ranging from teaching for improved problem solving skills to teaching content through a problem solving approach.

There is, not surprisingly, general agreement on the importance of problem solving as a learning activity essential in the mathematics curriculum, and on the difficulty of getting it to happen in typical classrooms. Teaching problem solving requires a radical change in teaching style, classroom organisation and management skills, all involving a change in the relationships, intellectual and pedagogical, between teacher and student. Teachers need confidence and experience in problem solving, suitable models for their teaching, and a wide range of curriculum materials for immediate use.

There is a strong call for a problem solving emphasis in preservice training, and for the general availability of inservice training support for teachers from consultants and education authorities. These general points were common to the reports of many groups. The more specific comments and recommendations from the groups on the nine different topics are summarised here.

PS1 *'Problem Solving — Getting Started'*

Teachers at all levels are concerned that their students have difficulty in solving problems outside the narrow range of illustrative examples used in class. The aim of these working groups was to explore ways of introducing some problem solving into the classroom, to identify obstacles and difficulties one may encounter in doing so, and to look for practical ways to overcome those difficulties whilst still meeting the constraints of the 'set syllabus'.

While recognising the variety of purposes that teaching problem solving might have in the classroom, the four working groups all focussed on its use as a means of introducing, developing and applying concepts and/or skills. In contrast to most other groups, they tended to concentrate on problem solving as an attitude to teaching and learning, rather than as a body of skills or procedures to be acquired. Problem solving is recognised as a valuable, enjoyable and pedagogically sound approach to teaching which may be initially difficult to implement, so the reports included considerable practical advice on how to do this.

Inservice training for principals and teachers was seen as an important aid to making curriculum innovations successful, and it was recommended that consultants and advisers be asked to provide practical assistance during the phasing in of a problem solving approach. All the Groups discussed the central need for good resource material and the potential contribution of different types of problems. The support that teachers can offer each other was also highly valued, especially in collecting or devising resources and comparing experiences. A change in the attitude of students (and some teachers) was seen as necessary before an environment conducive to teaching problem solving can develop in a school. Practical suggestions were made on how to make students willing to grapple with unfamiliar problems, to take time to think about a problem, to become actively involved, to accept a challenge without fear of failure, and so on.

PS2 *'Problem Solving in the Curriculum'*

The 'brief' said:

'There is considerable interest in establishing problem solving in the curriculum, but less understanding of how it may be done. Devise, explain and comment on various practical ways of introducing it.'

Broadly speaking, these four working groups defined problem solving as 'using knowledge, skills and insights of one's own in a new situation'. They addressed the following obstacles to implementing problem solving on a broad scale.

There is little consensus as to what problem solving means, and much chaos in consequence. There seems to be 'too much to teach and not enough time to do it in'. As problem solving is not typically assessed at the end of a mathematics course, many teachers are reluctant to devote much time to it. Suggestions for finding time for problem solving include periodically reviewing the curriculum for 'dead wood', giving students investigations or problems they can work on at home, and in their own time, and working problems in small bits over many days. It is stressed that work must be done at the 'system' level, providing teachers with authorisation to stress problem solving, and ammunition for political arguments. (See the NCTM's 'Agenda for Action' and the Cockcroft Report). At the classroom level the difficulty perceived is not that there is a lack of good problems, but rather that there is a lack of good support structure and materials embodying them. The need for identifying sources of 'instruction kits' (such as the Shell Centre's TSS Project, the RIME materials in Australia and others) was stressed, as was the need for much more extensive preservice and inservice training for teaching problem solving.

PS3 *'Problem Solving and the Real World'*
These two Groups were concerned with real problem solving in the classroom. In recent years there has been clearer recognition of the importance of developing children's skills in tackling practical problems of concern in their everyday life, using their mathematical and other skills. This complements the teaching of mathematical technique and of standard applications of mathematics.

The task suggested for the Groups was to discuss some good examples of real world problems, and to suggest how such problems may be used in the classroom and may be related to problems that are the students' own.

The criteria for good problems figure large in the reports. Such problems should be interesting and relevant to the student, thought provoking and open ended, accessible and adaptable, with various facets stimulating a lot of activity from the students and little from the teacher, leading on to enrichment, both mathematically and otherwise. Mathematically, they should help skill acquisition and consolidation as well as deployment; their practical outcomes should be able to be implemented and validated.

A range of exemplary problems was provided. There was emphasis on student-generated problems, with discussion of the various challenges to the teacher in handling them in the classroom. Different kinds of learning environments for real problem solving, and how teachers can be helped to create them in their classrooms, were described, as were various modes of class organisation. A range of suggestions for overcoming the difficulties of implementation of real problem solving in the classroom included the gradual introduction of applications, school support by consultants or visiting 'master teachers', teacher networks for individual support, timetabled opportunities for teachers to observe each other, collections of good materials, and a carefully generated climate of public support. The idea of exam-driven change was seen as powerful. The importance in teacher training of helping teachers broaden their traditional role was stressed.

PS4 *'Teacher Education for Problem Solving'*

Among the factors mitigating against the establishment of problem solving in schools are the lack of confidence and experience in problem solving of teachers, and the difficulties of creating a learning situation directed to a process based curriculum as opposed to one based on mathematical content. There are no clearly defined syllabi, nor even a well defined, well understood body of pedagogical skills for problem solving. In schools, there is a general lack of classroom support materials, and support personnel for problem solving, while teachers lose the security of existing textbooks, tests and specific structured sequences. Teachers must make a commitment to change — not only in what they teach, but also in their classroom organisation, teaching style and management techniques, and must cope with the reactions of parents and colleagues.

For the preservice training of teachers, the small number of teachers using a problem solving approach makes it difficult to provide a model during teaching practice. Most schools follow the pattern of instruction followed by practice, leaving a critical gap between any purely theoretical course and practice teaching 'in the real world'.

Both preservice and inservice teacher education courses require a balance between practical experience and reflection on that experience. The key components should include personal problem solving experience, preferably in a group, reflection on this experience, observation of a teacher involving students in problem solving activity, reflection on teacher behaviour which facilitates problem solving, opportunities to work with small groups of students, and investigation of the resources available. Several detailed outlines of possible courses were suggested in the reports.

Preservice teacher education is insufficient in itself. Teachers need continual inservice education, support networks for problem solving, resource materials, appropriate guidelines for assessment, and programs to inform school communities of the benefits of problem solving in the curriculum.

PS5 *'Research and Development in Problem Solving'*

This Working Group addressed issues related to research and development in understanding problem solving and its teaching. The Group did not seek to define a comprehensive program of research and development. It pointed instead to a need for initiatives in a number of areas, for some clarification of the methodologies and results appropriate to various aspects of teaching problem solving, and for the linking of more powerful research methods with imaginative and experience-based curriculum development procedures.

Some members took a cognitive science view with links to artificial intelligence; they saw a need to acquire a detailed understanding of problem solving processes before making an attempt to design teaching. Others saw the need for a more phenomenological approach, with instruction along the lines established by Polya and others. It was regarded as important that we seek to determine what types of knowledge are used in problem solving, how they interact and how effective they are.

Various approaches to enhancing the effectiveness of teaching were put forward. Three ingredients were common to these: practice in solving problems, some instruction in strategies, and the encouragement of reflection by students on their own methods. Recent studies have largely consisted of examining the effects of teaching packages, whereas there is a need for more analytical studies integrating both research and development.

The third main area of discussion was how teacher behaviour might be broadened to achieve problem solving objectives. While there is agreement on the elements of teaching style and strategies that are important to stimulating and supporting problem solving in the classroom, there is a need for a much better understanding of what ordinary teachers are likely to be able to achieve with realistic support.

PS6 *'Detailed Analysis of Problem Solving Performance'*
These working groups explored in detail the processes that take place during problem solving sessions. Video tapes of students working on a range of problems were provided. The goal was to discuss methods for analysing such tapes, indicating the uses and limitations of such analytic methods.

One Group focussed largely on methodological issues. It noted that while videotapes may capture some aspects of student behaviour, they may obscure or distort others. For a 'complete' view, such videotapes must be supplemented by other information such as tests and interviews. Moreover, the observer's analytical perspectives and frameworks often shape how taping sessions are structured, and thus affect what appears on the tape to be analysed. Other factors affecting the analysis include task variables, interpersonal variables in the experimental group, the social context (e.g. co-operative versus competitive sessions), observer interventions, and the nature of the instructions for verbalisation. Frameworks for explorations of such behaviour, by Balacheff, Schoenfeld, and Zimmerman, were discussed.

The second working group focussed on developing a process for problem solving to serve as a backdrop and context for the observation of problem solving behaviour. The process of solving a problem was seen as a progression through five stages: being confronted by the problem, accepting it, creating an internal representation, solving, and checking.

Progress through these five stages is not always linear; when difficulties are encountered, the problem solver may jump back to an earlier stage. At all stages, situational and personal variables may affect behaviour and result in shifts within the process. These variables include the instructions given, the social and physical context, the time allowed, past experience, relevant knowledge, and cognitive style.

This kind of approach, based on an analytic framework allows for detailed analysis of problem solving behaviour and the precise communication of the results obtained; two aspects essential for our increased understanding of such behaviour.

PS7 *'Technology, Teaching Style and Problem Solving'*
The potential contributions of technology are so many and varied that the members of the group did not expect to encompass more than a fraction of them. The suggested task also ranged particularly widely, reaching through teaching style to teacher development. The groups were asked to identify and discuss various ways of using technology to support problem solving activities.

One group presented a simple, clear model based on classifications of modes of 'teacher operation', 'student operation' and 'technology use'. Teacher operation was divided, using the six 'roles' from the Shell Centre research, into 'traditional' (manager, task setter and explainer), and 'innovative' (counsellor, fellow pupil and resource). Student operation was labelled 'passive', 'involved' or 'autonomous', while technology use was labelled 'tutor' ('reference mode'), 'tool' ('data manipulation' or 'formulation' modes) or 'tutee' ('creator/presenter', 'simulation' and 'programming' modes) using Anderson's classification. The classification was filled out with detailed descriptions, and references were given to materials which illustrate them.

The other group looked at aspects of teaching style for problem solving as the focus for their discussions, referring at appropriate points to the choice of suitable problems and the support that technology could provide. Their essential requirement for the teacher's style and strategies was that it put the children into an investigating role in a situation involving them in decision taking, and to treat their decisions and results with respect appropriate to these responsibilities. They described the positive outcomes that may be expected but warned against using problem solving as the main channel of learning mathematics, using a personal discovery approach — 'the long, steep hill which mankind has slowly scaled over the centuries'. They preferred problems matched to the student's independent capability, preferably discovering something that the teacher did not know beforehand , including problems from the local environment, mathematical or practical. They want technological support (calculators and microcomputer programs) selected to fit the planned learning program, not vice versa, with multi-media representations where possible. This in turn requires the teacher to be aware, and to think analytically about the range of materials and modes of use. The variety of uses was discussed and the need for quality was stressed.

PS8 'Problem Solving Competitions'
Problem Solving activities in the form of difficult exercises, puzzles and mathematical competitions have traditionally been used to enrich the education of highly gifted students. Three levels of competitions were identified: international and national olympiads and competitions related to these; broadly based huge entry national competitions, and local competitions, including magazine, newspaper, and radio contests.

Apart from the high level competitions, the main aim of contests relates to improving public attitudes and encouraging students and teachers, while secondary aims relate to the actual process of problem solving. The identification of top talent was not regarded as an important aim in itself. Publicity for competitions was regarded as very important, not only to attract sponsorship but also to raise community awareness and to produce materials for use in schools.

There was considerable discussion on question format and the difficulties of marking huge entry competitions. In particular, the value of multiple choice questions, the appropriateness of strict time limits and the conflicting demands of enabling all entrants to gain a high sense of achievement, while identifying and ranking top students were discussed. The Group reviewed a wide variety of competitions from all over the world.

PS9 'What is Problem Solving?'
This group worked largely in Japanese; contact with the rest of the Theme Group was maintained by the group's Australian co-ordinator, Tom

Frossinakis. The suggested task involved creating a classification scheme for problem solving: dimensions included aims and methods, problems 'real' or mathematical, student-generated or presented, complete or topic-illustration, trial and error or 'insight' approaches. The discussion covered a more detailed specification of the classification and of some methods and materials.

(b)The Mode of Operation of the Working Groups
The organisers of this theme took seriously the wish of the Congress planners that there should be opportunity for all participants to participate actively in the Theme Group sessions. It was assumed that every member of the Congress had valuable ideas and experiences to share with others, and that a summary of their discussions would be of interest to a wider group.

Each working group received a 'brief' that raised some specific issues and suggested a focussing task. The groups were designed to build on the collective and complementary experience of their members; the format aimed at wide ranging discussion, leading to the crystallisation of issues and the production of a report. The use of 'expert' leaders of the groups was deliberately avoided. The groups elected their own chairpersons and recorders, although each had an Australian co-ordinator who helped the group get started and facilitated its work throughout the four sessions.

Since this mode of organisation was most unusual for a large international Congress, the response of the participants was of particular interest to us. Although the evaluation was informal, we sought and received a great deal of comment. Both individual participants and individual working groups differed markedly in their response to this mode of working, ranging from great enthusiasm to sharp criticism. Though almost all had some detailed suggestions for change (some groups in particular would have liked more specific leadership), over 70% saw these working groups as a valuable element of the Congress. Attrition over the four sessions varied, but all the groups survived to the end of the Congress, attendance at the last session varying from 40% to 80% of numbers at the first session. The reports from the groups, with perhaps one exception, showed the serious involvement of at least some of the participants. We regard these as positive results, vindicating the Congress organisers' experiment. Several of the groups made explicit, ringing recommendations that such groups be a feature of future ICMEs.

Workshops and Invited Talks

Workshops at ICME 5 offered participants the experience of solving problems, opportunities to look at materials and techniques for classroom use, and chances to reflect on the processes involved. The workshops were given by Shmuel Avital (Israel), Steven Krulik and Jesse Rudnick (USA), John Mason (UK), and Vern Treilibs and John Gaffney (Australia).

Invited Talks covered a broad range of state-of-the-art research and development work in problem solving around the world. Brief abstracts of the talks delivered at the Congress follow:

Cognitive Versus Situational Analysis of Problem Solving Behaviours.
Nicholas Balacheff (France)
Most of the research on problem solving is centred on *heuristics*. Thus it

overlooks both the mathematical content, in terms of psychological complexity, and the situation in which the activity takes place. The aim of the paper was to contribute to a discussion on this point with special focus on dialectical relations between situational components and cognitive (mathematical) ones.

Implementing Problem Solving in the Curriculum. *Leonie Burton* (UK)
Three assumptions underlie attempts to incorporate problem solving into the normal curriculum: that there exists motivation on the part of both teachers and pupils; that there exists 'know-how'; that there exists materials. We now have sufficient research experience to state that problem solving establishes a confident and enthusiastic environment in which pupils of all ages demonstrate positive attitudes both to mathematics and their ability to learn it. The know-how which is developed supports flexible and creative learning in mathematics and it is to be hoped that the production of materials continues to support busy teachers in shifting their teaching style and affecting the consequent learning experiences of their pupils.

Production System Analysis of Mathematical Problem Solving. *Tom Cooper* (Australia) and *Rod Nason* (Australia)
Problem solving is a complex phenomenon to study. Attempts to characterise factors affecting problem solving performance are often thwarted by the sheer size of the problem. The approach (borrowed from artificial intelligence research) called 'production systems' can incorporate both the complexities of problem solving and the essential simplicity of effective teaching frameworks. By being able to supply the necessary detailed complex analysis of procedures and the insight from which global models emerge, production systems are an exciting alternative to more traditional studies of problem solving behaviour.

On Mathematical Thinking. *Tony Gardiner* (U.K.)
Tension between passive learning and active doing is a universal educational phenomenon. The 'Art of Doing Mathematics' is a collection of material which consists of carefully devised sequences of activities which are unfamiliar, elementary, yet intriguing and elusive, and which generate some interesting mathematics. Students find the experience of mathematical discovery so novel that, although it is technically very elementary, it is not perceived as trivial even by very bright undergraduates.

Developing Material and Implementing Curricula for Problem Solving. *Siegfried Grasser* (RSA)
This paper was a case study report on distance teaching at University level. It discussed the difficulties of problem solving in the distance learning environment, with its lack of easy feedback. Approaches in tackling these difficulties were described.

Using Microcomputers to Stimulate Problem Solving. *Mary Grace Kantowski* (USA)
The presence of a microcomputer in the classroom can stimulate interest in problem solving and support students in mathematical investigations that would otherwise be out of reach. An investigation into 'squumbers' was

carried out by students beginning computing. The process of writing a program to find 'squumbers' helped students clarify exactly what a squumber was, and the results offered many opportunities for exciting investigations of mathematical patterns and motivation for looking back.

Evidence for a Pattern of Problem Solving of a 'Matheracy' Type. *Tadasu Kawaguchi* (Japan)
An earlier analysis of the processes of problem solving was illustrated in detail by a specific example. The usefulness of a programmable calculator emerged as an important factor.

Generalisations of Moessner's Process: An Exercise in Problem Solving. *Calvin T. Long* (USA)
Apparently simple problems can be explored, modified and generalised to produce interesting and surprising results. Alfred Moessner observed that one could obtain the sequence of perfect squares by listing all integers, striking out every other one, and adding up partial sums. A modified procedure yields perfect cubes, and a generalisation yields kth powers! A variation on the procedure yields factorials and leads to some fairly deep mathematics. Ways the problem can be used in the classroom were discussed.

Teacher Education Towards Problem Solving. *Ian Lowe* and *Charles Lovitt* (Australia)
The RIME Teacher Development Project has developed a support for both inservice (on-the-job) and preservice education of mathematics teachers at secondary school level. This support consists of a pack of documented, tested, lesson plans, each a complete working model of a problem solving classroom in action, and a support network of demonstration lessons and videotapes. Some lessons introduce problem solving strategies, but most use problem solving methods to help pupils learn the regular content of secondary mathematics.

Problem Solving in Japanese School Mathematics. *Tatsuro Miwa* (Japan)
Trends and issues of problem solving in Japanese school mathematics were discussed. Teaching problem solving was considered in relation to the aim of fostering pupils' mathematical thinking, and features of mathematical problem solving were elucidated. Verbal problem solving in elementary school, including the methods of solution and structure of verbal problems, were described. In junior and senior high school, applications of mathematics present a serious difficulty as few pupils or teachers have experience in solving real problems by applying mathematics. Teachers have devised interesting means of motivating pupils.

Problem Solving in Australia. *Roland Mortlock* (Australia)
In recent years there has been a great upsurge of interest in problem solving in Australia and it has been included formally as a major thrust in the guidelines for mathematics curricula throughout the country. Teachers also recognise the need for problem solving and are now, to some extent, being provided with the training and resources necessary to implement it in their classrooms. Particularly helpful projects were surveyed.

Problem Solving Using 'Open-Ended Problems' in Mathematics Teaching. *Nobuhiko Nohda* (Japan)

One important aspect of ultimate objectives in mathematics education is to develop students' ability to identify the mathematical relations hidden in a concrete situation and to formulate them so as to make mathematical models. In a series of lessons, students were given open-ended problems and they were encouraged to form some propositions from different perspectives, then to investigate their types of quantities and properties so as to see if better mathematical activities and thinking could be fostered. The results were reported and discussed.

Trends in Problem Solving in the United States. *Sid Rachlin* (USA)

In the United States, although the last decade saw 'problem solving' become if not a household word, then a classroom word, there has been little change in problem solving performance. In many schools, problem solving is being taught by rote. There are some excellent programs which do, in fact, organise the mathematics curriculum around problem solving but they are generally at elementary school level. Curriculum materials need to be developed which help students develop problem solving behaviours as they learn the high school mathematics curriculum. One example of such an endeavour is the Algebra Learning Project currently in progress at the University of Hawaii.

Innovation in Mathematics Education — The TSS Approach. *Jim Ridgway, Malcolm Swan, Anne Haworth, Jon Coupland* (UK)

This sequence of four papers described various aspects of a new approach to large scale curriculum change in England. The TSS project is an ambitious attempt to reshape the balance of classroom activities in mathematics lessons. It depends upon collaboration between the Shell Centre for Mathematical Education and a Public Examination Board committed to curriculum innovation, which is aware of its responsibilities both to innovate and to support the educational community which it serves. All innovation is done slowly and its effects are examined empirically in 'representative' classrooms via purpose built, systematic observation techniques. The unit of innovation is a 'module' which corresponds to about 5% of the curriculum. The module offers material for classroom use, and extensive support material to help teachers respond positively to the new challenges they face.

This Module aims to develop the performance of children in tackling mathematical problems of a more varied, more open and less standardised kind than is normal on present examination papers. It emphasises a number of specific strategies which help such problem solving. These include the following:

• try some simple cases;
• find a helpful diagram;
• organise systematically;
• make a table;
• spot patterns;
• find a general rule;
• explain why the rule works;
• check regularly.

Such skills involve bringing into the classroom a rather different balance of classroom activities than is appropriate when teaching specific mathematical techniques: for the pupils, more independent work and more discussion in pairs or groups, or by the whole class; for the teacher, less emphasis on detailed explanation and knowing the answers, and more on encouragement and strategic guidance.

The microcomputer has two distinct roles. One is to support a problem solving session amongst the staff of the mathematics department, the purpose of which is to help the teachers identify with the feelings and needs of children facing unfamiliar tasks. Teachers are invited to think about the importance of discussion, and the level of support that they themselves would welcome as problem solvers. The second role is to support style change. In related research on the impact of microcomputers in class, it has been shown that investigative work can be facilitated by many of the programs of the ITMA Collaboration — the microcomputer can be programmed as a 'teaching assistant' to take over temporarily much of the detailed managing and task-setting that is usually performed by the teacher, thus liberating the teacher and leading to other styles of classroom behaviour (such as 'fellow pupil' or 'counsellor').

Curriculum Development in Problem Solving. *Kaye Stacey* and *Susie Groves* (Australia)
Many teachers who wish to improve the problem solving ability of their students feel insecure and uncertain how to proceed. To assist these teachers, material has been developed for schools, based on earlier experience with preservice problem solving teaching. For primary schools, it is recommended that, from the earliest grades, children be given regular and frequent experience of tackling non-routine problems in a supportive environment. In secondary schools a more systematic approach is both possible and desirable. With the co-operation of local teachers a two year course based on a few simple but important strategies and hints has been developed.

Evaluation in Problem Solving: An Implementation Problem. *Walter Szetela* (Canada)
Today, the increasing acceptance of the Polya four-stage problem solving model in the teaching of problem solving is making the task of evaluating students' attempts at solving problems more difficult. Instead of a right-wrong fixedness based upon a rule- and algorithm-oriented curriculum, evaluation procedures must become more flexible, reflecting for example, understanding, planning and result. Questions of reliability are paramount but the question of feasibility of use in the classroom should receive special attention.

Problem Solving in South Africa. *John Webb* (RSA)
Within the official syllabus there is ample scope for problem solving, but because the educational system is highly exam-oriented, the teaching and textbooks opt for a stereotyped, work-to-rule approach, opposed to the attitudes one wants to inculcate in a problem solver. In order to promote problem solving from outside the system, a magazine for high school pupils with prizes offered for problem solutions is produced. Other magazines are now being published and there are a number of local and national mathematics contests which promote enthusiasm for problem solving.

The Way Forward After ICME 5

It is clear from the whole range of activities at ICME 5 that problem solving is a central concern in mathematics education world-wide. It is equally clear that there is a long way to go before it is found in even a majority of mathematics classrooms. The work on the Problem Solving Theme clarified the need and the directions for work aimed at making it a reality.

The support structures available to help teachers interested in problem solving are, in general, woefully inadequate, both in terms of training and in terms of materials for classroom use. Moreover, much of what does exist is material developed on an intuitive basis, which needs to be better grounded in our understanding of students' thinking, more thoroughly researched, and better developed and evaluated.

Teaching problem solving demands a clear shift in teaching style, a change in the deeply ingrained responses of many classroom teachers. There are hopeful signs that this may be achieveable, but little firm evidence as yet. Indeed, little is known about the dynamics of teacher behaviour, and much less about modifying it for such difficult tasks as teaching problem solving. What works and why? There is a need for the careful documentation of successful programs of all kinds. Can such programs be packaged for others to use, and how?

There are some good materials available on problem solving, but the quality has often suffered because of an entrepreneurial approach and the very superficial view of problem solving popularly espoused. Materials are rushed into production or dissemination before they are adequately designed, developed and tested. The difficulty in producing adequate materials — which will work 'in the hands of those other than the ones who designed them', or their fellow enthusiasts — is usually grossly underestimated. There is a need for developmental studies (not just with volunteers) in realistic circumstances, with adequate observation for development and evaluation. Such work should be long- as well as short-term. Problem solving skills are complex and take time to develop and our work should reflect this obvious fact. The gimmicks and puzzles of many popular materials are not the best medium for it.

The research and development communities in problem solving have in general been quite isolated from each other, to great mutual disadvantage. A large fraction of the materials are produced in ignorance of research findings, often ignoring essential components of mathematical thinking. Conversely, the research community needs to consider far more seriously than it has the generality and potential utility of its studies, especially short-term laboratory studies that are claimed to have classroom relevance. In *Critical Variables in Mathematics Education*, Ed Begle found a survey of the empirical literature quite depressing — from one ICME to the next, he noted, there are new faces addressing the same old problems with new solutions that do not build on previous work, and will themselves be ignored in four years' time. It is our sincere hope that by ICME 6 substantial steps will have been taken in a coherent effort towards making problem solving a reality in most classrooms.

TOPIC AREA: EVALUATION, EXAMINATIONS AND ASSESSMENT

Organisers: Kenneth J. Travers (USA), David F. Robitaille (Canada), Robert Garden (New Zealand)

Twelve sessions were offered on this topic area, involving some thirty presenters from many regions of the world. Several major themes emerged from the presentations and discussions that followed.

International evaluation, as embodied in the Second IEA Mathematics Study, is currently being carried out in twenty-four countries* around the world. From an international perspective, the Study provides countries with an opportunity to view their curricula within the broader context of patterns of subject matter emphasis and classroom practice. This activity is proving to be of particular value to countries that are in the process of revising their programs. From a national perspective, examples were provided by several countries of how the IEA Study findings are of special interest as national issues are being faced.

National assessment activities, apart from those related to the IEA Mathematics Study, appeared to be characterised by a concern for finding appropriate ways for assessing the learning of students who are not 'mathematically-able'. Some examples were provided of non-standard examination practices in the United Kingdom that have promise in this regard. Clearly, as greater proportions of young people are retained in school, and greater numbers of students are enrolling in mathematics courses, this matter becomes increasingly urgent.

The cultural context within which learning, and in particular mathematical learning, takes place, was raised by Ubiratan D'Ambrosio in his plenary address to the Congress and continued in this topic area. As ways are sought to provide more and more students with the needed mathematical competencies for daily living in an increasingly technological society, it will be imperative to take into account cultural factors not only in assessment activities but in teaching and learning as well.

Theme 1 — Evaluation from an International Perspective: The Second IEA Mathematics Study (SIMS)

The First Mathematics Study of the International Association for the Evaluation of Educational Achievement (IEA) was carried out in the mid-

---

*These countries are: Australia; Belgium (French and Flemish); Canada (British Columbia and Ontario); Chile; Dominican Republic; England and Wales; Finland; France; Hong Kong; Hungary; Ireland; Israel; Japan; Luxembourg; Netherlands; New Zealand; Nigeria; Scotland; Swaziland; Sweden; Thailand; United States.

1960s. The Second Study is now reaching its conclusion. Some preliminary findings were shared at ICME 5. The target populations for the study were Population A: students in that year of school at which the majority of the students are 13 years of age by the middle of the school year, and Population B: students in the terminal year of secondary school and who are studying mathematics as a substantial part (approximately five hours per week) of their academic program.

SIMS Phase I: Curriculum Analysis (The Intended Curriculum)
The methodology of the SIMS called for ratings from each country as to the importance of each of a variety of topics from an 'international menu of school mathematics'. From these ratings, various patterns of emphasis on mathematical topics for each of the two target populations were identified. Examples of the findings include:
Algebra. Two clusters of countries can be identified for Population A as high algebra and low algebra. That is to say, at this level, approximately one-half of the countries are well into elementary algebra while the remaining countries have not yet reached this stage. For the latter countries, the mathematics curriculum appears to be dominated still by arithmetic.
Geometry. No patterns of importance were detected for geometry among the countries in the study. There appears to be little international agreement as to the content of school geometry.
Calculus. Two clusters of countries can be identified for Population B as high importance and low importance. Approximately one-third of the countries are in the high importance calculus group. Explanations for this finding were sought in contextual factors, such as the importance of algebra in Population A, proportion of age cohort in school (retentivity of the system), and proportion of grade cohort in Population B mathematics.

SIMS Phase II: Classroom Processes (Implemented Curriculum)
Data were obtained from each country on what mathematics is taught and how it is taught.
   Eight groups of teachers and students participated in the Classroom Processes component of SIMS: Belgium (Flemish), Canada (British Columbia), Canada (Ontario), France, Japan, New Zealand, Thailand, and the USA. Each teacher was asked to complete five comprehensive questionnaires dealing with specific aspects of the methodologies they employed in their teaching of the following five areas of the mathematics curriculum: arithmetic, algebra, measurement, geometry, ratio and proportion, and percent.
   One of the more significant findings to date has been the high degree of comparability among teachers from these eight disparate educational jurisdictions. Not only do they have fairly similar backgrounds, training, and years of experience, but they appear to share many opinions and a general orientation toward teaching. For example, there is widespread agreement among all these teachers that learning mathematics 'helps one to think logically,' and that mathematics is a field in which new discoveries are constantly being made. Teachers from all eight systems also agree that an emphasis on feedback to students and attention to matters of discipline and classroom management are two of the hallmarks of effective mathematics teaching.

The greatest differences among countries occur in the teaching of geometry, where there is no apparent consensus of opinion regarding either the content to be taught or the teaching approach to be emphasised. For example, while teachers in Belgium, France, Luxembourg, and Thailand emphasise a formal transformational approach in their teaching of geometry, those in other countries tend to use an informal Euclidean approach.

SIMS Phase III: Student Outcomes (Attained Curriculum)

Achievement results at the Population A level are available for students from twenty different countries. These involve 157 multiple-choice items divided among five major subtests: arithmetic (46 items), algebra (30 items), geometry (39 items), measurement (24 items), and descriptive statistics (18 items). Furthermore, data indicating the degree of appropriateness of each item to the curriculum in each country are available, as well as data from teachers indicating whether their students had been taught the mathematics required to respond correctly to a given item. These *appropriateness* ratings and *opportunity-to-learn* measures provide important contextual information that should be employed in any attempt to interpret the achievement results. For example, the fact that a given country considers a particular item to be inappropriate for its students and that the teachers in that country give the item a low opportunity-to-learn rating may account, in large part, for relatively poor performance by students in that country on that item.

The fifteen items that students found easiest dealt with several different topics. The easiest item, one which required students to estimate the length of a segment, was answered correctly by 84% of the students. Of the fifteen easiest items, five concerned the use of standard computational techniques; three, familiarity with SI metric units; and two, interpreting statistical graphs.

The fifteen most difficult items were answered correctly by less than 25% of the students. Of the fifteen, four derived from a fairly rigorous approach to transformational geometry and another four with topics from Euclidean geometry. In other words, slightly more than half of the most difficult items dealt with some aspect of geometry. This is likely a reflection of the fact that there is so little agreement among countries as to what constitutes appropriate geometric content, as well as of the fact that most of these items were indeed difficult and it is not at all surprising that students did not do well on them.

SIMS Classroom Processes: The 'Missing Link' Between What is Taught and What is Learned: Thomas J. Cooney (USA) and Curtis C. McKnight (USA)

Plans and progress were reported for data analysis on classroom practices in eight countries. Data came from questionnaires dealing with five topics: fractions; ratio, proportion and percent; algebra; geometry, and measurement. Data analysis to date has focussed on balance between abstract, symbolic instruction vs. concrete, perceptual instruction. Some topics (for example, measurement) were found to be taught in a highly symbolic way in all countries while other topics (for example, algebra) varied more from country to country. The full report will be available in about one year's time.

Theme 2 — International Evaluation from a National Perspective

Reports from the following countries taking part in SIMS were given by the individual named.

The Teaching and Learning of Mathematics in the Dominican Republic (TLMDR): Eduardo Luna, Rafael Yunen and Sarah Gonzalez

In 1978, the Interamerican Committee on Mathematics Education endorsed the participation of Latin American countries in the Second International Mathematics Study (SIMS).

The data of the TLMDR were gathered in 160 classrooms (117 in urban, and 43 in rural areas), belonging to 116 schools altogether (76 urban and 40 rural).

Types of Schools

The sample was selected from six different types of schools:

P: Primary and Intermediate School (Public, Urban)
T: Intermediate and High School (Public, Urban)
R: Schools in the Reform Program (Public, Urban)
F: Private Schools which are authorised to give examinations (Urban)
O: Private Schools without authorisation to give examinations (Urban)
Ru: Rural Schools (Public)

The mathematics teacher of a typical eighth grade class is a young person (32.6 years) with 11 years of teaching experience but only 5.7 years of mathematics teaching experience at this grade level. Mathematics training is poor (one-third have only a high school diploma). This teacher has a heavy teaching load (34 periods per week) and teaches several subjects. The teacher depends on the textbook as the main teaching resource, although only 24% of the students own textbooks.

Student achievement was very low in all the mathematics areas, even though the majority of the items (83%) were considered appropriate by the teachers. There are large differences in achievement between F schools and the other types of schools. Achievement in the pre-test of F schools was greater than the achievement of the other types of schools in the post-test in each one of the mathematics areas considered. The socio-economic status (SES) of students in F schools is higher than the SES of students in the other types of schools. It seems that the educational system emphasises and deepens social inequalities.

Japan: Toshio Sawada

Japan participated in the project in order to gain a better understanding of patterns of achievement and attitudes toward mathematics, and to make use of those findings to improve the quality of mathematics education.

At the Population A level (seventh grade), achievement results were highest in measurement and descriptive statistics. The lowest results occurred in geometry. The mean percent correct for all items was 62.5%.

The mean percent correct for Population B in Japan was 69.2%, and here too the lowest results were in geometry. Highest results were obtained in algebra, finite mathematics, and sets, relations and functions.

The topics that showed the greatest degree of growth in achievement from pre- to post-test at the Population A level were algebra and geometry.

On the whole there were no significant differences between the results obtained on the items which were used in both the first and second IEA mathematics studies for Population A students. For Population B, however, there was a significant increase in achievement in SIMS.

There were no significant differences in achievement between boys and girls for Population A. For Population B, however, boys outperformed the girls.

Population A students from schools in medium-sized cities and towns did better than other students. For Population B, schools that had higher proportions of students continuing on to institutions of higher learning showed better achievement than the others.

Boys and girls responded very differently to items assessing the extent of sex stereotyping in mathematics and to the 'Mathematics and Society' scale. A majority of students in both populations agreed that a knowledge of mathematics is necessary in everyday life, and that mathematics is used in most occupations.

Luxembourg: Robert Dieschbourg
Only Population A was included in the study with 2005 students (50.7% girls and 49.3% boys).

The national sample was drawn from the following school types:

Type	Percent of Population Tested	Examination for Admission	Language of Instruction
(1) Secondaire	27.3	Yes	French
(2) Moyen	11.4	Yes	French
(3) Professionelle	8.2	Yes	French/German
(4) Secondaire Technique	35.4	Yes	French/German
(5) Complementaire	17.7	No	German
	(2,3,4 — same exam)		

Note: At the time of testing, school type (4) was experimental, and has now replaced (2) and (3) with a modern program, which is a weaker version of the modern program of school type (1).

Normally, one would expect that the students in school type (1) would have the best results, since that school is most selective. However, this occurred only in algebra, statistics and measurement. School type (3) had the best scores in arithmetic and geometry. The lowest performance on all topics was in school type (5).

Schools (4) and (5), enrol immigrant students (Portuguese) in seemingly greater proportions each year. Upon comparing the performance of Luxembourg students with their foreign counterparts, one finds that the latter group performs less well on all verbal problems, regardless of whether the language of instruction is French or German. In general, the results for boys were better than the results for girls.

Netherlands: Hans Pelgrum and Henk Schuring
The Netherlands participated only in the Population A study. Four subgroups were included in the study involving those preparing for education requiring mathematics for at least four years. A second group in intermediate general education required mathematics for the first two years. In addition, students from two practically-oriented programs (lower technical and lower domestic science) were included.

Large differences in achievement between the four subgroups were found, corresponding to the ability levels of students and the implemented curricula in these programs. It was noteworthy that the subgroup from the lower domestic science program achieved rather well on the statistics portion of the international test, reflecting the emphasis placed on this topic in the program.

There have been changes in the school system since the first IEA study in 1964, hence school types are not directly comparable. Also, in the first study the First Form, rather then the Second Form, was tested. For all subgroups, decay in arithmetic was noted, but performance in algebra has improved over the 20 year period.

The findings of the second mathematics study (SIMS) will be used in both teacher training and in curriculum development.

In teacher training, students will compare SIMS findings with those observed as part of their classroom study activities. As curriculum changes are planned, the SIMS data will be important for information on what is currently attained by students. The findings are of use not only to government groups, but also for practising teachers.

Nigeria: Wole Falayajo
Participation of Nigeria in the Second International Mathematics Study (SIMS) was coordinated from the Institute of Education, University of Ibadan. Because the project did not have any special funding but had to rely on funds provided by the Institute, the scope of participation had to be curtailed. Only Population A and the cross-sectional component of SIMS were studied.

Results on the cognitive tests show very low performance. Mean score on the core test was 14.7 (which represents 36.8% of the total) with a standard deviation of 5.9.

This performance seems to be closely related to the opportunity to learn (OTL) rating. In arithmetic, which had the highest mean score of 38.5% of the total, the mean OTL rating was also the highest — 80.7%; while geometry had the lowest OTL rating of 69.7%.

Results from the affective measures show that students think that mathematics is important, and they like it, but find it difficult.

Correlational analyses have been done in a parallel study that involved some 1200 students in 40 schools at the end of the secondary school in one of the states. The results of these analyses show some quite substantial correlations between cognitive achievement on the one hand, and some affective variables like mathematics in school, attitude to mathematics, and also such teacher variables as qualification, pedagogic training, and experience in teaching mathematics at this particular level.

Thailand: Patrakoon Jariyavidyanont
For Population A, the national sample included 4015 students in the eighth grade in 99 schools from 63 provinces. For Population B, the sample consisted of students in the twelfth grade who were studying in the mathematics and science track. A total of 4150 students in 107 classrooms in 64 schools from 33 provinces was included.

Characteristics of Students: The fathers of the majority of students were skilled workers. Father's education was largely at primary level, although more fathers of Population B students had completed secondary education or higher, than had those of Population A students. Population A students spent

8 hours per week on homework with about 4.5 hours devoted to mathematics. Population B students spent about 13 hours per week on homework, with about 6.5 hours devoted to mathematics.

Characteristics of Teachers: After secondary school, teachers of Population A studied an average of 5 semesters of mathematics, 2 semesters of methods of teaching mathematics and 2 semesters of general pedagogy. For Population B, teachers had 8 semesters of mathematics, and the same amount of methods and pedagogy as Population A.

Student outcomes: The mean achievement for Population A was approximately 40 percent of the items on the international test. For Population B, achievement ranged from about 35 percent on the calculus to about 53 percent on sets, relations and functions. For Population A, little difference in achievement was found between students in the metropolitan, urban and rural areas. For Population B, students from metropolitan schools performed better than did those in urban schools, and they in turn performed better than students from rural schools.

United States: Curtis McKnight
Both Populations A and B took part in the Study, and classroom processes were studied for both Populations.

For Population A (Grade eight) four class types were identified with rather distinctive content coverage and achievement patterns, namely remedial, typical, enriched and algebra classes. On average, 145 hours of instruction in mathematics is offered during the school year.

The fit of the international test was good for arithmetic, modest for algebra and measurement, and poor for geometry. Instruction tended to be symbolic and formal, and dependency on the textbook was high.

Student achievement in arithmetic and algebra was close to the international average, while achievement in measurement and geometry was relatively lower on the international tests.

For Population B (Grade twelve) two class types were identified, namely pre-calculus and calculus. The calculus classes followed the detailed syllabus of the Advanced Placement program. On average, the content coverage for the Population B classes was rather low for many parts of the SIMS test.

The achievement of the pre-calculus classes, the majority of the US Population B mathematics students, was well below the international average, while the calculus classes achieved at or near the international average in most content areas.

Theme 3: National Assessment Projects

Burma: Chit Swe
Burma is a socialist country where education is a major responsibility of the State. The Burmese school system is composed of primary schools (kindergarten, plus four years), middle schools (4 years), and high schools (2 years). System-wide examinations are conducted by the State at the end of middle school and at the end of high school. These examinations play an important role in determining the academic and professional futures of students.

In 1980, the Government approved a plan calling for the reform of the entire school curriculum at all levels and in all subjects. Recognising the importance of enlisting the support of the teachers in such a venture, the Government has taken steps to involve as many good teachers as they can in the process of change. Although the initial plans called for the whole system to implement the new curriculum at the same time, these have now been revised to allow two years for the implementation to take place. The need for a comprehensive program of in-service education has been foreseen, and plans for such a program are currently being developed.

Curriculum guides and textbooks, both in Burmese and English, are being prepared by committees of teachers and other educators. In order to ensure that curricular materials adequately serve the needs of the people, surveys of government departments, teachers, representatives of industry, and other professionals were conducted. The results of these surveys, together with exemplary materials obtained from other countries, were made available to the committees working in the project.

The Caribbean: Ian Isaacs

The Caribbean Examinations Council was established in 1972 to develop examinations to replace the Cambridge Overseas G.C.E. examinations with ones which catered to a wider range of abilities and career intentions. Three mathematics syllabi were planned to correspond to three levels of student ability: the specialist level, the general level, and the basic level. The first two are directed at the top 40% of students, and the basic level at the remaining 60%.

The general syllabus was an innovation in the Caribbean since it included topics such as consumer mathematics, statistics, and computation which were deemed more relevant for this sector of the school population than trigonometry, vectors, or matrices. The first two syllabi included a statement of the goals and objectives of the program as well as sample test items. The latter were intended to illustrate, in a concrete fashion, what was to be covered in some of the newer topics.

Examinations based on the new syllabi were set for the first time in 1979, and the number of students sitting these examinations has risen rapidly over the past five years.

The examinations will probably continue to undergo modifications to ensure that they reflect more accurately the abilities, interests, and needs of Caribbean students. It is hoped that they will help reduce the number of students who see mathematics as a terrifying trial in their rite of passage to adulthood, and, at the same time, increase substantially the number who perceive and use mathematics as a tool for effective living.

England, Wales And Northern Ireland: Derek Foxman

Regular surveys of the mathematical performance of 11 and 15 year old pupils are carried out by the National Foundation for Education Research on behalf of The Assessment of Performance Unit (APU) of the Department of Education and Science. The surveys are designed to provide a national picture of pupil performance and to monitor changes in performance over time. They were conducted annually from 1978 to 1982 but are now to take place every five years, with the next scheduled for 1987.

The APU surveys attempt to obtain a general picture of pupil performance which reflects the variety in the curriculum. Use is made of both written and practical tests. The innovative practical tests are administered in a one to one interactive interview situation by teachers trained as testers. The testers work from a 'script' of questions which includes prompts and probes, so that the methods used by pupils to obtain their answers are highlighted.

A feature of successive surveys has been the development of items to explore or extend particular findings obtained in previous surveys; these findings relate to the understanding of concepts and the effects of wording, presentation and context on the mathematics. Extensive coding of error responses to individual items has been developed, so that the contrasts in performance on different items relate to error and omission rates as well as facility.

The most important aspects of the project have been the development of new assessment techniques and the breadth and richness of the information provided.

Kenya: George Eshiwani

The Kenyan system of education is fairly typical of a number of educational systems in Africa, which share many of the same kinds of problems. Kenyan schools use a 7—4—2 structure, with formal examinations at the three break-points. The majority of students leave school by the end of the primary level, that is, Year 7. Of the entire population of the country, 60% are below the age of 20.

The examination that is held at the end of Year 9 serves primarily as a selection mechanism, and mathematics is regarded as one of the more important subjects in this selection process. There are three examinations in all, mathematics, English, and a third which contains sections from other areas of the curriculum. In order to be successful, that is, to be admitted to the next level of the school system, a student must do well in mathematics. Excellent performance in both of the other examinations will not compensate for poor performance on the mathematics examination.

Since 70% of students do not continue their formal education beyond the primary level, there is a great deal of concern about their ability to cope with the mathematical requirements of adult life. In order to assess the state of affairs, the examination at the end of primary school focuses on two areas of mathematics, namely arithmetic and applied arithmetic. Results to date indicate a satisfactory level of performance in the former, but much weaker performance in the latter. This is currently a matter of considerable concern.

Scotland: Gerard Pollock

Over the past thirty years, the Scottish Council for Research in Education has carried out six assessment projects in Scottish schools related to various aspects of mathematics achievement at several different levels. During this period, thanks to progress in the technology of large scale assessment, a number of improvements have been made to the process, not only with regard to the selection and use of appropriate samples of students, but also with regard to the kinds and number of test items employed.

The use of a multi-matrix sampling design for test booklet design, and of scientific sample selection has meant that the 1983 project produced a number of important benefits. In addition to making it possible to examine results on individual items, it is now possible to examine subtest scores, to produce national profiles of student and school performance, and to conduct analysis of patterns of errors made by students. Since 1978, moreover, the design of the assessments has made it possible to track changes in patterns of student achievement over time.

The most recent development in this area has been the introduction of *practical work*, involving the use of 'stations' as a part of the assessment. An initial set of such items was utilised in the most recent project (June, 1984) and, although there was not time to pretest the materials before they were used in the field, it is expected that a good deal of valuable information will be derived from their use.

Singapore: Wong Kooi Sim

The origins of the current system of national assessment in Singapore may be traced to the 'Report of the Ministry of Education, 1978' by Goh Keng Swee. The report highlighted a number of problems in the educational system (high levels of wastage, low rates of literacy, ineffective bilingual programs) as well as a number of factors which contributed to those problems (the rigidity of the system, lack of effective programs to deal with individual differences, and a lack of overall planning including evaluation and feedback). As a compulsory component of the curriculum at both the primary and secondary levels, mathematics was a source of concern, especially as failure rates in it were generally higher than in other subjects, and there were wide variations in performance levels in mathematics between students and schools.

The report recommended that the entire educational system be restructured. In particular, it recommended that a program of streaming of students be introduced at all levels and that the organisational effectiveness of the system be improved by monitoring the effectiveness of pupils, teachers, and schools. This has resulted in the introduction of several new assessment programs at various levels in the system, as well as in the development of diagnostic tests for use by teachers.

Along with the restructuring of the system, a centralised educational data base has been developed. This permits the Ministry of Education to monitor not only patterns of achievement, but also to track patterns of student mobility, to investigate variations in student and school performance and the underlying factors associated with such variations.

Theme 4: Emerging Practices in Examination and Assessment Techniques

Graduated Testing: Susan Pirie (England)

Graduated testing has been developed in response to radical changes that are taking place in the English public examination system. These changes have been motivated by several factors, including the need for a less divisive, more comprehensive examination system for the majority of the pupils and the search for appropriate ways to assess the abilities of low achievers (the bottom 40%).

The Cockcroft Report has had an important role to play in these developments. For example, it recommended the design of a 'curriculum which is suitable for low attaining pupils' and suggested a 'foundation list of mathematical topics which should form part of the mathematics syllabus for all pupils'.

Graduated tests, in which all pupils can demonstrate success at one or more of a series of progressive levels, go part way towards offering a means of effecting the desired changes.

Two projects on graduated tests are the Oxford Certificate of Educational Achievement (OCEA) and the Graded Assessment in Mathematics (GAIM). Both aim to produce tests for a progressive series of levels of achievement appropriate for all students from eleven to sixteen years of age. The lowest levels are to be based on the Cockcroft Foundation List and achievable by 90-95% of the school population by sixteen years of age.

The central idea of OCEA is that it should record a pupil's success, publicly record what the pupil *can* do, and demonstrate where the pupil's competencies lie. To do this, each level will demand a high degree of mastery in the areas under assessment, and judgments of mathematical performance must be criterion referenced.

Probably the most revolutionary feature of OCEA is its intention to assess not mathematical content but mathematical processes. At times such as the present, characterised by rapid technological development, specific skills frequently become redundant. In the future, pupils will need to be able to interpret new situations, adapt existing knowledge and develop new strategies for dealing with problems. Problem solving is at the heart of all mathematical activity. Eight processes, ways of thinking and acting mathematically, have been designated by OCEA as suitable and appropriate for assessment.

It is expected that the certificate will be available nationally in 1987. The great strength of a graduated examination system is that it can provide the opportunity for all students to build through their schooling a detailed and accurate record of their mathematical achievement.

Practical Mathematics Assessment: Aileen Duncan (Scotland)

The project was funded by the Scottish Education Department for three years and involved teachers of children aged seven to eleven. The major purposes of the project were to develop practical assessment materials, and to consider assessment as a means of curriculum development.

The two questions of why and what to assess were addressed. The major purpose of assessment is as an essential part of the teaching process that provides feedback so that future teaching can be effectively planned. It is generally agreed that what is taught should be assessed. If concepts, language, understanding and problem solving are emphasised in teaching, these aspects should be assessed. In assessment the process of finding an answer should be focused on as well as the answer itself.

The assessment activities are short and concentrate on only one or two specific mathematical behaviours. Each child attempts the activity individually. He or she responds by doing and/or by talking. Writing is minimised.

There are no specific requirements for classroom organisation for using the activities. Organisation should fit the teacher's own needs (individual, small groups, *etc*) and teaching style. The activities chosen can all be on the same topic or be based on a range of mathematics. They can be used before, during or near the end of some teaching. The assessment should take place whenever feedback is required. This is likely to mean assessment is carried out with only one or two groups at a time, and different activities used for different groups.

Discussion with the group of pupils about the assessment activities is essential. In this way, the teacher can probe deeper into aspects of difficulty experienced by the children who also can learn from each other in determining acceptable and unacceptable responses.

Feedback from the activities has highlighted several types of learning gaps or difficulties. These include examples of concepts related only to a specific context, limited understanding and use of mathematical vocabulary, inability to use materials and equipment effectively, trial and error as the main strategy for problem solving.

Many teachers who use the assessment activities receive feedback which has far reaching effects on their future teaching. Assessment can be a powerful means to achieving innovation in the curriculum.

The Use of Grade Related Criteria in Assessing 16 Year Old Pupils Taking a New Mathematics Program in Scotland: Edward S. Kelly (Scotland)
In an effort to broaden the mathematical experience of students, a new program has been developed based on problem solving for which four main processes have been identified: interpreting the information presented in a problem, selecting a strategy, processing data, and communicating the solution of a problem. Making use of applications, and providing a context in the teaching and learning of mathematics, are also major features of the new courses. The program is intended to cover the whole ability range in the secondary school during the final 2 years of compulsory education (15 and 16 year olds).

Throughout the course, teachers are expected to award grades to students on selected pieces of classwork and tests. Towards the end of the course, teachers send to the national Examination Board, a profile of each pupil's grades on five 'assessable elements' — interpretation/communication, selecting a strategy, processing data, problem solving and practical investigations. The assessable element, problem solving, is concerned with the pupil's attainment in solving problems in the round, that is, problems which require the use of two or more of the elements interpretation/communication, processing data, and selecting a strategy. Practical investigations are extended assignments carried out by the pupils and may include measurement, or the use of real-life data (possibly drawn from a survey), or may be a purely mathematical investigation.

The pupils sit an external examination that also produces grades for interpretation/communication, processing data, selecting a strategy and problem solving. Practical investigations, however, are only assessed by the school. On the basis of two sets of grades — one set internally derived, the other set externally derived — the pupil is awarded a certificate in the form of a profile of performance on each of the five assessable elements.

Levels of performance have been prepared by subject specialists for each of the assessable elements at six grade points so that pupils and teachers as well as others know what targets they have to reach to obtain a particular grade.

External Examinations in Mathematics: Gila Hanna and Howard Russell (Canada)
The data gathered by the Second International Mathematics Study have raised some issues related to the use or misuse of external examinations. Of necessity, the SIMS tests are very comprehensive, since they are devised in an attempt to

provide useful measures of achievement across a large number of countries. But such tests could well result in being 'unfair' for particular countries, or for schools within countries.

The SIMS methodology, dealing with concepts of various aspects of the curriculum, — intended, implemented and attained — provides a means of taking into account variations in curricula as data from examinations are used. An example from Canada (Ontario) was provided.

External Examinations in New Zealand: Issues and Trends: Ian Livingstone (New Zealand)
The present secondary school system offers comprehensive, external examinations at the end of the first three years. Most pupils remain in school and attempt the School Certificate examination in Form Five at the age of 15–16 years. They may take up to six subjects from more than 30 options offered. Mathematics became an increasingly popular option during the 1970s; it is currently taken by four-fifths of all pupils sitting the examination.

At the Form Six level, two certificates are available. The University Entrance qualification caters specifically for pupils wishing to proceed to university. The other, the Sixth Form Certificate, is a general leaving certificate, designed to allow assessment over a wider range of objectives and subject options. Both are internally assessed. The moderating device for University Entrance is a national 'back-up' examination. For the Sixth Form Certificate group, results in the School Certificate Examination from the previous year are used. The Government has indicated that, in the near future, a process of absorption will occur. The University Entrance Examination as such will disappear, and an award will be granted on the basis of results from the Sixth Form Certificate.

Some pupils, mainly those intending to enter university, choose to spend a further year at secondary school, in Form Seven, giving them the opportunity to qualify for more generous bursary or scholarship aid. At both the Form Six and Form Seven levels, there has been widespread consultation between education authorities and teachers, with the aim of making prescribed mathematics courses more relevant for the increased numbers of pupils who are remaining at school to these higher levels.

Technology of Educational Measurement

Latent Trait Application in Cross-national Comparisions of Subtest Scores: Patrick Griffin (Australia) and Mo Ching Mok (Hong Kong)
SIMS data from Population A students in Finland and Hong Kong from the 40-item core test were examined using the Location Dispersion Rasch Model. The items were grouped into nine subtests covering fractions, arithmetic, verbal problems, formulas, sets, geometry (lower cognitive level), geometry (higher cognitive level), statistics and measurement. Comparison of item p-values (proportions of students passing) within sub-tests showed considerable inconsistency across populations. At the sub-test level, the tests relating to fractions and lower level geometry skills, were transposed in relative positions on the underlying trait. Curriculum influence was identified at both the item and sub-test level indicating a treatment by instrument interaction. This effect may invalidate comparisons at the total score, and subtest score, and perhaps

at even an item level. Some subtests were identified however, for which valid comparisions might be made.

Problem-solving and Assessment Problems: Vinicio Vallani (Italy)
In recent years, much has been said about 'problem-solving' strategies, but not enough has been done in order to adapt assessment tests and written examination papers to this teaching method. In the opinion of the author, assessment procedures should not neglect relevant student activities such as verbal description of simple but mathematically rich and interesting situations, construction of examples and counter examples, and ability to choose suitable methods for solving problems that are stated in an 'open' formulation.

Assessment problems of this kind should not be too 'routine' but neither should they be too tricky nor too difficult. Some examples appropriate for students at the upper secondary level are:
- Cutting a cube by a plane, is it possible to obtain triangles, quadrilaterals, pentagons, hexagons, heptagons? Which of these polygons may be regular?
- In a given coordinate system in the plane, suppose the coordinates of the vertices of two triangles ABC, A'B'C' are given. Describe an algorithm in order to check whether the second triangle is completely contained in the first.
- It should be noted that each of these problems may also be stated in more vague, more provocative and more difficult formulations. Finally, it should also be noted that evaluating student solutions of problems of this type requires accurate, intelligent, personalised and time-consuming correction.

Issues in Item Development: Aurum Weinzweig (USA)
The development of items for use in international contexts poses unique problems,and several selection criteria need to be observed.
- Items should be real. When items refer to a given context or situation, these should be real to the test taker. The use of money (currency) may provide a natural context for problems involving decimals — but in Japan, for example, the yen does not have decimal subdivisions. Moreover, conversions to yen may result in numbers that are too large.
- Items should ask one question. Frequently, questions posed demand much more knowledge and sophistication than intended precisely because they do *not* ask only one question. For example, an item on measurement that uses a diagram showing a rod against a rule may require knowledge of how to use a ruler as well as the ability to visualise a three dimensional object.
- When using multiple choice items distractors should be plausible. Unreasonable distractors are in effect non-distractors. While the multiple choice format may limit the kind of questions that can be asked, it can also result in making certain questions reasonable. For example, to determine all the values of n for which '2 to the 2nth power minus 3 to the nth power' is divisible by 13 is difficult. But in a multiple choice format, the problem is more reasonable and still very nice.

Cultural Dimensions of Evaluation, Examinations and Assessment

Two sessions, chaired by Ubiratan D'Ambrosio, were devoted to this topic. In these sessions consideration was given to viewing, from a broader perspective, the cultural context in which evaluation, examination and assessment take place. There is growing concern that evaluation procedures have been based upon factors which are internal to the disciplines and to the school systems, both at the local classroom and at the national and international levels, without taking into account cultural components. On the other hand, assessment and evaluation procedures that do not take into account socio-cultural influences on teaching and learning may strongly bias both the assessment procedures used and the interpretation of the findings obtained.

Following are summaries of the presentations given.

Effects of Schooling on Arithmetic Understanding: Studies of Oksapmin Children in Papua New Guinea: Geoffrey Saxe (USA)

The objective of this research was to gain insight into the way in which Western schooling affects the development of basic arithmetical understanding of the Oksapmin, who live in a remote section of Papua New Guinea and have been contacted by outsiders only recently. The number system of the Oksapmin consists of 27 conventionally defined points on the human body. The research revealed that the Oksapmin are not normally required by their society to perform arithmetic operations. However, when asked to solve arithmetic problems in school, they make use of their indigeneous number system and devise independently novel procedures for performing operations such as addition and subtraction. The paper provides details of these novel conceptual and symbolic developments.

Factors Affecting Evaluation Procedures: Paulus Gerdes (Mozambique)

The analyses reported in this paper were at six 'levels of interference': teacher/student; classroom/school; classroom/local community; local context (classroom, school and community)/provincial or national context (ministries or secretaries of education, examination boards and inspectors of schools); local context/employer; national context (local context, national institutions plus employers)/international context (funding agencies, professional community, 'legitimation', recognition).

For each of these levels of interaction, examples were provided drawn from reports, examination results, and so forth.

Socio-cultural Context of the SIMS Study in the Dominican Republic: Eduardo Luna — presented in his absence

The analysis presented was based upon the question, 'Can this kind of analysis result in relevant changes in a society's educational system?' The paper began with an examination of the concept of curriculum and of the relation between the social space (where-when) and the curricular space (objectives, content and methods). The conclusions of the paper point in the direction of proposing curricular changes that may have an effect in creating educational opportunities in the system.

Contextual Dimensions of the Second IEA International Mathematics Study (SIMS): Elizabeth Oldham (Ireland)

The Curriculum Analysis of SIMS has two aspects; the study of *content*, and the study of *context*.

It should be noted first that the Study is limited by the fact that it chose to examine just two populations; the grade containing the modal number of thirteen-year-olds, and a senior specialist mathematics group. Also some incidental variation was brought into the Study because of the different ways that countries interpreted these definitions for their own use. (The need for such interpretation, of course, reflects real differences between education systems.)

Bearing this in mind, we consider the students in the two populations, the way in which their courses are organised, the sources from which their curricula derive, and the mechanism by which courses are monitored and standards enforced. In the junior population; mathematics is taken by almost all students; at senior level, there is a tendency for classes to be male and of higher socio-economic status, especially in countries retaining only a small cohort in mathematics courses at that level. Widespread mixed ability teaching, even at junior levels, seems to be the exception. Selection, or differentiation of some kind, usually takes place sooner or later, and many countries offer a choice of courses (as well as the option of dropping mathematics) in senior grades. The content of the various courses is typically decreed at national level, though some countries offer guidelines rather than a prescribed curriculum, and a couple allow curriculum to be determined at local level. Most countries, again with a few exceptions (only one of which determines its curriculum locally), have a national examination or assessment system which monitors the implementation of courses, and sets national standards for student achievement. Interesting ways of involving teachers in the certification process have developed in some countries.

The foregoing summary may give an impression of greater uniformity than actually exists. Behind these general patterns lie very varied ways of organising mathematics education in the countries of the Second International Mathematics Study.

TOPIC AREA: COMPETITIONS

Organiser: Peter O'Halloran (Australia)

In recent years the popularity of competitions at regional, national, and international levels has increased greatly. Competitions can be an instrument for highlighting the importance of learning mathematics both at the school level and within the community. A series of four panel sessions reflected the variety and flavour of competitions that are offered around the world. Competitions aimed at primary students to university students, from small local contests to prestigious international events, and from a radio quiz to a mathematical relay, feature in the reports that follow.

As well as the four designated sessions for competitions, a brief report is included on a new competition, a meeting to form a world federation of national mathematics competitions, and a special poster display of journals for students that focus on problem solving.

1. The Creation of Competition Questions

The Creation of Mathematical Olympiad Problems. Arthur Engel (FRG)
It is far more difficult to create a problem than to solve it. There exist very few routine methods of problem creation. As far as is known, there has not been a Polya among problem creators who wrote a book with the title 'How To Create It'.

By a problem we mean an Olympiad type non-routine problem. There are, of course, routine methods of creating routine problems. One of these is based on the inversion of a universal problem solving paradigm as described below:
'You have a difficult problem; transform it to make it an easier problem; solve this easier problem; invert it to get a solution to the difficult problem.'

Problem creators often use the inverse procedure: start with an easy problem; transform it to make it a difficult problem; pose the transformed problem to challenge the problem solvers.

Engel described in detail the development of a problem for the 1979 IMO that is not included here.

Problem Proposing and Mathematical Creativity. Murray Klamkin (USA)
There are very many mathematical competitions being given around the world each year and at many levels. As a consequence of these competitions, there have been quite a number of duplications in the problems set, either inadvertently, or in some cases by direct copying.

Since these competitions have apparently become more important, there are training sessions which help prepare for a number of them. In view of all this, competition examination committees now have to be much more vigilant than ever before in setting their competitions. They must now keep abreast of problems set in other competitions and they have to be very careful about duplicating problems from known books and journal problem sections. Even if a book or journal used is not too well known, the problems used could have been duplicated in other books and journals which are well known. To play safe, problems should either be created which are new (nouveau as well as nouvelle) or else based on some nice result from some non-recent mathematical paper. In this session, we will be concerned with the creative aspects of problem solving and proposing.

It is certainly quite helpful to have a good memory and to be observant. George Polya makes the analogy to finding a precious uncut stone on the shore and tossing it away since it is not recognised as being valuable. Here one has to do a certain amount of cutting and polishing before the value of the stone is recognised. An expert usually gets away with just a careful examination.

With respect to a problem which has just been solved or whose solution has been looked up, we should not immediately pass on to something else. Rather, we should stand back and re-examine the problem in the light of its solution and be asking ourselves: does the solution really get to the 'heart' of the problem? Mathematically, one of the points being made here is to check whether or not the hypotheses of the problem are necessary for the result. Additionally, although our solution may be valid, there may be and usually are, better ways of looking at the problem which makes the result and the proof more transparent. Consequently, it should be easier to understand the result and the proof as well as to give a non-trivial extension.

Further discussion considered suitable results in the areas of equilateral and equiangular polygons; intersecting curves on a surface; maximum number of terms in a sequence; isoperimetric inequality; maximum volume of a tetrahedron; a two triangle inequality; the Weierstrass product inequality and several others.

2. National Mathematics Competitions

Convenor: Walter Mientka (USA)

This report represents a summary of the various competitions discussed by the seven panelists during the session on National Mathematics Competitions. Dean Dunkley (Canada) served as moderator and organiser of the panel.

The South African Mathematics Olympiad.
Speaker: John Webb, member of the SAMOC; Editor, Mathematical Digest.
Origins: Began in 1966 as a consequence of a recommendation of the South African Mathematical·Association and Academy of the Arts and Science.
Objectives: To identify and encourage exceptional mathematical talent in the last two years of high school.
Format: The contest consists of two rounds. The first round, written in May is a 20–25 question multiple-choice test to be completed in three hours with a total participation of 4000 to 5000 students in South Africa and Namibia. The second round, offered to the best 100 students of the first round, is written in

September and consists of six essay-type problems to be solved in three hours. The papers for both rounds are taken in the individual schools of the contestants, but the answer books are centrally marked.

Awards: The students who qualify for the second round receive a certificate and a small cash reward. The top ten students in the second round are awarded medals.

Sponsors: The Old Mutual Life Sciences Research Council.

Publications: The problem sets in the two rounds are usually published in two South African Magazines: *Spectrum* and *Mathematical Digest.* The problems and solutions of papers set from 1966 to 1974 have been published in the book *The South African Mathematics Olympiad* by A. P. Malan (Editor), Nasou, Capetown, 1976.

Olimpiadas Colombianas de Matematicas
Speaker: Ricardo Losada.

Origins: Founded in 1980 by Maria and Ricardo Losada.

Objectives: To create a genuine interest in formative education in mathematics and in the solution of ingenious and original problems.

Format: Pre-Olympiad examinations (Junior and Senior levels) are formulated from the American High School Mathematics Examination (AHSME) questions. The rules for the implementation of the examinations are similar to those associated with the AHSME. Approximately 2700 participate at the Junior level and 4500 at the Senior level. Olympiad participants travel to the capital to take the examination and receive awards.

Awards: The winners are presented with gold medals.

Sponsors: The Ministry of Education, Universidad Antonio Narino.

Publications: Problem solving materials are prepared in Spanish; a weekly newspaper column is distributed, reaching over 300 000 homes; a bimonthly magazine for students is published.

Mathematical Challenge: A Scottish Mathematical Council Competition.
Speaker: Edward Patterson, Chairman of the Scottish Mathematical Council and of Mathematical Challenge.

Origins: Founded in 1976 by the Scottish Mathematical Council.

Objectives: To stimulate interest in mathematics generally and thereby to attract pupils to the study of mathematics and its applications; to encourage pupils to think for themselves and develop their powers of mathematical and logical reasoning; to discover and foster talent in the art of problem solving.

The competition is not directed solely at the 'high-flyers'! The intention is to encourage as many people as possible rather than to single out those who have exceptional talent.

Format: Each year four sets of problems are sent by the Council to participating schools and four weeks are allowed for their solution. Students are trusted to submit only individual work. Scotland is divided into four sections, each based on one or more universities. Each section is controlled by a committee consisting of university and school teachers with responsibilities which include the distribution of problems, solutions, grading of entries, and selection of prize winners. The same problems are used in all four sections at the same time. The average number of participants is 424 students from 155 schools.

Awards: Over 100 prize winners are chosen in the different sections on the basis of overall attainments. The prizes consist of cash awards (20 pounds) which are presented during an award ceremony held in each of the examination sections.

Sponsors: IBM, Scottish Life, BP, London and Edinburgh Mathematical Societies, Science Ltd., as well as the universities from the four sections.

Publications: Copies of reports, problems and solutions from previous years are provided on request.

Canadian Mathematics Competition

Speaker: Ronald Dunkley, Director.

Origins: Started in 1963 as a local Waterloo competition.

Objectives: To provide stimulus for students to test their mathematical prowess at various levels against a national standard and against other schools. To provide a good test for the very best students, but also to cater to students who work conscientiously and consistently in school. To instil a desire in students to do further study in mathematics.

Format: The following examinations are available, with the grade level, number of questions and minutes allowed indicated in parentheses: Gauss (7–8; 26; 60), Pascal (9; 25; 60), Cayley (10; 25; 60), Fermat (11; 25; 60), Euclid (12; 21; 150) and Descartes (13; 10; 150). The first four contests are multiple choice; the Euclid contest consists of 15 multiple choice and 6 essay questions; the Descartes contest consists of 10 essay questions and is compulsory for students applying for scholarships in the faculty of Mathematics at the University of Waterloo.

With the exception of the Gauss Contest, the competitions are centrally marked at Waterloo. In 1983, the total number of entries was 80 000.

Awards: Awards are given to top ranking students and schools in each of the Contests except the Gauss Contest. These awards include certificates, plaques, championship shields, gold, silver, and bronze medals, and cash prizes ranging from $250 to $50. The top 60 students are invited to attend a one week seminar held at the University of Waterloo in June.

Sponsors: The Waterloo Mathematics Foundation, The University of Waterloo, The Grand Valley Mathematics Association, Mutual Life of Canada, Dominion Life Assurance Company, IBM, Canadian Imperial Bank of Commerce, and Canon Canada, Inc.

Publications: Reprints of previous contests are available, as are solution booklets.

The Australian Mathematics Competition in French Polynesia

Speaker: Pierre–Olivier Legrand, Regional Director of the Australian Mathematics Competition, French Polynesia.

Origins: The French edition of the Australian Mathematics Competition (AMC) was first administered in French Polynesia in 1981. The AMC became a national competition in 1978 and was extended to countries in the South Pacific in 1980.

Objectives: To stimulate the teaching and learning of mathematics at the high school level, to encourage good average students, and provide a bank of interesting questions for classroom discussion.

Format: The following examinations are available with the grade level, number of questions, and minutes allowed indicated in parentheses. Junior division (7–8, 30, 75); Intermediate division (9–10, 30, 75); Senior division (11–12, 30, 75). These examinations are centrally scored in Canberra, along with all Australian entries. The total number of entries is 300 000.

In addition, two Australian Mathematical Olympiads are administered. A preliminary examination called the AMO Interstate Finals consisting of five questions to be solved in four hours is administered to top students in the Senior Division of the competition. This examination is followed by the Australian Mathematical Olympiad which consists of two papers, each containing three questions and lasting four hours. The AMO is administered to 40–50 students. The top nine AMO participants attend the AMO IBM one-week Mathematical Olympiad Training School held at the University of Sydney.

Awards: Outstanding students are awarded medals, about 1000 receive cash prizes, the top 15% qualify for a distinction certificate, and the next 30% for a credit certificate.

Sponsors: Canberra College of Advanced Education, Westpac Banking Corporation and the Canberra Mathematical Association.

Publications: Past Solutions and Statistics booklets, Past Competition Papers, Individual School/Regional Question Analyses, and the Australian Mathematical Olympiads.

American Mathematics Competition

Speaker: Walter Mientka, Executive Director, AJHSME, AHSME, AIME, USAMO.

Origins: AHSME–1950; USAMO–1972; AIME–1983; AJHSME, offered for the first time in December 1985.

Objectives: AJHSME (American Junior High School Mathematics Examination): To promote the development of problem solving skills and positive attitudes towards mathematics by providing challenging problem solving activities; stimulate interest in continuing the study of mathematics; identify students with mathematical problem solving ability; reward mathematical excellence by recognising high achievement on a national examination; encourage the improvement of junior high school mathematics curricula nationwide by increasing academic expectations beyond the level of basic skills and minimum competencies.

AHSME (American High School Mathematics Examination): To spur interest in mathematics and develop talent through the excitement of friendly competition at solving intriguing problems in a timed format.

AIME (American Invitational Mathematics Examination): To provide further challenge and recognition beyond that provided by the AHSME to the many high school students in North America who have exceptional mathematical ability. It is also the qualifying examination for the USAMO.

USAMO (USA Mathematical Olympiad): To discover and challenge mathematical talent, superior mathematical knowledge, ingenuity and computational competence.

Format: AJHSME — Grades 7–8; 30 questions; 60 minutes; to be centrally scored in Lincoln.

AHSME: Grades 9–12; 30 questions; 90 minutes, generally scored by the school examination manager — centrally scored in some states. Administered by 69 Coordinators in the USA and Canada, with 400 000 participants.

AIME: Administered to 1000 to 2000 participants who achieve a score of 95 or above on the AHSME (this cut-off will change to 100 in 1986). Fifteen question 'essay' examination; answers are integers from 0 to 999; 180 minutes; centrally scored in Lincoln.

USAMO: Limited to the top 50–100 students selected on the basis of AIME scores. Five question 'essay' examination; 210 minutes. Papers are graded in Lincoln.

Awards: AHSME: Ten outstanding students in each of the ten examination regions receive awards consisting of medals, plaques, books, subscriptions to the Mathematics Magazine, and certificates. The top scoring student in each of the participating schools receives an award pin. In each of the ten regions, the five schools with the highest team scores (sum of the highest three scores by contestants) are recognised by Donor or Committee Awards which include silver and bronze cups, New Mathematical Library sets, Freeman books, and the Mathematical Magazine.

AIME: Each participant receives a Certificate of Participation.

USAMO: The top eight students are honoured at ceremonies held in Washington, D.C. which are made possible by a generous grant from IBM. They receive awards from IBM, Hewlett–Packard, MAA, NCTM, Mu Alpha Theta, and several publishers. Twenty-four of the top students are invited to participate at the Military Academy at West Point or the Naval Academy at Annapolis (alternating training sites).

Sponsors: Mathematical Association of America, Society of Actuaries, Mu Alpha Theta, National Council of Teachers of Mathematics, Casualty Actuarial Society, American Statistical Association, and American Mathematical Association of Two-Year Colleges.

Publications: Specimen sets (examinations and solution pamphlet) of previous examination are available for the Competitions as well as the National Summary of Results and Awards. Four Contest Problem Books containing AHSME from 1950 to 1982 are also available, as is the International Mathematical Olympiads book containing the 1959 to 1977 Olympiads. *The Arbelos*, a journal (5 issues per year) containing short articles and challenging problems for gifted students may be obtained by subscription. This journal was founded by Samuel Greitzer in 1982. Further details about the competitions may be found in *AHSME, AIME, USAMO: The Examinations of the Committee on High School Contests*, by Stephen Maurer and Walter Mientka in The Mathematics Teacher, vol. 75, no. 7, October 1982.

George Berzsenyi gave some further details about AIME. There was very little time during the panel presentations to discuss the impact of national competitions on school curriculum or to discuss such questions as:

What effect have national competitions had on the mathematics curriculum?

What has been student and teacher reaction to the competitions?

Have the objectives of the competitions been achieved?

It is hoped that these and other questions will be addressed at future ICME sessions.

3. A Kaleidoscope of Competitions

This session provided an opportunity to look at some of the variety of competitions that are organised at a local level in six different regions of the world.

The Canberra College of Advanced Education Mathematics Day
Place: Canberra, ACT, Australia.
Aim: To encourage the talented mathematics students; to foster closer links between CCAE and its feeder high schools and junior colleges.
Participants: 24 teams of 5 final year students representing their schools.
Format: A day of mathematical activities at the College.
Contact Address: The Convenor, CCAE Maths Day, School of Information Science, Canberra College of Advanced Education, P.O.Box 1, Belconnen, ACT 2606 Australia.

Cantamath
Place: Christchurch, New Zealand.
Aim: To encourage interest in mathematics among children at all levels of achievement.
Participants: About 10 000 children, ages 11 to 15. Teams from about 50 schools at each level.
Format: Displays; Posters, designs, models, written work, class projects. Race competition between teams of four.
Contact Address: Cantamath Secretary, Canterbury Mathematical Association, P.O.Box 31–014, Christchurch, New Zealand.

Challenge on Radio
Place: Merseyside, around Liverpool, England.
Aim: To popularise mental arithmetic and problem solving, and through the preliminary rounds, to foster mathematical links between neighbouring schools.
Participants: Teams of three (age under 14.5) from 40 schools.
Format: Preliminary leagues produce the seven challengers to last year's champions. Remaining seven matches broadcast on local independent radio.
Contact Address: The Chairman, Mathematical Association of Merseyside, The Mathematics Building, University of Liverpool, P.O.Box 147, Liverpool L69 3BX. England.

Mathematical Olympiads for Elementary Schools
Place: Based in Long Island New York, USA, but taken now by schools all over the US and in other countries.
Aim: To develop children's enthusiasm for problem solving; to contribute to their cognitive development; to provide the initial foundation for an intellectually stimulating and pleasurable activity.
Participants: Capable, talented and gifted children in elementary schools (age 12 or less). 26 500 children from 956 schools in 1984. School teams limited to 35, each team has a teacher sponsor.
Format: Five Olympiads are administered each school year. Each contains five problems, each with a time limit. Focus is on individual rather than team achievement.
Contact Address: Dr. George Lenchner, Executive Director, Math Olympiads, Forest Road School, Valley Stream, NY 11582, USA.

American Regions Mathematics League
Place: USA, especially the eastern states.
Aim: To encourage and recognise student achievement in mathematics, and to bring the best students together.
Participants: Teams of 15 representing states of the US (some states have up to six teams).
Format: Teams meet at a host university. Problems for solution by teams and by individuals. Teams selected by local and statewide competitions.
Contact: Professor Al Kalfus, President, American Regions Mathematics League, Babylon High School, Babylon, NY 11702, USA.

William Lowell Putman Competition
Place: USA and Canada
Aim: To allow undergraduates to win honour for their colleges by intellectual activity.
Participants: Teams of 3 from universities and colleges in the USA and Canada, also individual groups. A total of 2024 from 348 institutions last year.
Format: All contestants work individually on problems; two 3-hour sessions of six questions each on one day.
Contact: Len Klosinski, Director, Putnam Competition, Mathematics Department, University of Santa Clara, CA 95053, USA.

4. The International Mathematical Olympiad
Reporter: Henry Alder (USA)

At ICME IV, an informal session on the IMOs was held for the first time. It had four purposes: to review the origin and history of the IMO; to explain the current operation of the IMO; to inform those interested in participating in future IMOs how to be invited by host countries; and to solicit the views of participants on what type of organisation to establish to assure continuity in the operations of the IMO.

In 1980 there were serious concerns about the future of the IMOs. No IMO was held that year as no country had extended an invitation to host it and no organisation existed to make this fact known to possible host countries or participants.

In May 1980, the Finnish National Committee on Mathematics Instruction proposed that ICMI set up a Site Committee for the IMO. This was one of the possibilities discussed at the informal session at Berkeley and was favourably received by the meeting. As a result, the IMO Site Committee was appointed by ICMI in April, 1981.

The IMO session at ICME 5 (the first formal session on the IMO at an ICME) had the same first three purposes as the Berkeley meeting, and in addition, included a report on the changes which had taken place in the operation of the IMOs since the appointment of the IMO SC.

The first panelist, John Hersee (UK), chief organiser of the 1979 IMO in London and secretary of the IMO SC since 1981, provided a brief history of the IMO up to 1980. He recalled that the IMOs had their beginning in 1959 when Romania invited six other countries from Eastern Europe to send a team of eight students to the first IMO. In 1965, with 10 countries now participating, Finland became the first country outside Eastern Europe to join the IMO. England, France, Italy and Sweden entered in 1967, and the United States in 1974, bringing the number of participating countries to 18. In 1976, Austria became the first Western country to host an IMO. In 1979, the United Kingdom was host to 22 countries.

Although great interest is taken in the team positions, the IMO is an individual competition. Participating countries may see the IMO as a standard and a target to work towards as they strive to improve the teaching of mathematics in their schools. Some continue to see the IMO and the competitions which lead up to team selection as a way of identifying future mathematicians and scientists. For all countries, the competitions that lead up to the IMO are a way of stimulating interest in mathematics in all schools. The meeting of able young people from many nations at an IMO is a major contribution to international understanding.

The next panelist, Claude Deschamps (France), has been leader of the French delegation to the IMO every year since 1978 and was chief organiser for the 1983 IMO in Paris. He described in detail what is involved in the organisation of an IMO for the host country. In particular, he explained the procedures used to determine which countries to invite, the method used to decide on the questions to be used in an IMO, and the grading procedures.

Finally, Jan van de Craats (Netherlands), who has attended all IMOs since 1973, led the Dutch delegation since 1975, and has been a member of the IMO SC since its establishment, spoke on the IMO since 1981 and its future.

He noted the 'spectacular revival' of the IMO beginning in 1981 when it was held in the USA with 27 participating countries. The number of participating countries has steadily increased from that time and reached 34 in 1984 when it was held in Czechoslovakia. The traditional team size of 8 (up to 1981) had to be abandoned both for financial and organisational reasons. This does not endanger the spirit of the Olympiad however, as it will remain a competition between individuals. Nevertheless, it is hoped that a maximum number of 6 students per country will continue to be possible in future teams.

The 1985 IMO will be held in Finland, and the 1986 IMO in Poland. Two countries have expressed interest in hosting the 1987 IMO and the IMO SC is presently negotiating with those two countries to persuade one of them to defer their invitation to a later date. This position is a sharp reversal from the situation that saw the need for the appointment of an IMO SC.

These presentations were followed by a lively question and answer period. Some 80 persons present at the session included representatives of countries which have not previously participated in an IMO. Several indicated that their countries were planning to join the IMO soon. It is anticipated that by 1988 some 50 countries will be participants. Many members indicated that they had found the session most interesting and helpful, and expressed the hope that a formal session on the IMO would be a regular feature of future ICMEs.

5. Why 1 in 50 Australians do a Competition
Presenters: Warren Atkins (Australia), Peter Taylor (Australia)

In 1972/73, Peter O'Halloran of the mathematics staff of the Canberra College of Advanced Education spent twelve months' special study leave in

Canada, including six months at the University of Waterloo. While there he studied the operations of the Junior Mathematics Contest (now known as the Canadian Mathematics Competition) and the similar Annual High Schools Mathematics Examination held in the USA by the Mathematical Association of America.

These competitions clearly had a more significant impact on their respective countries than any of the competitions being run in Australia at that time, the latter being run on a state basis and directed only towards more gifted students. As there appeared to be a need for a more broadly based competition in Australian schools, O'Halloran formed a committee to organise such a competition in 1976.

The aims of this competition were to ask questions which all high school students could relate to, either from the syllabus of their school or other experiences. By providing an appropriate award structure (certificates for the top 45% of all entered students), there was an opportunity for the average student to gain a sense of achievement in mathematics. In view of the number of entries anticipated, it was decided that all questions would be multiple choice to enable computer marking and reporting.

Entries from the 1976 competition were so encouraging (1300 students from 33 schools) that it was decided to run a pilot national competition in 1977. Schools from the other states were invited to enter, and some 4500 students from 73 schools took part.

In 1978 the competition became fully national and known as the Australian Mathematics Competition, with an invitation issed to every Australian high school to participate. While still centrally organised by the staff of the CCAE, the competition received a boost with new commercial sponsorship from the Bank of New South Wales (now Westpac Banking Corporation).

The competition has grown at a healthy rate since 1978 and has been extended to New Zealand and a number of other South Pacific countries, with a French language version available for French speaking countries in the Pacific and a braille version for blind students.

Three separate papers are set with some questions appearing on more than one paper. The basic aims of the AMC, as stated earlier, are to give the average student a sense of achievement in mathematics, and to highlight the importance of mathematics as a subject in the high school curriculum. These aims have been enhanced by awarding certificates of distinction and credit to the top 45% of all entrants within a state or region. As well, outstanding students receive cash prizes ranging from $20 to $140, while up to 15 special medals are awarded to the most outstanding students on an national basis. The medals are presented at a special ceremony in one of the state capitals, and students are flown to the city especially for the award ceremony. With over 300 000 participants, the AMC appears to have met its aim of providing a competition for a wide range of student ability.

6. World Federation of National Mathematics Competitions Meeting

Twenty mathematicians from ten countries attended the inaugural meeting of the World Federation of National Mathematics Competitions. It was agreed that there was a general need to establish an organisation which would be a focal point and resource for those who are concerned with national mathematics competitions. The meeting elected a committee with the following members.

President: Peter O'Halloran (Australia)
Vice-Presidents: Ron Dunkley (Canada); Walter Mientka (USA); Pierre–Olivier Legrand (Tahiti)
Secretary: Sally Bakker (Australia)
Editor: Warren Atkins (Australia)
Committee: Maria de Losada (Colombia); Erica Keogh (Zimbabwe).
    The meeting decided to prepare and circulate a newsletter.

7. Posters for Student Mathematical Journals

The following student journals were represented in a special poster display. The display was organised by the Competitions Topic Area. The journals represented on the posters all had an emphasis on problem solving for students.

Alpha (German Democratic Republic)
American Mathematical Monthly (USA)
Berita Mathematik (Malaysia)
Bulletin AMQ (Canada)
College Mathematics Journal (Canada)
Crux Mathematica (Canada)
Delta (Poland)
Fun With Mathematics (Canada)
Function (Australia)
Kozepiskololai Matematika I Lapok (Hungary)
L'Educazione Matematica (Italy)
Mathematical Digest (RSA)
Mathematical Digest (New Zealand)
Mathematical Spectrum (England)
Mathematics Magazine (USA)
Mathematik in der Schule (German Democratic Republic)
Ontario Secondary School Mathematics Bulletin (Canada)
Parabola (Australia)
Problem Solving Newsletter (USA)
Pythagoras (Netherlands)
The Problem Solver (England)
Trigon (Australia)

TOPIC AREA: THE TEACHING OF GEOMETRY

Organiser: Walter Bloom (Australia)

As there were few positive responses from individuals invited to present papers for this Topic Area, only one of the four sessions allotted was used. Despite this apparent lack of interest and concern prior to the Congress, this session was well attended by over 160 participants. It should also be noted that the teaching of geometry received considerable attention in several of the Action Groups.

The three invited presentations were:
Walter Bloom (Australia), A Review of Developments in the Teaching of Geometry since ICME IV;
Nathan Hoffman (Australia), Geometry for the Mathematically More Able with Emphasis on the Lower Secondary;
Aurum Weinzweig (USA), A New Look at the Child's Conception of Space — Piaget Revisited.

Bloom introduced the session with the observation that, notwithstanding the obvious interest in this particular session, there seemed to have been a general decline in interest in geometry at the secondary school level. This had followed several years of attack on the traditional approach to the teaching of Euclidean geometry in favour of a more modern approach (transformation, coordinate, vector geometries in particular). The decline had become apparent in Australia during this decade, and was matched by a decrease of journal articles appearing on the subject. He pointed out that this coincided with the growing influence of computing and statistics on the secondary school curriculum, a point taken up in later discussion when it was remarked that the microcomputer could be a very valuable aid in the teaching of many geometric concepts.

The second speaker, Hoffman, gave an outline of a suggested geometry course for mathematically more able students at the lower secondary school level. He defined the audience to be those students between 13 and 15 years of age and in the top 15% of their peer group in terms of mathematical ability. His main argument was that it is inappropriate for goemetry alone to carry the full burden within school mathematics for the development of axiom systems and ideas of proof. Furthermore, a globally axiomatic approach to the teaching of geometry is too sophisticated for school students. Hoffman advocated a course in geometry that combines the traditional, vector, transformation, and coordinate approaches. This should be a course in which the proofs of surprising and/or structurally fundamental results are the only proofs presented, and emphasis is on work in three dimensional space.

The following is an outline of the proposed course:

1. Year 8 (Age 13).
- Review the main two-dimensional and three-dimensional geometric ideas that students have already met.
- Develop simple ruler and compass construction work.
- Derive conditions for congruency of triangles from construction work.
- Use derived congruency conditions to prove simple geometric results.
- Derive properties of parallel lines by measurement.
- Use properties of parallel lines to prove simple geometric results.

2. Year 9 (Age 14)
- Introduce directed line segments and lead on to vectors (coordinate free).
- Introduce vector addition and subtraction. Use these concepts to prove selected geometric results.
- Introduce the notion of translation as a mathematical description of a slide. Study the properties of translations.
- Introduce the notion of reflection and study its properties.
- Introduce the notion of rotation, and study its properties.
- Introduce the notion of enlargement,and study its properties.
- Consider the effects of combining pairs of transformations and use transformations to prove selected geometric results.

3. Year 10 (Age 15)
- Study similarity transformations, conditions for similarity of triangles, and the theorem of Pythagoras.
- Study coordinate geometry, including midpoints, distances and gradients. Use coordinate methods to prove selected results.
- Study circles and tangency. Use previous approaches to prove key results in this area.

The point of studying geometry in three-dimensional space was raised also by the third speaker. Weinzweig chose for his topic a study of the child's conception of space. He reviewed the influence of Piaget in this area and observed that some of the 'simple' concepts that are now used, for example *line* and *translation*, have present meanings far removed from those originally intended. In these two cases he referred to Euclid's concept of a line as being more related to an extendable line segment, and the relationship between the common notion of translation as applied to movement of rapid bodies and that of a mapping of the plane (or three-dimensional space) onto itself. Weinzweig suggested that commencing the study of geometry with three-dimensional work, rather than as at present almost exclusively with the study of the plane, would relate better to the child's intuition and experience with the surrounding environment.

In the discussions following the presentations, it was quite clear that those present considered the acquisition of concepts and skills in geometry important enough to ensure a continuing place in the school curriculum at all levels.

TOPIC AREA:
RELATIONSHIP BETWEEN THE HISTORY
AND PEDAGOGY OF MATHEMATICS

Organisers: Bruce Meserve (USA), George Booker (Australia)

'Its use is not just that history may give everyone its due and that others may look forward to similar praise, but also that the art of discovery may be promoted and its method known through illustrious examples'.

Leibniz

In contrast to these sentiments of Liebniz, mathematics is predominantly presented in a polished, logical form showing none of the difficulties, errors, guesses, or stumbling that went into its creation. Such presentation is prepared from hindsight, a-historically, largely as a consequence of a stress on symbolism and abstraction, and has led to an emphasis of content over method; the justification of mathematical truths rather than analysis of the processes by which they were created. Yet there exist means of demystifying mathematics, of showing its origins as well as its results and applications, of revealing its history. The International Study Group for the Relationship between the History and Pedagogy of Mathematics (HPM) exists as an informal group to encourage colleagues throughout the world to use aspects of mathematical history in the teaching of mathematics to motivate interests, develop positive attitudes, and encourage appreciation of the nature and role of mathematics.

A knowledge of the history of the development of the mathematics curriculum can support its teaching by demonstrating how modern mathematics has its roots in the past, and by revealing the improvements in mathematical rigour. Such calls to teach mathematics relative to its history also extend to investigations of national mathematics history, to the uses which might be made of this particular knowledge, and to the light it might shed on prior assumptions about the manner of development, the persons involved, and their roles.

Meetings held in conjunction with ICME and ICMI, and on a regional basis with NCTM in the United States of America, provide the major forum for discussion and dissemination. During ICME 5, a series of four meetings was held with the intention of introducing mathematics educators to the group and its aims. Specific examples of such ideas in practice, from a variety of countries, and across all levels of education were presented. In the first session George Booker (Australia) provided a framework of suggestions for using the history of mathematics supported by examples demonstrating the extent of such use in Australia. Rina Hershkowitz (Israel) and Amy Dahan (France) examined uses which might be made of the history of mathematics by teachers and talented students in the second session. In the third session, historical documents suited to secondary school pupils were discussed by Jacques Borowcyz (France), Amy Dahan (France), and Lucia Grengetti (Italy). For the final session, Florence Fassanell (USA) examined the interplay between the history of art and mathematics. The session concluded with a summary of the objectives being fostered by the HPM group and an account of the way these were being implemented in Canada by Israel Kleiner (Canada).

In the first session, two contrasting positions on the role of the history of mathematics in mathematics education were given and the discussions over the four days of the Congress can be summarised in these terms. The first position was that for students, mathematics evolves from a problem that interests them, thus focussing on the process of mathematical thinking rather than the end-product of the mathematical thoughts of earlier mathematicians. On one hand this could involve problems and their solution from the past such as those suggested in the recent book *Geometrical Investigations* by John Pottage. However, there is also the view provocatively described by Roland Stowasser as 'ransacking the history for teaching mathematics' whereby critical incidents or examples are taken to illustrate a particular point or to generate a technique or method. An example of this presented in the second session examined the developments in mathematics by the Arabs as they took mathematics from the orbit of Greek geometrical form to a theoretical discipline with its own methods and precise objects. Two currents were involved in this renovation; one from the geometrical construction of the roots of equations of degree higher than two, the other building from a deep dialectical movement between arithmetic and algebra. This led to the definition of the null power, $x^0 = 1$, and the use of a board

4	3	2	1	0	1	2	3	4
x^4	x^3	x^2	x	1	$1/x$	$1/x^2$	$1/x^3$	$1/x^4$

to represent the multiplication of x^m by x^n *(if n > 0)* by taking n steps to the left from m. The practical basis for this invention with origins in very real problems provides a valuable means of accessing this area of algebra which often seems very artificial to children and consequently a stumbling ground for their learning. Further details of this are available in Amy Dahans recently published book, *Routes et Dedales* (1982).

The second proposal was to examine the development of mathematical ideas within an historical context, bringing out the relationship between mathematics and the rest of human thought and invention. In this way, students would gain a sense of perspective on mathematical discoveries and be able to tie mathematics to the problems from which it arose as well as to its discoverers. This setting of mathematics within the broader social and cultural fields has been addressed by R. Wilder, most recently, in *Mathematics as a Cultural System*. It was also the theme of a paper presented by Ubi D'Ambrosio at the pre-ICME meeting of the group and of his plenary address to the Congress where he raised the issue of 'ethno mathematics' as opposed to 'learned' mathematics.

Having identified these somewhat different positions, the discussion then focussed on ways in which they might be effectively used in teaching. The first suggestion was to use an anecdotal approach, one frequently used in both elementary and high schools when brief biographies of relevant mathematicians are given, or some context for the use of particular systems of numeration or computation is provided. While this would seem to be the easiest and most basic level of involvement of an historical orientation, it is important to bear in mind the distinction between history and story-telling. People who will not present a mathematical theorem or statement without checking it will often copy or rely on loose recall of history giving rise to inaccuracies such as, 'the integral was discovered by Riemann', 'common fractions' can be ascribed to the Babylonians, 'Egyptian 'rope-stretchers' made use of a 3,4,5, triangle to determine right angles', while the description of Fermat as 'the prince of amateurs' led to his being considered an amateur mathematician. None the less, the provision of background material in this way can give some perspective on mathematical discoveries and inventions and can also provide interest and motivation to the subject. Such an approach can also provide insightful, more intuitive or alternative teaching procedures. For instance, the early Greek proofs that the series

$$1 + \tfrac{1}{2} + \tfrac{1}{4} + \ + \dots \text{ sums to } 2$$

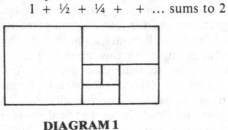

DIAGRAM 1

or for Pythagoras' theorem:

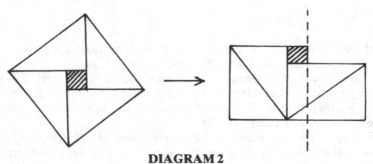

DIAGRAM 2

A second way of using an historical approach has been through an application of the genetic principle in teaching mathematics, that is, effective learning requires each learner to retrace the main steps in the historical evolution of the subject under study. While this has been interpreted in the past in terms of 'a discovery approach', the frame of reference increasingly used to describe mathematics learning has been a constructivist one, and a similar approach to this use of the history of mathematics is warranted. That is, attention is focussed on the process of reconstructing mathematics rather than rediscovering it. In contrast, the notion that effective mathematics learning requires rediscovery is derived from an emphasis on the completed form of mathematics rather than the process by which it is formed. A constructive perspective is likely to highlight inherent difficulties more clearly (such as the zero concept or the notions of algebra), whose existence or formulation were long in gaining acceptance and understanding. It would also highlight the pedogogically more valuable order of presentation from the intuitive to the formal and add to motivation through student recognition of the origin and developmental paths of problems, concepts and proofs. An historical approach of this kind would tend to lead to understanding rather than memorisation.

One example of this approach is the recent publication *Una Historia Breve del Algebra* written by the Norwegian mathematics historian Otto Bekken. Themes from this history of algebra were presented in the pre-ICME conference of HPM, and at a similar meeting prior to the 1984 NCTM conference. A source book of documents suitable for both children and teachers has been developed through the IREMS, in France, by groups of mathematics teachers at both secondary and university level, and addresses the broad issue of 'object and utility in mathematics' as well as more subject specific areas such as arithmetic, algebra, analysis and astronomy.

A similar approach has been taken in developing in-service programs for mathematics teachers in Israel at the Weizmann Institute. Each topic is created as a series of worksheets that begin with a brief introduction to set the historical scene, together with an historical source. From this leading questions are posed which are pursued in active learning through workshops or correspondence courses. This is followed up by a discussion leaflet with detailed solutions to any problems and further source materials.

Topics treated so far include negative numbers, irrational numbers, and the story of equations, with materials developed around primary sources and designed for active learning.

One unfortunate outgrowth of the 'genetic principle' was an emphasis on mathematical structures as a unifying idea for mathematics. Again, this evolved from a focus on the content of mathematics and its form rather than its processes of construction. Because its focus has to be constructivist, the history of mathematics is more likely to provide a basis of unification, showing the interdependence of the various parts of mathematics through their common origins, evolution or response to similar problems or cultural forces.

Mathematics has grown in part because of those working in the subject as professionals, but men and women working in other fields have also contributed greatly to the subject, for example, the philosopher Pascal, the clergyman Bolzano, the poet Omar Khayam and the artist Durer. Indeed, the writings of Durer as a theorist of art provided one of the first mathematical textbooks in the German language, *The Teaching of Measurement with the Compass and Ruler* (1525), giving birth to German scientific prose and replacing traditional formulas with analysis and new creative constructions. Examining the interplay between subjects such as the history of art and the history of mathematics provides another entry to mathematics topics which otherwise appear dry and devoid of practical application or interest.

An examination of the distinction between content and form provides a third way in which the history of mathematics can be utilised in the teaching of mathematics. Content refers to the methods and results of mathematics, while form refers to the symbolic notation in which these results are usually expressed, and the chains of logical reasoning through which a proof is given. The two are inextricably, but not inevitably linked, for much of the content of mathematics would not have been discovered if it were not for advances in form. New results have often become possible because of a new mode of writing, such as the introduction of Hindu-Arabic numerals or the notation Liebniz provides for the calculus. 'An adequate notation reflects reality better than a poor one, and appears endowed with a life of its own which in turn creates new life' (Struik). Advances in form have often made it easier to learn mathematics, although they can also give rise to learning difficulties and bar the way to understanding. As Ginsburg notes, 'Children's informal arithmetic is powerful, their understanding of written symbolism is weak'. An examination of the evolution of the expression or proof of particular results, of the derivation of other results from these, and the ways in which ideas have been inhibited could provide valuable insights into mathematics.

A more explicit development of this approach is the notion of *Proofs and Refutations* proposed by Lakatos. While both content and form are examined, it is the limitations and the processes by which mathematics is formalised that come under examination. Lakatos contends that it is formalisaton that divorces mathematics from its history and only an awareness of the mathematical disputes and errors that went into the formulation of mathematics will provide real understanding of the content.

If these four frameworks for utilising the history of mathematics are so well known, what then is inhibiting their adoption and use in the classroom? The first reason must be the sparsity of teaching materials at the level of the students involved or for the content they are to be taught. The difficulties associated with the mathematical form used in the student texts or the teachers' own education also limit the capacity to see opportunities for incorporating historical material or methods. Teachers may feel inadequate with the different teaching methods needed to present and discuss history as opposed to mathematics. Changing emphasis in teacher education can help, but the existence of these inhibiting factors is one of the major reasons for the establishment of HPM. It provides a venue for sharing teaching ideas and materials, a network of support and contacts through the HPM Newsletter, and by its existence raises for discussion the issue of including an historical approach in all levels of mathematics education.

The sessions concluded with a business meeting of the group at which future activities were discussed.

Chairpersons of H.P.M.:
Ubiratan D'Ambrosio (Brazil), *Christian Houzel* (France)

Members of the International Committee: Otto Bekken (Norway) *George Booker* (Australia) *Sergei Demidov* (USSR) *Paulus Gerdes* (Mozambique) *Maassouma Kazim* (Qatar) *Bruce Meserve* (USA) *David Pimm* (UK) *Roland Stowasser* (FRG) *Lee Peng Yee* (Singapore) *David Wheeler* (Canada)

Initial contact with the group and requests to be placed on the mailing list for the HPM Newsletter: Charles Jones (USA)

TOPIC AREA: LANGUAGE AND MATHEMATICS

Organisers: Ed. Jacobsen (UNESCO), Lloyd Dawe (Australia)

Structure

The program was organised around three themes: Oral Mathematics, Reading and Writing Mathematics, and Teaching and Learning Mathematics in a Second Language. A simplified version of the program is shown on the accompanying page. Each session was of 90 minutes duration and presentations were generally restricted to about 20 minutes to allow for questions and discussion. A leader was selected in advance from the presenters in each session to coordinate reporting, prepare summaries, and to sit on the panel in the closing plenary session. Chairpersons for each session introduced the speakers, kept presentations to the appropriate time limit, and led question time. Attendances were very encouraging and showed active interest in all three themes. An international group of over one hundred people has been formed to encourage cooperation in research and correspondence between members with a view to making real progress in the four years to ICME 6.

Presentations

Papers presented in the different sessions were generally of high quality and provoked much discussion. The following is a summary of important issues raised and research interests being followed. There is considerable overlap between the various themes.

At the opening plenary session, Ed Jacobsen introduced the program for the four days. There were speakers from every continent, reflecting the commonality of the problems of language and mathematics learning.

Jacobsen reminded us of some outcomes of the UNESCO/ICMI Nairobi Symposium on the interaction between linguistics and mathematics education, of the clear statement that every African language possessed the logical structure to enable the learning of mathematical concepts, but that, as in all languages, there was a need to enlarge the technical vocabulary. Although it is best that instruction in mathematics be given in the mother tongue (L1), there may be political and economic reasons why this is not always possible. He then described how one country, Lesotho, is implementing the Nairobi Symposium recommendations to ameliorate its mathematics language problems. In preparing for the switchover from L1 (Lesotho) to L2 (English) teaching, the L2 is adjusted in order to develop needed language skills. A glossary of mathematical terms was prepared for the L2 teachers, as well as suggestions for a simplified mathematics vocabulary and sentence

ICME5: Language and Mathematics Topic Group

Session / Group	Saturday, 25 August 1930–2100	Sunday, 26 August 1030–1200	Tuesday, 28 August 1030–1200	Wednesday, 29 August 1030–1200
	Plenary Session			*Plenary Session*
	*Introduction to the program * Report on progress made since Nairobi Symposium and ICME 4. *Speakers* Ed Jacobsen (UNESCO) David Pimm (UK) *Chair:* Lloyd Dawe (Australia)		*The sociolinguistics of Mathematics Classrooms*	Summaries from groups and a discussion of research areas: planning of possible joint research in preparation for ICME 6.
Oral Mathematics		*The Relationship between spoken Language and the Development of Mathematical Concepts in children* Elmar Cohors-Fresenborg (FRG) Colette Laborde (France) Francis Lowenthal (Belgium) John Conroy (Australia) *Chair:* Dora Helen Skypek (USA)	Hilary Shuard (UK) Pam Harris (Australia) Marilyn Nickson (UK) *Chair:* Lloyd Dawe (Australia)	
Written Mathematics		*The Move from Natural Language and Symbolism (and vice versa)* Bernard Parzysz (France) Glen Lean (PNG) *Chair:* Peter Jones (Australia)	*Problems in the Reading and Writing of Mathematics* Daphne Kerslake (UK) Alan Larkin (Australia) *Chair:* Gilbert Cuevas (USA)	
Teaching & Learning Mathematics in a Second Language		*Bilingualism and Learning Mathematics* Lloyd Dawe (Australia) Gilbert Cuevas (USA) Helen Watson (Nigeria) *Chair:* Ed Jacobsen (UNESCO)	*Teaching/Learning Mathematics in a Second Language and the change over from a second language to mother tongue and vice versa* James Taole (Lesotho) Peter Jones (Australia) Dilip Sinha (India) Jan Thomas (Australia) *Chair:* Glen Lean (Papua New Guinea)	

construction. Teachers are thus sensitised to switchover problems the learner must face. Two sets of resource booklets have been prepared; one offering mathematics topics to use in language classes, the other providing language skill activities for classes. He invited participants to examine their national mathematics language situation and then consider appropriate steps at national, local, school and classroom levels.

David Pimm (UK) provided a 'second' summary on *Mathematics and Language: the state-of-the-art*

There is a vague, often impenetrable, no-man's-land between the discourse of poets, philosophers or people in general, and the discourse of mathematicians in general. People say things multiply when there is increase. Mathematicians also say they multiply when there is decrease (times a half) or when neither increase nor decrease is in question (times a matrix).

Despite claims to the contrary (Austin and Howson), Pimm believes that sense can be made of the claim that mathematics is a language, provided that statement is seen as a structuring metaphor (Lakoff and Johnson, Pimm). The Cockcroft Report (1982) stated, 'Mathematics provides a means of communication which is powerful, concise and 'unambiguous'. While many take exception to the word 'unambiguous', this claim serves the purpose of introducing communication as one of the central concerns of anyone interested in mathematics education.

One of the fundamental divisions is between the spoken and written channels as different realisations of language. Both of these modes of language have different qualities and perform different functions in our society (Stubbs). Among the characteristics of unrefined spoken utterances in mathematics is a widespread use of pronouns such as *it*, together with many indefinite or vague expressions. However, communication does take place under such circumstances. In particular, the purpose and context of the utterance determines criteria for adequacy and hence acceptability. The move toward context-free utterances, necessary for successful utilisation of written mathematics, needs to be justified in terms of the demands of the situation.

Numerous researchers have devised constraints on the communicative situation in order to focus attention on the form of the language. For instance, Booth found consistent employment of brackets by pupils when she made it a requirement of the calculating machine she was using. Similarly, the microcomputer requires scope information to function. Machine communication in fact may provide the first genuine situation in mathematics classes where formal language and precise attention to syntax is *required* rather than artificially imposed. Mason et al. suggest the sequence seeing — saying — recording as the fundamental cycle in doing and learning mathematics. Recent calls for increased discussion in mathematics, while not new, are providing a new impetus for attending to the middle of these three terms. The general premise is that articulation externalises thought and provides greater access to it for refinement and modification both to the speaker and others.

Finally, work which has been carried out on Halliday's notion of the mathematics register was reported, in particular the place of metaphor, both structural and extra-mathematical, in the creation and extension of meaning for mathematical terms and interference with the customary register of English.

Oral—Mathematics

Two sessions were devoted to specific consideration of spoken language in mathematics classrooms. The first was concerned with the relationship between language and cognitive development. John Conroy Australia) examined the many verbal and written ways in which the equivalent horizontal or vertical symbolic forms of the four operations on natural numbers may be expressed. Some simple investigations have shown that children at different levels of schooling find some of the verbal expressions much more difficult than others. They can however, carry out the operations quite successfully if the identical items are expressed in symbols. There appears to be a hierarchy of ease or difficulty) for various verbal expressions: intuition and spontaneity; extension and refinement; symbolisation; consolidation and transformation; and finally, mathematical discourse. Conroy thus described a general framework in which specific examples could be studied.

Francis Lowenthal (Belgium) provided an example of the arguments of 6-year olds learning to describe a triangle on a geoboard in terms of coordinates. He studied the way these children verbalised their mathematical experience, and talked about their strategies for solving the problem. The technical constraints built into the material encouraged the children, without any adult intervention, to reach a common understanding, to build sensible arguments and *then* to solve the mathematical task. He claims that such a 'non-verbal communication device' favours the development of structured communication, allowing very young children to learn how to build a convincing verbal argument.

While Lowenthal uses a mathematical situation as a context and lets children talk about it in the framework of natural conversation, Colette Laborde France) uses a mathematical situation as an object which has to be described. She focussed on the kind of language used by 12 and 13 year olds who were asked to describe geometrical figures to other pupils in such a way that the second group would be able to reconstruct them exactly. She noticed that if a mathematical notion is not well assimilated by pupils, they may have considerable language problems when they try to speak about it. Further, it was apparent that they found it difficult to decide by themselves to use a symbolic code when this was more appropriate. LaBorde stressed the need for teachers to create mathematical situations such as these where the social dimension of communication is essential. She feels that this is a positive way in which teachers can allow pupils to develop cognitively, by exposing them to the use of mathematical language in a social context.

In the final paper relating language to cognitive development, Elmar Cohors–Fresenborg (FRG) suggested that the way children use a programming language in creating a computer program can be used to describe different thinking strategies. For example, some children, using a strategy which he calls 'conceptual', have a detailed plan of action before making the first step. Other children , using a strategy which he calls 'sequential', have a global aim at first and fill in the details later. These latter children use a more algorithmic procedure. They imagine the first steps of their program and, by conversing with it and de-bugging it, eventually succeed in building it up step by step. These different preferred ways of programming are of importance for classroom teachers and could influence the choice of a teaching strategy. Cohors–Fresenborg described a method of teaching deaf children to write a program by using his invention of a dynamic maze. This should be looked at closely by teachers of the deaf as a significant breakthrough in non-verbal mathematical communication.

The second session on oral mathematics focussed on the sociolinguistics of mathematics classrooms. Three presentations were given:

The observation, description and analysis of teacher/pupil interaction in primary mathematics classrooms.
The influence of language and culture in the mathematical education of Aboriginal Australians.
The relationship between the perceptions teachers have of the nature of mathematics and the kinds of social interaction which take place in their mathematics lessons.

In the first presentation, Hilary Shuard (UK) described the styles of classroom talk in the classes of four teachers who worked in similar socio-economic settings with 7–9 year olds. These teachers, all women in their late twenties, were confident and willing to be recorded. Shuard found that the first priority of the children was survival, personal psychological survival, that is, 'getting through the day without anything too awful happening.' In the mathematics lesson it is very easy for awful things to happen. The way to survival is to try to make the system roll as smoothly as possible. For both teacher and children this is by getting it right. The tapes revealed a number of strategies used by teachers and pupils in their efforts to achieve this. For example the 'guess what I'm thinking of' strategy. Here a teacher prompts a child for a word or phrase of which she is thinking. The child is to guess what the teacher has in mind and then to go along with it. Shuard often found a mismatch between the teacher's thinking and the child's thinking. She found that meaningful communication often floundered on the rocks of unknown mathematical terms. It was clear that children really tried to make sense of a problem, often using their own methods, and making very good sense even if it was not quite as intended. Such sense was often made on the basis of limited information — many wrong answers were correct answers to slightly different questions. On the other hand there were examples of interaction where the child was simply asked 'what are you going to do?', a style which encouraged and guided the child's own thinking:
T: So what are you going to do first?
P: Got to get 7.
T: Right.
P: ...five, six, seven.
T: Now what have you got to do?
P: Take 3 away.
T: Mmm.
P: Four.
T: Yes. You've taken away 3. And it leaves you with 4. So what are you going to put here?
P: Four.
T: Right. So 7 take away 3 equals 4.
Shuard's paper made it clear that we need to know much more about detail.

In the second presentation, Pam Harris Australia) provided us with a picture of life in a traditional Aboriginal community in Central Australia. In particular, she highlighted the language difficulties involved in teaching mathematics to Aboriginal children at Yuendumu. The children take part in a bilingual, bicultural mathematics program which has to take account of cultural differences between traditional Warlbiri and modern Western society. For example, the Warlbiri people have no standard measure of length, so Cuisenaire rods are quite meaningless in the teaching of number. Other strategies have to be found related to their lives outside the classroom. Language difficulties often occur with words which have a range of meaning in Warlbiri yet are specific in English. For example, there is no equivalent for the English word 'full' — the nearest Warlbiri word simply means 'bodily presence' and would be used if a glass had, for example, any beer in it at all, full or nearly empty. The conflict between traditional Aboriginal and Western ways demands a very special approach to the teaching of mathematics, an approach which recognises the importance of the child's first language and culture while opening up new horizons for them. No easy solutions are expected yet the question of what to do, in both the short and long term, needs urgent support.

In the third presentation, Marilyn Nickson (UK) reminded us of the general need to recognise that the kinds of social interactions which occur in a given teacher's classroom are dependent upon the beliefs and perceptions of teachers with regard to mathematics and the nature of its foundations as a kind of knowledge. She made the point that for mathematics teachers to be able to meet the demands currently being made upon them in terms of curriculum development, an effective rationale is needed for their subject, supportive of the changes they are being asked to make. Recent research in the United Kingdom suggests that teachers at both primary and secondary levels see mathematics as a linear, somewhat mechanistic subject concerned with facts and skills, but above all, individual in terms of pupil activity. It is an authoritarian discipline tinged with a bit of mysticism, with emphasis on rules of operation and formal examples rather than on practical problems.

Nickson argues cogently for a view of mathematics founded in terms of growth and change theory. In this view social factors cannot be ignored. Mathematics is seen as a shared human activity based on inquiry, questioning and critical discussion in exploring mathematical ideas. The need for clarity of purpose, expression, and the inherent logic of mathematics is stressed. Learned facts and skills are applied in problem solving situations and pupils are encouraged to make hypotheses and test them. Thus a social context is identified which results in the sharing of ideas and problems amongst pupils and between pupil and teacher. As Nickson so clearly points out, the old theoretical interpretation of mathematical knowledge is still being adhered to in spite of the recent efforts in many countries to bring about such changes. Nickson argues that a rationale for these changes exists and we must not wait any longer to acknowledge it. It is important, therefore, that every teacher and teacher educator clarify and articulate their perceptions of the subject since it is these perceptions that support our beliefs and values with respect to mathematical knowledge and which will propagate in turn those beliefs and values in the minds of our pupils.

Reading and Writing Mathematics
The second major theme of the program was directed towards the reading and

writing of mathematics and the move from natural language to symbolism. As it was considered important that reading and writing mathematics be addressed in a classroom setting, Daphne Kerslake (UK) and Alan Larkin (Australia) made presentations which linked reading and writing with oral expression. Larkin placed emphasis on the participation of learners in talking, thinking and communicating. The language used by the learners should be monitored by the astute teacher as the ideas and language evolve towards mathematically acceptable and precise forms. The oral work should be supplemented then with written work as a forerunner to reading and writing formal mathematics. Thus Larkin sees the teacher's role as one of facilitator and resource person, who, with the assistance of the children, establishes a positive, caring, supportive learning environment.

Analysis of the nature of language used in the classroom suggests that, for most of the time, the talking is done by the teacher. In her research with Shuard and others in the United Kingdom, Kerslake found that anything that approaches real mathematically valuable discussion, or that encourages children to generate their own ideas, was rare. It seems likely that the use of an investigative approach to the learning of mathematics might call for different verbal approaches from the teacher. With the help of a transcript of a classroom investigation recently recorded in England, Kerslake drew attention to communication between teacher and pupils. This provided a practical endorsement of Larkin's presentation and enabled participants to share their reactions and ideas in a practical way. Discussion of a wide range of issues involving talking, reading and writing followed the presentations.

The move from natural language to symbolism, and vice versa, is an issue of world wide concern in mathematics education. Two presentations were given on this topic. The first, by Bernard Parzysz (France), highlighted problems of the interaction of natural language with mathematical metalanguage and symbolism in French secondary schools. Throughout the secondary course, mathematics is taught in French, that is, in natural language. However, the language of the text is mostly 'special' French — mathematical metalanguage. This metalanguage presents peculiarities in vocabulary and syntax and varies with the age of the student, being 'dynamic' in the beginning, but growing progressively 'static' under the influence of the teaching received, and tending to become increasingly closer to symbolic language structures. The most utilised written or verbal language is therefore an intermediate one, with the purpose of bridging the gap between truly natural language and symbolism with various possible levels in between.

Parzysz made two important points. Firstly, symbolic language is exclusively written. When one wishes to read it, one is obliged to 'translate' the symbols into natural language. Secondly, this process is introduced very late in the mathematics courses of French secondary schools. Consequently, teachers need to constantly translate symbolic language in order to be understood. Problems of reading comprehension, the use of quantifiers, and interference with natural usage for example, 'negation' were all discussed. There is clearly a need for teachers to think about the language they use in the classroom and for pupils to ponder on the texts they are given. In this way, correct manipulation of metalanguage can be progressively developed, as well as that of symbols. At the Paris VII IREM, cooperative teaching between mathematics and French language teachers has brought fruitful results. Other countries could well follow this example.

The second presentation by Glen Lean (PNG) explored the cognitive processing of written mathematical tasks, illustrated by reference to recent research in Papua New Guinea. Once encoded, a written mathematical task may be subject to both verbal and visual symbolic processing. Children tend to show preferences for a particular processing mode. The mode or modes employed will depend, first, on the child's predisposition to use a particular mode, which is affected by genetic and socio-cultural factors, including the nature of his mathematical education. Second, the nature of the task may also affect the mode or modes used. A novel or unfamiliar task makes greater demands on processing capability and requires flexibility in the selection of the processing mode.

In terms of mathematical development, it is likely that a child's early learning relies heavily on visual coding and processing but that this ceases to be the dominant mode as language competence increases. Lean argues that the mathematical cast of mind of a child appears to be influenced by, in addition to other factors, the tendency for mathematics teachers to over-emphasise the verbal — symbolic — analytic approach to the detriment of the visual. In the solution of routine problems, the analytic approach appears to be quite sufficient. However, in the solution of non-routine problems and in the area of mathematical creativity, it may well be that the visual mode of processing plays a role of at least equal importance. It is time that any imbalance in classroom teaching is brought to the attention of teachers.

Teaching and Learning Mathematics in English as a Second Language
The third theme in the series of presentations focussed on the problems being experienced in different countries around the world on teaching and learning mathematics in a second language.

In the first session, issues concerning bilingualism in relation to teaching and learning mathematics were discussed. Lloyd Dawe (Australia) outlined a recently developed theoretical framework for research into the interrelations of language and thought in bilingual children. This theory asserts that a cognitively and academically beneficial form of bilingualism can only be achieved on the basis of adequately developed first language skills. He discussed the results of his recent study in the United Kingdom which set out to test this theory with respect to the ability of bilingual Punjabi, Miripuri, Italian and Jamaican children to reason deductively in mathematics in English as a second language. In general, the results supported the theory. In particular, knowledge of logical connectives in English (words which join propositions in reasoned argument, such as 'because', 'and', 'or', *etc.*) was an important predictor of scores on the reasoning test in mathematics for both bilingual and English monolingual children. The results suggest that published weaknesses in mathematics found among certain Asian and West Indian pupils may well be due to language factors. Furthermore, there are strong cultural factors which predispose differential performance among boys and girls (see Dawe, L. (1983) *Educational Studies in Mathematics*, 14, 325–353).

The second speaker, Gilbert Cuevas (USA) addressed the teaching of language skills in the mathematics classrooms of the United States. He described a current project in Miami, The Second Language Approach to Mathematical Skills (SLAMS), which follows a diagnostic-prescriptive approach to teaching, incorporating strategies for dealing with language skills to assist limited English proficient students in mastering mathematical content. The teacher training model contains an incentive component, to sensitise teachers to the language problems of students; a procedural component, which describes strategies for carrying out the training activities; and a content component which includes cultural, linguistic, mathematical and pedagogical topics. This innovative work is highly recommended reading (Ceuvas, G. (1984) *Journal for Research in Mathematics Education*, 152, 134–144.)

The third speaker in this session, Helen Watson (Nigeria) spoke of her research with Yoruba — English bilingual children in Nigeria. In formal schooling in Nigeria, children are taught in English after the age of seven. Thus, to receive an adequate education, they must be effectively bilingual. However, English and Yoruba are radically different languages. For example, Watson has found that in reporting the physical world, English and Yoruba speakers refer to differently situated objects. Thus the conceptual processes of quantification in the two languages are basically different. In English, quantification processes are counting and measuring. In counting, the qualities of 'thingness' and 'numberness' must be cognised. In measuring, various context dependent qualities must be construed as a scalar continuum. By contrast, in the Yoruba language, the quantification processes are 'ikaye' and 'iwon'. For both these processes, the quality of unicity must be cognised, for ikaye its spatial aspect and for iwon its temporal aspect. Thus to 'measure' is not the same as to 'won'. This fascinating research draws attention once again to the debate begun by Whorf some thirty years ago as to whether people in a given culture are at the mercy of the language they speak in understanding and reporting the physical world.

The second session on this topic explored the teaching and learning of mathematics in a second language with specific emphasis on the changeover from a second language to mother tongue and vice versa. D.K.Sinha (India) reported on some recent developments in India aimed at studying the interaction of language and mathematics education in a wide variety of contexts. In particular, Sinha reported on two common Indian contexts, namely, learning mathematics in a language other than one's mother tongue, and learning mathematics in one's mother tongue but using texts written in another language and then being examined in that other language. The issues of concern were the effect of the 'changeover' from one language to the other on the growth and development of mathematical ideas.

Jan Thomas (Australia) identified some challenges facing mathematics educators in a multi-cultural society. In particular, she referred to the problems in Victoria, an Australian state with approximately 25% of children from non-English speaking backgrounds (NESB). While such children perform adequately in the computational aspects of mathematics, they fall well below their English speaking peers in their ability to apply their mathematics in real world situations. The need for appropriate teaching strategies in mathematics, based on recent developments in the methodology for teaching English as a second language, may offer a way ahead for schools with a large proportion of NESB students. Such approaches are currently being developed in Victoria through the Child Migrant Education Service.

James Taole (Lesotho) highlighted the specific problem of lack of congruence between the learner's mother tongue and the language of instruction, and the adverse effect this may have on learners who still have to translate between the two when decoding mathematical language. This was clearly demonstrated with a group of Lesotho children in the first year of secondary school. When presented with a series of tasks based on the polarised comparatives 'more' and 'less', a high proportion (80%) tested in the vernacular successfully completed the task, while for those tested in English the success rate was considerably lower about (30%). However, when the tasks were presented in English, but in a form that was syntactically congruent to the vernacular, there was a considerable improvement in the success rate about (50%) suggesting that some second language learners may benefit if the mathematics teacher is able to use English language forms that more closely parallel those used in the vernacular. This idea is currently being developed in Lesotho and it is hoped that learning materials based on this principle will soon be available.

Peter Jones (Australia) reported on the language based problems faced by children in Papua New Guinea where learning mathematics in English means learning a distant culture's mathematics in a distant culture's language whilst growing up in a society that is unable to provide the everyday experiences necessary to support this cross-cultural learning. Three major problem areas were identified and illustrated in the Papua New Guinea context. These were:
• the mismatch between the language demands of the mathematics classroom and the learner's English language competency;
• the nature of mathematics and the methods of reasoning in the learner's culture; and
• the difficulty of translating English language terms and expressions used in mathematics into the learner's mother tongue.
While it is difficult to establish causality, it would appear that these language based problems considerably influence the relatively poor performance of Papua New Guinea children on any mathematics that goes beyond the routine performance of standard algorithms. For this reason, a considerable effort has been made in Papua New Guinea to develop a mathematics curriculum that takes into account some of the language based problems of Papua New Guinea children, although there is still much to be done before the problem is solved.

Research Issues

At the final plenary session, group leaders presented summaries of contributions made in the three areas of oral mathematics, written mathematics, and teaching and learning mathematics in a second language. In preparation for ICME 6, it was decided to make a list of research questions of immediate importance and to form groups to tackle them. The areas identified for research were:

Oral Mathematics
• The description and classification of teacher/pupil interactions — general and mathematical — in mathematics classrooms;
• Differential interaction patterns: small group/large group/individual; high/low ability; male/female; problem solving/exposition;
• Gaps between teacher's language and child's language; ambiguity; different

types of language used by different teachers, for example, primary and secondary).

Reading Mathematics
- Reading mathematical text; balance of text to visuals; the cognitive and linguistic demands of text;
- The effect of symbolism on order of reading text;
- The reactions of students to different types of text.

Writing Mathematics
- A detailed study of the range of functions of written language and how far children are aware of these;
- The acceptability of written records to teachers;
- The move from natural language to symbolism.

Language And Cognitive Development
There is continued research interest in the ways children verbalise mathematical experience and their strategies for problem solving. Ongoing work in relation to the importance of classroom discourse and cognitive development is anticipated, in particular, the importance of the social context of mathematical situations and the use of computer programming to study thinking strategies. There is also interest in the role of the left and right hemispheres of the brain in relation to preferred cognitive styles for processing mathematical information. The impact of such research on the choice of programming languages to use with young children is important.

Bilingualism, Cross-Cultural Studies And Teaching And Learning Mathematics In a Second Language
The following issues and questions were identified as needing further research:
- The documentation and analysis of mathematical language in various vernaculars, so that it is recorded for posterity and so that different conceptual systems built into the languages are identified;
- What kind of instructional models/strategies may be developed given the knowledge we have regarding the roles language and culture play in the learning/teaching of mathematics in English as a Second Language;
- What are the skills that teachers need to implement these strategies?
- How can teachers be motivated to particpate in this training?
- There is a need to develop appropriate courses to train teacher educators from third world countries. Closely related to this is the need to document cultural differences so as to produce 'culture oriented' materials for teaching mathematics in schools;
- There is a need to develop culturally appropriate materials for students outside 'Indo-European' culture, for example, North American Indians and Aboriginal Australians;
- There is a need to document and understand concepts of mathematics in conceptual systems remote from those implicit in Indo-European languages;
- Very precise linguistic studies to find out how different languages express mathematical ideas;
- Studies of language switching in bilingual classrooms by children learning mathematics in a second language;

PROCEEDINGS OF ICME 5

- The desirability of bilingual teaching: under what circumstances is it of positive benefit for both L1 and L2 to be used in the mathematics lesson?
- The study of linguistic interference between L1 and L2 when mathematics is learned in L2 — is it possible to translate?
- The study of speech in bilingual classrooms, including the 'translation' of L2 to match more closely the structure of L1;
- Can instruction be improved by using texts where L2 is consciously modified so as to match, where possible, the syntax of L1?
- The problems of the changeover from the mother tongue to L2 (or vice versa) in the mathematical education of children in many countries needs continued effort.

Conclusion

The quality of the papers presented and the enthusiasm of those who attended hold out real hope for progress in this important area of mathematics education. In his plenary address to the ICME 5 Congress, Jeremy Kilpatrick asked, 'What do we know in 1984 that we did not know in 1980?' It is our goal for ICME 6 to be able to review progress made in the years 1984–88 so that in 1988 we will indeed know a great deal more than we do now about the interactions of language with mathematics teaching and learning in different cultural, social and psychological contexts. We commend the set of papers presented at ICME 5 to you and record our thanks to UNESCO for making their publication possible.

References

Austin, J.L. and Howson, A.G. (1979) 'Language and Mathematical Education', *Educational Studies in Mathematics*, 102), (pp) 161–197.

Booth, L.R. (1984) *Algebra: Children's Strategies and Errors*, NFER- Nelson.

Halliday, M. (1974) 'Some Aspects of Sociolinquistics' in *Interactions between Linguistics and Mathematical Education*, Nairobi, Unesco.

Lakoff, G. and Johnson, M. (1980) *Metaphors We Live By*, U. of Chicago Press.

Mason et al. (1985) *Routes to Algebra*, Open University Educational Enterprises.

Pimm, D. (1985) 'Seeing Mathematics Linquistically', in Unesco.

Stubbs, M. (1980) *Language and Literacy*, Routledge and Kegan Paul.

TOPIC AREA: PSYCHOLOGY OF MATHEMATICS EDUCATION

Organiser: Gerard Vergnaud (France)

At Exeter in 1972 (ICME 2), Professor Efraim Fischbein instituted a working group bringing together people working in the area of the psychology of mathematics education. At Karlsruhe in 1976 (ICME 3), this group became affiliated with the International Commission for Mathematical Instruction (ICMI).

The International Study Group for the Psychology of Mathematics Education (PME) has met every year since 1976 at Utrecht (Netherlands), Osnabruck (FRG), Warwick (England), Berkeley (USA), Grenoble (France), Antwerp (Belgium), Shoresh (Israel), Sydney (Australia).

PME has a constitution that provides for the election of a President and an International Committee. As President of PME, Gerard Vergnaud (France) had the responsibility of organising PME's contribution to ICME 5.

The major goals of the group are:
- to promote international contacts and exchange of scientific information;
- to promote and stimulate interdisciplinary research with the cooperation of psychologists, mathematicians and mathematics teachers;
- to further a deeper and more correct understanding of the psychological aspects of teaching and learning mathematics and the implications thereof.

There are several hundred members in the group. Membership is open to persons involved in active research to further the group's aims, or those professionally interested in the results of such research.

Session 1: What Do We Learn from Analysing Students' Work from Interviews with Students.

Presenters: Nicolas Balacheff (France), Kath Hart (UK), Robert Hunting (Australia), Leen Streefland (Netherlands)

One of the aims of PME research is to elucidate pupils' conceptions about mathematics concepts. Such work necessitates the use of special means to investigate what is usually hidden, that is, the intellectual constructs and processes of children.

PME researchers use, or have designed, experimental settings in which to make explicit witness of pupils' intellectual representations and understandings. The methodology used seeks to unfold processes, unlike questionnaires which allow only access to results and performance. Interviews, clinical investigations, and social interaction between small groups of pupils have a contribution to make to the development of research since they illuminate the answers and performance of pupils in broader enquiries carried out using paper and pencil means.

In introducing teachers to this methodology it is hoped they will consider their pupils' behaviour in a different light, going beyond immediate performance to diagnosis of conceptions which may be wrong or incomplete. The collection of data is through interviews with one (or two) children. The discussion is focussed on a particular area of mathematics learning, tape recorded or video recorded, and then the discourse is written. The presentation included four protocols illustrating different kinds of interview investigation.

At the Start of the Research
The first protocol illustrated the registration of spontaneous utterances of children and the related mathematical behaviour. This kind of protocol not only helps the researcher become aware of what may be tender germs of hypotheses for future research but also helps him/her otherwise. The behaviour of a single child may not be generalisable but the problem, phenomenon or event and the child's reaction to it may be of general interest to other children. The analysis and subsequent results in turn may be of general interest within the research context.

Case Study of One Child
These protocols were used to illustrate the interview method over a long period of time in which the researcher was seeking to discover the child's concept of fractions. It is from the interviews that the conflicting nature of Rebecca's two views of fractions was discovered.

Using Interviews to Collect Data from a Large Sample
In order to discover the views/use children have of practical work with tangible objects after they have moved to a formalisation/generalisation, interviews in different contexts with different children are used. The protocols are of children aged 8–13 learning mathematics in their normal classrooms but being asked three months after using the materials, to use them again to demonstrate a mathematical statement.

Interactions Between Two Children
For some research (such as research on proving processes) it appears to be necessary to minimise the pressure of the observer for verbalisation, such as 'What are you doing?'; 'Why are you sure?'. For this purpose an experimental setting has been designed within which pupils have to interact and cooperate in order to solve a common problem. In such a situation, by means of the socio–cognitive conflict which arises and has to be overcome, the researcher has access to the debates between the pupils. These reveal — within the limits specific to social interaction — the background of the decisions taken both in terms of knowledge, rationality and the meaning of knowledge.

Session 2: Long Term Evolution of Student's Conceptions: An Example from Addition and Subtraction

Presenters: Jacques C. Bergeron (Canada), Nicolas Herscovics (Canada), James Moser (USA)

Much too often in the past, research on early arithmetic has been limited to the acquisition of skills, that is, essentially the memorisation of basic facts. However, with today's emphasis on problem solving, it is necessary to question the old belief that children have to first master these basic number facts before they can be introduced to problems. But we cannot discuss this topic without examining it in the broader context of the long term development of conceptual understanding of addition and subtraction.

To study this long term development, we need an appropriate frame of reference. Vergnaud, has suggested the notion of a *conceptual field*, a primary example of which is that of additive structures, as an organisation of knowledge that develops over a very long period of time through experience, maturation and learning. To him, a conceptual field is an informal and heterogeneous set of problems, situations, concepts, procedures, relationships, structures and operations of thought, connected to one another and likely to be interwoven during the process of acquisition. He suggests that progressive understanding of the conceptual field of additive structures develops over a long period of time beginning at age 3 or 4. Let us examine the period that extends from 3 to 8 years of age.

If we look at the conceptual field of additive structures we find that it is much richer and more complex than the one associated with purely arithmetical skills. In fact, following a semantic analysis of word problems, three classes of problems have been identified: the transformation type, the combination type and the comparison type (Carpenter and Moser; Nesher; Riley, Greeno and Heller; Vergnaud). And these give rise to 14 types of problems. So the question is raised as to how children solve these problems.

Earlier research tended to focus more on correctness of response and reports tended to characterise difficulty of tasks or of learning of so-called number facts. Happily, recent research emphasis has changed to an interest in the child. Through observation and recording pupils' performance on a variety of arithmetic tasks, investigators attempt to determine the procedures and strategies children use. There is overwhelming evidence indicating that young children's early conceptual understanding of addition and subtraction has its roots in and develops from their early conceptions of number and the counting process.

Carpenter and Moser (1984), for instance, have reported a three year longitudinal study of children's solutions to simple addition and subtraction word problems. Over this period of time, 88 children were individually interviewed eight times (three times in grade 1, three times in grade 2, and twice in grade 3) to identify the processes used in solving six types of problems.

For addition, three basic levels of strategies have been identified: strategies based on direct modelling with fingers or physical objects, strategies based on the use of counting sequences, and strategies based on recalled number facts. For sums between 10 and 16, and with manipulative objects available, the following results were obtained.

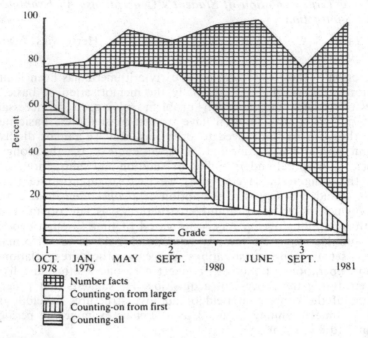

Figure 1. Strategy use over time on join addition problems with larger number facts (sums between 10 and 16) and manipulative objects available.

We first note that close to 80% of the children succeeded in the very first interview before any instruction was given. By the third interview, in May 1979, the success rate was close to 95%. These results clearly indicate that children initially solve the problem with a counting-all strategy, and that this strategy gradually gives way to counting-on and the use of number facts.

It was found that in the case of a much more difficult addition problem, the join-missing addend, about 55% of the first graders succeeded without any prior training by using two counting strategies. The same pattern was observed concerning the gradual replacement of the counting strategies by the number facts. Moreover, even after three years, about 40% of the children were still relying on these counting strategies! Hence, the counting strategies proved to be very important for the young child. But these in turn necessitate the mastery of different counting procedures whose acquisition also takes a long time, starting from the use of the number word sequence, the mastery of which is certainly not spontaneous.

Fuson, Richards and Briars have data showing how the length of the memorised sequence evolves from the age of 3 or 4 to 9, and how the different uses of the counting word sequence can be related to the various counting procedures, thus proposing a learning hierarchy.

Bessot and Comiti have shown that it is by asking first grade pupils to handle many different tasks for which they need different procedures, including counting, that the number word sequence and counting procedures are mastered.

It is important to distinguish between the recitation of the number word sequence, which is a pre-requisite skill, and the act of enumeration, which is a procedure setting a 1-1 correspondence between the number word sequence and a set of objects to be counted. In fact, a child may often know how to recite the number word sequence without necessarily being able to coordinate it correctly in his enumeration procedure.

We can identify six procedures for enumerating objects corresponding to the different skills described by Fuson and her colleagues:

- counting by starting from 1;
- counting from 1 in a row of objects, up to a given number;
- counting of objects starting from a given number;
- counting of objects from a number A to a number B, A < B;
- counting backwards;
- counting backwards and stopping at a given number.

These are the counting procedures on which children have to rely before they can use their number facts.

Implications for Instruction

The studies reported suggest that, considering the rich informal mathematics that children bring to instruction, word problems can be integrated into the curriculum even before computational skills have been mastered. The progression of skills identified is generally not reflected in instruction, thus ignoring the fact that there is an extended period of time during which children count on and count back to solve addition and subtraction problems. Since a variety of counting procedures are used to solve addition and subtraction problems, it could be useful to incorporate the various forms of counting into the curriculum, thus enabling students to develop a greater mastery of the less common procedures such as counting backwards. Finally, in order to help children extend their knowledge to the whole conceptual field of addition and subtraction, a great variety of contexts and problem types should be used.

Session 3: How Students' Conceptions Conflict and Change in the Process of Learning

Presenters: Leonie Burton (UK), Klaus Hasemann (FRG), Colette Laborde (France), Francis Lowenthal (Netherlands).

Videotapes provide a unique opportunity to observe behaviour which under normal classroom conditions might pass without remark. Information relating both to pupils' cognitive processes and to the pedagogical situation in which these develop can be deduced from the tapes. As a research method, this is particularly valuable to those interested in the psychological problems associated with mathematics learning.

The following hypotheses provided the basis for viewing the videotaped sequences to be shown:

- the view of mathematical concepts in teachers and in pupils is frequently framed differently;
- provoking conflict is one way of handling this mismatch;
- the learner requires a challenge in order to build understanding;
- the learner has powerful mathematical thinking processes which are rarely used in traditional communications with the teacher.

The following sequences have been used to exemplify *cognitive processes*, particularly:
- formulating problems;
- constructing logical arguments to explain observed phenomena;
- using these arguments to convince oneself and others;
- relating logical argument and action;
- moving from problem to solution and vice versa; and

pedagogical points, including:
- the importance of concrete representation to the student's mental images;
- the necessity for precision in communication;
- the impact of group interaction;
- the role of the teacher in challenging the students;
- the teacher's recognition that the responsibility for accommodating understanding to observation rests with the student.

The sequences span the age range 6 to 13 years, and different aspects of mathematical learning.

The main purpose of the session was to relate current theoretical approaches to classroom practice. Each tape was viewed with a focus on the following major components:
- the content of the tape;
- the moments where cognitive changes can be observed;
- the cognitive processes involved;
- the pedagogical points;
- additional remarks, if any.

Emphasis on each of the five components differed from tape to tape. An analysis was made of each of the tapes in terms of these five components and is given here.

'Coordinates' (five 6 year olds and the teacher)

Content: This tape showed the gap existing between numbers as counting objects and numbers representing distances. It also exemplified the power of an image to influence conceptions: representations are needed. Finally, it showed that changes in conception depend upon group interaction.

Moments of change: Firstly, Julie hesitated and lost confidence although she referred back to a previous lesson. She then used her intuition, watched David's demonstration and used that method to construct an explanation to justify her intuition: Julie matched counting with distance. Later, another girl, Anne, went from distance to counting. In a third moment of cognitive change, all the pupils (except David) adopted this technique which they had rejected at the start.

Cognitive processes: Two processes were observed, namely, constructing logical arguments to explain observed phenomena and convincing others by using these arguments.

Pedagogical points: The teacher appeared to have an important role by asking some questions, by being neutral and only *repeating* the children's responses, by throwing the responsibility back to the children, by challenging them, by reversing the procedure to enable all responses to be acceptable, and also by demonstrating the 'symmetry' of the argument. The need for representation and the role of group interactions also seemed to be relevant.

Additional remarks: Julie was not as popular as David. At first her proof was rejected and she hesitated, which showed the effect of social pressure in group learning. But Anne, who was better accepted, moved the class to agree with Julie.

'Awkward Shapes' (six 10 year olds and the teacher)

Content: This tape demonstrated first the use of conversation; second, a misplaced generalisation, 'area is calculated by multiplying side by side'; and third, the power of the 'side by side' image to influence conceptions.

Moments of change: There were three moments of cognitive change. First, Katy, shifted by a tangram reference, used visualisation; second, David partially shifted from a situation where he was blocked to a misconceptualisation of the way to transform a triangle into a rectangle by cutting; and third, Andrew compared a slant height with the actual 'perpendicular' height and noted that they were different although he assumed them to be equal.

Cognitive processes: There were four important cognitive processes to be observed: constructing logical arguments to explain observed phenomena; convincing others by using these arguments; using past experience to unravel present puzzlement; relating action and argument.

Pedagogical points: The teacher played an important role by *reinterpreting* and *clarifying* what the children observed, by asking questions, by remaining neutral, by throwing the responsibility to the children, and by challenging them. The need for representations, and the role of group interactions as 'conception changers' seemed also to be relevant.

Additional remarks: A psychological remark: Katy's lack of confidence: she cut, folded, cut and ... kept quiet.

'Write a message' (two 13 year olds, no teacher)

Content: This tape demonstrated the difficulty in choosing relevant from irrelevant data in order to describe a picture. It also exemplified the decentring process from a child's own point of view to that of another, as well as showing the power of the image 'triangle' to influence conceptions.

Moments of change: Pierre changed his point of view. He first described things two-dimensionally and then one-dimensionally; he moved from triangles to edges, and then to points. Both children changed also in another way: from ego-centring to communicating so that others could reproduce their diagram.

Cognitive processes: Three cognitive processes were observed: communicating in a very precise fashion, reducing from a two-dimensional presentation to a one-dimensional one after identification of the necessary pieces of information and convincing by action.

Pedagogical points: In this case there was no teacher but the task was challenging and put the responsibility onto the children. It must also be mentioned that the choice of the task demanded interpretation and communication. Finally, as seen in other tapes, the solution was dependent upon interactions and discussions.

Additional remarks: There are no additional remarks.

'There are no more' (two 9 year olds with the teacher)

Content: This tape demonstrated the difficulty of proving. It also gave an example of proof by exhaustion. It also showed the power of the image 'ladder' to influence conceptions.

Moments of change: There were three main cognitive changes: first, both children changed from observation to proof, second, they extended their system of representation to cover all the cases, and finally, the *teacher's idea* of 'correct proof' for the given problem expanded to include the children's proof.

Cognitive processes: Four cognitive processes should be mentioned here: formulating a problem, realising that convincing means are needed, constructing logical arguments to explain observed phenomena, and reversing the movement from building to nets, from nets to constructions by matching.

Pedagogical points: A concrete apparatus was here central and the children soon realised that a representation was needed. The teacher had an important role in being neutral and in challenging the children. As noted before, the changes in conception depended upon the interactions between the children. This tape was unique in showing an example where the teacher accepted the need to expand his own conception in the light of the children's argument.

Additional remarks: There are no additional remarks.

A lively discussion followed the viewing of each sequence. The discussion focussed on the use of videotapes for the purpose of teaching and research.

Session 4: The Nature of Mathematical Thinking: Intuition, Operation, Representation, Discovery and Proof.

Panel: Gerard Vergnaud (France), Leslie Booth (UK), Rina Herskowitz (Israel), Carolyn Kieran (USA), Hartwig Meissner (FRG), Elmar Cohors-Fresenborg (FRG), Alan Bell (UK), Richard Lesh (USA).

Introductory comments on the previous three sessions were presented by Kieran, Herskowitz, Booth, and Meissner. Other panelists and members from the floor contributed to the discussion which followed. Many points of interest were raised on understanding mathematics and the nature of thinking involved.

Case study methodologies have allowed researchers to make significant contributions to our knowledge of how children learn to add and subtract, interpret fractions, ratios and algebraic expressions, and how they construct different mathematical structures and processes. Though there is considerable variety in the methods of data collection, most studies typically involve the gathering of data by means of audio and video tape, with the data ultimately represented by written protocols. These protocols are then used as a basis for the analysis of underlying processes. Such methods allow researchers to go beyond the kind of inferences which are possible with the more traditional agricultural methods. Some of these methods could be learned and used by teachers in their training. They could also be used in their classroom when they need them, and have the possibility of using them. Some of these methods, however, may appear quite difficult and time-consuming.

A case study with one child illustrated the conflict of two different ways of thinking for an extended period of time and showed that cognitive conflicts are not resolved easily. Another study showed the communality of students' ideas before and after a period of instruction. The teacher was informed of the data and could therefore appreciate the benefit that his students had gained from instruction and the way in which the pre-instructional methods had been transformed. His assumptions were quite different from the findings.

In order to overcome the bias of interviewing by an adult, a method has been developed which consists of recording the behaviour and arguments of two students solving a problem, without the presence of any adult. By providing only one pencil with which to write a statement, they are forced to come to an agreement. One can observe the interaction and its effects, although this interaction may fail to produce any effect for some period of time when one student cannot follow the other.

From the findings of research on the long-term evolution of students' conceptions and procedures, several needs can be identified:
* the need to identify more clearly the informal mathematical understandings that children bring to instruction;
* the need to help children develop and extend their own procedures and to enrich their own conceptual field;
* the need to capitalise on these understandings in helping children to construct more formal procedures;
* the need to allow time for this construction to take place.

But this is only part of the story: we must not only understand where the child is starting from, but also what the goals of mathematical instruction are. It is important to help children develop the mathematical processes that underlie the procedures they 'spontaneously' tend to use. There may be more mathematical thinking in those procedures and processes than in some formal algorithms that we teach them. Iterative procedures and curtailment processes are good examples of this. Children make extensive use of iterations, and spontaneously shorten their initial procedures, for instance, going from 'counting all' to 'counting on', then to 'counting on from larger', then using 'number facts' or 'counting by fives or tens'.

The reasons why conceptions conflict and change do not lie solely in the interactions of students with teachers and other students. Reasons can also be found in the cognitive development of individuals. For instance, children often go from a global perception of situations that may be sufficient in certain uses, to some kind of 'chaos' when they try to analyse the components of those situations, until they reach a new structured understanding. U-shaped curves have been observed on many notions. There are also incorrect or partial interpretations of instruction by children (subtraction algorithm, for instance), and also incomplete frames in the development of new concepts and new procedures (the width of a parallelogram in session 3 provides a good example). Finally, there are many other causes, such as the emotions of the child, his or her personal experience and social background, and the relationship to the teacher's authority. There are also 'black-outs' and perseverations.

Altogether, the nature of mathematical thinking, when it is viewed through the psychology of student behaviour, appears to be much more comprehensive and varied than some usual definitions in terms of abstraction, formalisation, rigor and proof tend to convey. It would be misleading to confine mathematical thinking to what is most specific of it. Poincare tended to see its essential component in mathematical instruction and recursive reasoning, Piaget in abstraction from action, as opposed to abstraction from objects. The very process by which many mathematical objects are created (create a set and then take this set as an object and examine its relationships to other objects) is certainly a very important feature of mathematical thinking, although it may not be specific to it. Curtailment is also important, and not specific to mathematical processes.

One can envisage mathematical thinking as focussing on specific objects like numbers, space, functions, symbolic expressions ... or as consisting of specific ways of reasoning. Rigine Donady has provided a very interesting distinction between 'tools' and 'objects', seeing new invariants as tools for understanding new classes of situations or new aspects; and then the reflexive process would take them as objects. Although this distinction looks like Piaget's theory of reflexive abstraction, it does not say that reflexive activity focusses on action only. Moreover, there is a dialectical interaction between tools and objects, and this process is endless.

Invariants are certainly the core of mathematical concepts, but one must not forget that they refer to situations (a large variety of them), and that they can be expressed by different symbolic representations. Therefore, syntactic activities are important in mathematics, although they have no value if there is no permanence in it. As students bring their own interpretation in the learning of these syntactic activities (algebra, for instance), it is essential to understand what metaphors they have in mind when practising them.

Intuition is important. It often consists of 'theorems-in-action' that students spontaneously use in dealing with new situations. These theorems-in-action often go far beyond the level teachers think their students have reached. But students' intuition can also be far behind the level of understanding that teachers expect. In proof, for instance, each sentence must be accepted either because it is obvious, or because it is obviously related to the preceding ones. These jumps consist of implicit theorems that some students may totally fail to accept because they conflict with their own conceptions.

In summary, mathematical thinking as inferred from the analysis of students' behaviour and explanations appears to be a very composite type of activity. Mathematical concepts certainly develop over a long period of time. But essential aspects of mathematics are present at the very beginning of children's cognitive development.

It is clear from the work presently available that theories of general stages of intelligence have been relinquished, although a developmental approach remains essential and individual differences are important. The use of computers by children helps them to become more aware of the relationship they have to deal with; it certainly does not permit them to skip over the conceptual difficulties lying in mathematical concepts and processes.

The present officers of PME are:
> *President:* Kevin Collis (Australia)
> *Vice-President:* Alfred Vermandel (Belgium)
> *Secretary:* Klaus Hasemann (FRG)
> *Treasurer:* Gerhard Becker (FRG)

References

Bessot and Comiti, *Recherches en Didactique des Mathematiques*, Vol 1.2, and Educational Studies in Mathematics, Vol. 1.3.

Carpenter, T. and Moser, J., (May, 1984) The Acquisition of Addition and

Subtraction Concepts in Grades One through Three in *Journal for Research in Mathematics Education*, Vol. 15, No. 3.

Fuson, K.C., Richards, J., Briars, D.J. (1982), The Acquisition and Elaboration of the Number Word Sequence in Brainerd, C. (Ed.) *Progress in Cognitive Development*, Vol. 1. *Children's Logical and Mathematical Cognition*. New York: Springer-Verlag.

Steffe, L., Von Glasersfeld, E., Richard, J., Cobb, P. (1983), *Children's Counting Types: Philosophy, Theory and Applications*. Praeger Scientific, New York.

Van Den Brink, J. (1983), Proceedings of the Fifth Annual Meeting of the North American Branch of PME, J.C. Bergeron and N. Herscovics (Eds.), Montreal.

Vergnaud, G., (1982) A Classification of Cognitive Tasks and Operations of Thought Involved in Addition and Subtraction Problems, in Carpenter, Moser, Romberg (eds.). *Addition and Subtraction: A Cognitive Perspective*. Lawrence Erlbaum Associates, New Jersey.

TOPIC AREA: RESEARCH AND TEACHING

Systematic Cooperation between Theory and Practice in Mathematics Education'.

Organiser: Bent Christiansen (Denmark)
Coorganisers: Guy Brousseau (France), Gerd von Harten (FRG),
Mary Grace Kantowski (USA)

The overall purpose of the meetings on Research and Teaching was to contribute to the exploration of frames, forms, and content of systematic cooperation and interaction between the researcher and the teacher in the classroom. The four sixty minute sessions essentially formed a miniconference in which three of the sessions were devoted to presentations of cases of cooperation, and the fourth provided the opportunity for a concluding debate.

Theory, research, and practice belong together. Thus, didactical theory is, due to its very nature, concerned with the problem field of teaching and learning in the system of schools. The objects of didactical research (such as investigating, comparing, developing or applying theoretical constructs) belong to, or are constructed in close relation to that same field.

Accordingly, both the initiation of research projects in mathematics education and the research processes are conditioned by, and in varying degrees determined by, the flow of the teaching/learning process in the classroom and in the school as a social institution. Finally, the importance of the relationship to school practice becomes evident in the evaluation of the research results, when questions are raised about their significance for the teacher and implementation into practice.

However, the relationships between theory and practice are of an indirect and complementary nature, and this makes mediation between these domains impossible by straightforward pragmatic means. Hence, there is an increasing interest in the identification of links between theory and practice which can be utilised and exploited through new and constructive forms of interaction and cooperation between the practitioner and the researcher. It was the overall purpose of the miniconference to contribute to the exploration of frames, forms and content of such systematic cooperative efforts.

Relation to Other Congress Activities.

The program of this Congress component was developed in coordination with the preparation for the Topic Area 'Theory of Mathematics Education', and is closely related to Theme Group 4, 'Theory, Research and Practice'. It was also linked to a Post-Congress Miniconference held in Adelaide on August 30-31, directed towards the establishment of structures and plans for future work related to the theory of mathematics education.

Report of the Sessions

In the lead paper of the first session, Erich Wittmann (FRG) took as his opening point the teacher's need for models of the more or less open situations with which he/she is confronted in daily practice. The teacher develops such a teaching model of a given situation by means of his/her theoretical repertoire supported by experience and existing teaching aids. Such a teaching model tells the teacher what to look for, what to do, what to expect. It may also provide reasons for the teacher's observations, decisions, and prognoses. Personal experiences with such models (which are being revised during the teaching process) will, over time, influence the teacher's theoretical repertoire, which may be seen as a collection of subjective theories. Using Popper's conception of 'the three worlds', the teacher's *practice* belongs to world 1, *theoretical repertoire* to world 2, whereas *theory* in the sense of didactical theories about teaching, is part of world 3.

Against this background, Wittmann identified the central issue of teacher education as the question: 'What is the best way to build up an effective theoretical repertoire for teaching?' He commented briefly on two well-known answers, that of the practitioner — through practice, and that of the theorist — through theory, and then gave a more detailed account of John Dewey's answer of 1904, — *through vital theoretical instruction*. Wittmann summarised and rephrased Dewey's position as follows. 'The main task of a teacher is to stimulate and to develop the mental activity and interaction of his or her pupils. The best way for a student teacher to acquire the necessary competence is to develop methods of mathematical thinking, to reflect upon them, to observe and analyse his or her own learning in interaction with other students, and to study mathematical thinking in children and groups of children.' In this perspective, Wittmann introduced his own position: 'This kind of doing mathematics and doing psychology shares the essential aspects of learning and teaching mathematics in the classroom, so it represents some sort of practice, which I would like to call *intermediate practice*. To me, intermediate practice is the keystone of relating theory and practice in education to each other'.

Wittmann next gave a brief description of the concept of *teaching unit* which he had treated in more detail in a recent publication (ESM, 1984). He proposed that didactical research and development, and also teacher education, should be related to groups of teaching units which allow for mathematical and psychological activities in the sense of intermediate practice and which are representative of the curriculum.

Finally, he described ways in which clinical interviews have been used in teacher education programs in Dortmund during the last decade. The themes are now taken from teaching units which are subjected to field-based exploration, and the interviews are performed in cooperation with kindergartens and primary schools. Themes, content and the form of interviews performed by student teachers were exemplified, and Wittmann concluded that, in his opinion, clinical interviews, under the conditions mentioned, constitute the easiest way of doing psychology, foster attitudes and skills of good teaching, and relate psychology to other domains of mathematics teacher education.

In the second presentation, Mary Grace Kantowski (USA) described a study investigating the role of instruction in the development of student ability to solve non-routine problems. The research was part of a teaching experiment in which teachers in service cooperated with a team of researchers. Two major aspects of the study were to promote teacher understanding of the educational problem field being investigated through their cooperative efforts, and to enable the findings of the study to become internalised by the teachers as useful tools.

The theoretical basis consisted of a theory of problem solving (information processing theory and Polya's four phases), and a theory of instruction for problem solving (concerning the relation between the teacher's role and functions and the student's personal activity). To become good problem solvers, students must first learn what constitutes the process of problem solving and how content and process interact in producing solutions to non-routine mathematical tasks. Furthermore, instruction must be planned so that students are provided with the tools or 'heuristics' for effective problem solving without having creativity impaired. Kantowski described how teachers changed their role during the project from 'model' via 'supporter' to 'facilitator', and how students took on a more active role in finding solutions as problem solving ability increased. Clinical interviews with the students were conducted by the team of researchers to provide data for the teachers.

The form, extent, and content of the cooperation depended heavily on the teacher's background. All teachers participated in observation and collection of data. Teachers with limited experience made no input into the selection and investigation of problems. Teachers with moderate experience were involved in the coding and interpretation of data and had provided some input to the planning for instruction. Experienced teachers were active in the selection of problems, in the planning of their own teaching, and in the gathering, coding, and interpretation of data.

Note was made of some general results of the study. The great variety of student solutions was remarkable; it was clear that there is a need at school for instruction in heuristics; there was a marked, general need for an 'incubation' period; and the experimental course had produced greatest gain in 'average' students. In working with teachers, there is a need to provide progressive types of teacher experiences as well as a general need for continuing experiences in research of this cooperative kind.

In opening the second session, Piet Verstappen (Netherlands) outlined the specific conditions for didactical research in the Netherlands where curriculum development has been centred in one institute while research and teacher education have been located elsewhere. He reported on the project 'Mathematics 12 — 16', based at the National Institute for Curriculum Development (SLO). Six members of the staff of SLO, three university researchers and twelve teachers and teacher trainers cooperate in the project which has been in operation since 1981. The major goal is the development of an overall curriculum design for mathematics teaching for 12 to 16 year olds. Verstappen characterised the perspectives and starting points in a series of brief statements: mathematics for all; relating mathematics to everyday life; presenting mathematics from contexts corresponding to the pupil's intuitive knowledge; relating mathematics to the individual's own approach; respecting the right of all for learning with insight. The theoretical objectives of the group of researchers were summarised as being to contribute to the development and improvement of learning materials, and to clarify the concept 'mathematics for all'.

Verstappen commented upon different types of research and characterised the research of the project as *education developmental research* (referring to the systematics of Terwel). This can be described as an informal, incremental, developing approach, with strong emphasis on the improvement of educational practice and the development of learning materials. Researchers have many roles (developing, training, supporting, investigating), and may interfere and influence events spontaneously. They play an important part in the dissemination of results and in forming the conviction of each group.

Developer's tasks are markedly creative, designing new learning materials and testing these for usefulness. Accordingly, they work in schools where they give lessons and participate in the teaching process. On the one hand the role is to analyse, to improve, to report findings concerning the arrangement of the mathematics lesson, and, on the other hand, to be an inspiration for teachers, teacher educators and authors of textbooks.

The problem field in consideration at the sessions was narrowed to the systematic cooperation between theory and practice. But, in addition to external relationships between theory, research and practice, there are internal relationships in research, particularly the relation between developer and researcher. Verstappen described the experience obtained in the project concerning this relation and the relations developer/teacher, and researcher/ teacher. For each relation the discussion focussed on differences in viewpoints and conceptions, in what is seen as problems, and in communication. Gaps had been experienced, such as those emanating from different opinions about theory, but Verstappen proposed that teachers, developers, and researchers understand each other in the light of practice, and that communication is promoted in relation to the child in the classroom. He emphasised the need for further exploration of the relations belonging to the triangle of teacher, developer, and researcher.

Leonie Burton (UK) then described a project which had investigated the feasibility of a problem-based approach to mathematics teaching of 9 to 13 year old pupils. Two related concerns formed the basis of the project. The mathematical concern was to address the relationship between content and process, and to attempt to improve the definition of mathematics as a problem-based discipline. The corresponding pedagogical concern was to find means whereby teachers and students could experience problem solving as part of their mathematics curriculum. For teachers, the overall aim was to identify a structure which would support changes in pedagogical role whereby insight into pupils' mathematical experiences could be gained, and to encourage the teacher as a researcher in the classroom.

For pupils, the aim was to build confidence, reinforce problem solving skills and procedures, and change perceptions of mathematics. Here the intention was to investigate whether problem solving processes already existed in young learners and if encouraging their use changed the learning context. The acquisition of processes was also to be tested.

The project team developed a bank of thirty problems for the pupils and supporting materials for the teachers. Teachers in five areas of England were invited to try out this pack of materials in collaboration with the research team, and twenty-seven classes became involved. Each teacher was invited to devote at least one hour per week for one term to problem solving, and to perceive their role in this work as a resource rather than as an informer. Close cooperation developed between the teachers and the team of researchers. In the term preceding the initiation of the project, a one day meeting was held in each area. The activities included group work on problems from the bank followed by discussions about the project materials, background philosophy, classroom organisation, and so on. The intention was that pupils would work in small groups and have free choice of problems from the bank. Teachers were encouraged to withdraw problems that they considered unsuitable for their class. At all times, responsibility for the class rested with the teacher (although team members regularly visited all classes), while responsibility for the work rested with the pupils (if practicable).

Burton described how results were recorded and evaluated. For the pupils, a substantial change from early dependency on the teacher towards a flexible and independent approach was apparent by the end of the term. As well, enjoyment of the work was evident. For the teachers, the change from accepting problem solving activity as a teaching experiment to seeing it as a necessary component of the teaching/learning process was observed. The study confirmed that such programs can be implemented by non-specialist teachers and that their feasibility is not affected by type of school, location, or ability range of pupils. However, considerable support had been provided in personnel and materials, and teachers should not be expected to undertake such radical shifts in style of work on their own without support, even when the materials are available.

Suggestions for future research included investigating a developmental approach to the recognition and use of problem solving processes, and exploring ways to simplify the theoretical background of the project.

The third session was opened by Gerd von Harten (FRG) who reported on the project LEDIS (Lehrplan und Differenziehrung in der Sekundarstufe). The overall aims of the project were to obtain general information about appropriate forms of didactical materials for teachers, and to develop teaching materials and assistance for teachers of stochastics at grade levels 5 and 6 of a comprehensive school. The project, which combines further education in the subject with classroom experience, seeks to attain its aims through cooperation with teachers. Earlier research projects in West Germany had come to more or less negative conclusions about the value of linking curriculum development and inservice education. Von Harten proposed that reforming the curriculum is of less concern for teachers than the daily problems of school and classroom. Overcoming the separation of the subject matter from the social aspects of teaching and learning is of central importance for the development of a more independent school mathematics. However, many projects have shown that interaction between researchers and teachers is very difficult. From this perspective, LEDIS was to be seen not as a solution, but as a possible new approach, attempting to integrate the mathematical subject matter, the organisation of the teaching, the theoretical and educational background, and the form of the teaching materials.

The major ideas and theses on which the project was based were presented. For example, the type of cooperation between researchers and teachers should depend on the content in question; the didactical analysis of the mathematical content must prepare for relating the educational/mathematical subject matter aspects with the interactive/pedagogical/organisational social aspects of teaching; the core materials for mathematics teachers should consist of tasks because teachers at this level plan lessons by means of tasks and not concepts; inservice education must link subject matter with classroom experience; projects such as these should not be evaluated in the classical way. It is much more important to establish groups of teachers working continuously with the problem field and that the project materials allow for continual adjustments according to the individual teacher's needs and experience.

The problem of how to relate the theoretical and innovative intentions of the project to the teacher's experience was a central one. A first version of the materials was developed by von Harten and Steinbring, and consisted of a didactical analysis describing the didactical intentions of the lesson and providing an overview of the concepts and definitions of stochastics relevant to grade 5/6. This system constitutes an offer of opportunities for the teacher. All the tasks contained ideas and requirements that could be altered easily by the teacher. For instance, they could be simplified, or made more difficult, or omitted altogether, thus providing different possible paths through the system. This first version was used by all teachers of mathematics at level 5/6 in five schools. The work of all the twenty-six teachers was carefully observed and studied in the project. The experience gained confirmed the usefulness of materials of this form, and it is now used by teachers at other comprehensive schools. A final version is not intended since the material is to remain a teacher's 'working book' to be continually altered and improved by teaching experience.

The final paper in the series was given by Guy Brousseau (France). His contribution was to exemplify how the two theoretical constructs of *the didactical contract* and *didactical situations* provide support for practice. A didactical situation is a set of relations established explicitly and/or implicitly between a pupil or group of pupils, some part of the environment (including instruments or materials), and the teacher, with the educational purpose of providing for the pupil's acquisition of some definite knowledge. A didactical contract is a set of rules which organise the relation within the classroom between the content taught, the pupils and the teacher, rules which are not normally stated explicitly. He chose counting as his example. This mathematical process appears at all levels of mathematics teaching from school through university, and seems to be involved in numerous difficulties for the individual. However, at no time is counting (l'enumeration) treated as an object for teaching in the sense that it becomes the core of intensive negotiations in the context of a complex didactical contract. Thus, two fundamental problems of school practice were to be considered, namely, the identification of didactical situations which can serve to generate appropriate activity and questions to be used as vehicles for the intended learning in the process of interaction and negotiation in the classroom, and communication to the teacher of such situations, together with appropriate knowledge and know-how about their potentials as well as of the didactical contracts involved.

The solution proposed in the project at the IREM, Bordeaux, consisted of presenting an identified fundamental didactical situation to the teacher on the computer screen. This provided access to each moment of the ongoing process of the child's activity and interaction with the objects displayed on the screen. Details were given of a computer program which provides a scene to motivate pre-school children to activity with high potential for learning related to counting, in this case involving actions with rabbits and ducks. The child using the light pencil was in continual interaction with the animals, and the teacher was able to observe and realise the role and function of the child in the developmental process. The teacher was able to reflect upon and participate in this dynamic situation. The program provided opportunity for the teacher to change the nature of the didactical contract in order to counteract the classical danger of destroying or deleting (by his/her very teaching actions) vital meaning from the child's understanding of the situation. Brousseau identified the didactical variables of this fundamental situation and pointed to the capacity of this type of didactical analysis for improving the contribution of didactical theory to practice.

As Chair of the fourth and final working session, Bent Christiansen (Denmark) emphasised that the theme of the series was to be interpreted in a broad sense, and the common field of interest for the panel included aspects of personal interaction between a didactician and a group of teachers. In this context he introduced two questions to the panel, the first being:

'What is the role of didactical theory, for the researcher in his/her interaction with the teacher, and for the teacher in his/her interaction with the students?'

In reply to this question the following points were made.

Wittmann asked if there was a gap between theory and practice, and pointed to the existence of an intermediate relationship between codified and subjective theory and proposed that it could be strengthened through intermediate practice.

Burton maintained that the teacher was, in fact, acting in the classroom from a theoretical position. There was a danger that it could become frozen, and therefore we should not try to transmit a specific theory to teachers, but rather aim at providing them with a *generative* theory by means of which the individual teacher could build theories from interactions with pupils.

Kantowski pointed out the difficulty of speaking in the abstract about the teacher and the researcher. In their first years, teachers pick up tasks produced by others, but as they grow and develop, theory becomes an increasing part of their professional life. The purpose of this cooperation, and the responsibility of the researcher, was to assist the teacher in a growing understanding of what mathematics teaching is about, to assist in the teacher's development from being a doer of tasks to being an understander of tasks and of their potentials.

Von Harten emphasised that practice should be seen not only as making lessons in the classroom, but also in preparing lessons, with different tools being needed for each of these functions. He found 'intermediate concepts' valuable, but pointed also to the problems met in a large competency-based teacher education project in the United States. In that project, intermediate concepts for use in all kinds of teacher functions increased to a number which was impossible to handle. The major problem is to identify intermediate concepts which are clearly related to the mathematical content, are practical without being trivial, and connect these with the social aspects of instruction.

Brousseau discussed the problem concerning the possibilities inherent in the various didactical contracts. In his opinion it is impossible for a participant in such cooperative projects to make decisions (in a given classroom situation) which are appropriate both for the ongoing research and the ongoing teaching. Clearly, the same person can do research and teach, but not at the same time. The researcher acts on the background of didactical theory and is responsible to that theory, whereas the teacher has responsibility and obligations concerning the learner's acquisition of mathematical meaning and must act accordingly.

Christiansen agreed that it would be important in all such interactive projects to distinguish clearly between expectations of the teacher and of the researcher. He then posed the second question:

'What is the potential of such 'interactive' projects for teacher education (pre-service and inservice) and in curriculum development?'

Kantowski recalled that teacher preparation often dealt with rather external things (how you speak and write, how you pass out papers, and so on). Her work had shown that when we have teachers deal with how students are learning and how they are responding to mathematics, the teachers learn much more about their role and functions and become more able to learn and benefit from theory. She suggested that a 'grey area' exists between research and curriculum development. We begin by searching for answers to some question, and for this we prepare materials and work with students. At some point as we continue to refine these materials, the project turns into curriculum development.

Burton saw the importance of research in the searching. The role of the researcher in these projects was to raise questions, to assist teachers in generating their own questions, and finding answers which were manageable in the classroom. She agreed with Kantowski that you could not distinguish the researcher from the curriculum developer.

Verstappen saw great potential in interactive projects with respect to teacher education and curriculum development. He saw the teacher as assisted by the developer in the teaching process and with materials, but in their cooperation in the classroom they are influenced by their separate personal goals and attitudes. So a third person, the researcher, is needed to evaluate what is going on. Perhaps someone else is needed to evaluate the researcher's evaluation.

Christiansen described courses in Copenhagen for groups of teachers working at the same school, meeting weekly to discuss and plan their teaching, and later to discuss their experiences. The course instructor interacts with the groups both as their consultant and as their teacher (setting tasks). This was a very promising model for in-service courses aimed at providing support from theory to practice.

Brousseau pointed out that it is impossible to explain for the teacher in his/her language, the available theoretical knowledge concerning a given classroom situation. Mediation and support must occur through other means. In this connection he described four different functions of the researcher during the interaction, and emphasised again that the same person in each given situation could perform only one function.

Nicholas Balacheff (France) was invited to contribute to the discussion. He said that clearly, the teacher is supported in his decision-making by a background of practical, professional conceptions. But to understand and support the teacher's activity, it is necessary to take into consideration the role of the didactical contracts as described by Brousseau. The researcher produces knowledge, but the teacher has to produce decisions and actions within his/her professional activity. Bourdieux has dealt in detail with these problems and differences in his 'theory of practice'.

Balacheff agreed with Wittmann that interviewing children should be part of the content of pre-service teacher education. However, it might sound as if an approach used by psychology and mathematics constituted the basic background for coping with the problems of teaching/learning mathematics. It is within the classroom that the teacher has to support the learner's construction of mathematical meaning and concepts.

TOPIC AREAS: THEORY OF MATHEMATICS EDUCATION (TME)

Organiser: Hans–Georg Steiner (FRG)

Sub-Title: A comprehensive approach to basic problems in the orientation, foundation, methodology and organisation of mathematics education as an interactive system comprising research, development and practice.

Background: Mathematics education is a field whose domains of reference and actions are characterised by an *extreme complexity*: the complex phenomenon 'mathematics' in its historical and actual development, and its interrelation with other sciences, areas of practice, technology and culture; the complex structure of teaching and schooling within our society; the highly differentiated factors in the learner's individual cognitive and social development, and so on. In this connection, the great variety of different groups of people involved in the total process plays an important role and represents another specific aspect of the given complexity. Within the whole system, several *sub-systems* have evolved. These do not always operate proficiently; in particular, they often lack mutual interconnection and cooperation. With respect to certain aspects and tasks, mathematics education itself as a discipline and a professional field is one of these sub-systems. On the other hand, it is also the only scientific field to be concerned with the total system. A *systems approach* with its self-referent tasks can be understood as an organising paradigm for mathematics education. It seems to be necessary in order to cope with complexity at large, but also because the *systems character* shows up in each particular problem in the field. Up to the present, mathematics education has often reacted to complexity either by non-scientific short-range pragmatism or heavy reduction of complexity favouring special aspects and research paradigms, frequently stemming from neighbouring disciplines. In this way, it comes close to 'normal science' with its strengths and weaknesses. However, it has also neglected many important tasks and problems concerning the orientation, foundation, methodology and organisation of the total enterprise. Some of these problems are:
- lack of unity, coherence, and stability;
- high susceptibility in basic orientations to short-lived issues and 'waves of fashion' moving between the extremes of polarised positions;
- one-sidedness and neglect of important problem domains and research areas;
- deficiencies in the relation between theory and practice;
- insufficient clarification of the epistemological status of mathematics education as a discipline and its relation to the various referential

disciplines, especially mathematics, but also pedagogy and psychology;
- uncertainty about the status and role of mathematics education within a university.

The program for this Congress component was developed in coordination with the Topic Area, 'Research and Teaching', and is closely related to Theme Group 4 on 'Theory, Research and Practice'. It was directly linked with a Post-Congress Miniconference on TME held in Adelaide on August 30–31, directed towards the establishment of working structures and proposals for future work related to TME. The papers prepared for the TME Congress sessions will be published by the Institute for the Didactics of Mathematics (IDM).

Report of Sessions:
In his introductory talk, Hans–Georg Steiner (FRG) concentrated on four major clusters of ideas:
- Systemsview — Interdisciplinarity — Transdisciplinarity;
- Towards normal science? — Micro vs. Macro Models — Metaparadigms;
- Place of Mathematics Education as a Discipline at the University;
- Strategies for a Developmental Program of TME.

With respect to *interdisciplinarity*, he argued that the straightforward borrowing of theory from other disciplines with 'homegrown theories' might not be an appropriate alternative. There are examples of very fruitful interdisciplinary approaches to problems in mathematics education which should be studied and analysed carefully both from a cognitive and social point of view in order to encourage more work of this kind. Steiner emphasised that the role of the didactics of mathematics in interdisciplinary research efforts is more than just at the level at which the various referential disciplines are or should be cooperating and interrelated. For problem identification and the coordination of interdisciplinary work, the didactics of mathematics also has an essential regulating and organising function located at another level which might be described as *transdisciplinarity*. Piaget has related 'transdisciplinarity' to a systems approach and this might be best understood in terms of C.W. Churchman's 'management philosophy as the true foundation of science'.

With respect to a *developmental program* for TME, a variety of activities on both national and international levels was suggested. It was considered important to hold a sequence of special conferences devoted to particular aspects of TME followed by related reports, as well as broad and open discussions at other international meetings such as the conferences of the International Study Group on the Psychology of Mathematics Education (PME).

In a report on some French research activities in the didactics of mathematics, Nicolas Balacheff (France), referred to an overall description of the field given by Brousseau in 1980: the didactics of mathematics is concerned with the pupils' acquisition of mathematical knowledge. The objects of its study are the situations and the processes which have been invoked with the intention of giving the pupils a sound knowledge of mathematics. It aims, therefore, at providing scientific insight which will allow us to reproduce these situations and processes, to conceive new ones, to foresee them and to control their development and results. However, both the definition and the reproduction of teaching-learning situations refer to a broader system which necessarily has to be taken into account and made a matter of investigation, since its actions and characteristics play a decisive role in the process by which knowledge is constituted and changed under didactical transpositions. The concept of *didactical transposition* is one of the key concepts being used and elaborated in many recent French research activities. The unavoidability of transpositions is evidenced by the difference which exists between the scholarly and didactical functioning of an explicitly labelled element of knowledge. Other such fundamental key concepts are 'didactical situation', 'didactical contract', and 'didactical — epistemological — ontogenetic obstacles'. The discussion of these concepts and ideas played an important role in several debates throughout the whole Congress component.

John Mason's (UK) way of viewing mathematics education had a very personal flavour. He used the image of a forest as a background metaphor. Mathematics education was seen as an ancient subject with recurrent themes that have to be researched and re-expressed by each generation, by each teacher. Just as the forest seems always the same, yet is always changing, so the essential questions are understood as being the same ones addressed by Plato, yet the context and the way of addressing the questions is in constant flux. The majority of his remarks addressed the establishment of a possible discipline for mathematics education. Such a discipline would provide both the justification and a self consistent and integrated methodology for the teacher (both as researcher and practitioner). The discipline can be thought of in terms of layers, each layer being concerned with the four headings of:
• methodology — learning and thinking about what is happening in a classroom;
• epistemology — criticising and modifying our knowing;
• discipline — ways of proceeding that allow for actions to take place;
• objections — critical statements that might be coped with by the discipline.

The theme of Mason's first layer was that of developing frameworks. As examples, such frameworks might enable one to express observations succinctly (methodology), find a mechanism for saying what I know (epistemology), and act as if a framework were valid (discipline). The theme of the second layer was that of invariance, but time did not permit a full discussion of this.

The discussion during Session 2 was initiated by two contributions. Rod Nason and Tom Cooper (Australia) gave a short introduction to their *information processing model of the learner*. Balacheff and Guy Brousseau (France) explained the fundamental role of the concept *didactical situations* as they use it in describing and analysing what is going on in the mathematics classroom.

The Cooper–Nason model of human information processing uses five distinguishable components: long term memory, semantic short term memory, external memory, motor interfaces, and sensory interfaces. Indications of some production system exemplars of the model in action were given and some implications for mathematics education drawn. Balacheff and Brousseau described didactical situations as a set of relations explicitly and/or implicitly established between a pupil or a group of pupils, the learning environment (including instruments or materials) and the teacher. These didactical situations are aimed at enhancing (that is, constructing) pupil's learning of some specific knowledge. They pointed out that several types of didactical situations have to be distinguished such as situations 'for actions', 'for communication leading to formulation', 'for validation', and 'for institutionalisation'.

The ensuing discussion was basically concerned with a comparison of the theoretical positions represented by the two contributions concentrating on the commonalities and the differences between 'information' and 'knowledge'. From the group, other models were brought into the debate. One briefly described his model of classroom interactions as being embedded in a system of interacting social institutions. In general, it was agreed that confronting and comparing different methods for the interpretation and analysis of phenomena and problems in mathematics education is a worthwhile task and one to be worked at more intensely in future activities of TME.

In his paper, Heinz Steinbring (FRG), gave an impressive example of how the systems character of mathematics education shows up in all problems in the field, and in particular in the understanding of mathematical content: 'If the subject matter treated in the classroom is seen more or less as a ready-made product provided by the mathematical discipline, it cannot function as a genuine systems element in the total system of mathematics education; it will remain an independent and closed component unrelated to the other elements of the system'. Therefore, Steinbring proposed that school mathematics not be confined to special techniques of pure mathematics as is still the case with many curricula in use, but rather extended to the development of the mathematical activity of knowing as a general social activity on the basis of a diversity of different aspects related to the mathematical concepts and working methods. A particular example was given with respect to the concept of probability which consists, on the one hand, of a mathematical structure (for example, an axiomatic system, implicitly defining equations, or rules of calculation) and refers, on the other hand, to the most diverse and specific interpretations regarding its contents. Its meaning cannot be the exclusive results of intramathematical consideration, its interpretation is rather an independent problem which is an additional task as compared to the purely mathematical components.

In the second paper for this session, Leslie Steffe (USA), began with the image of a mathematics education researcher as a personal union of a teacher, an observer and a theoretician, departing from Hawkin's statement that 'the teacher himself is potentially the best researcher'. While questioning whether every teacher should be a researcher, he gave four reasons why educational scientists should act as teachers: the first reason being the insufficiency of theoretical analysis; the second, the observation of and participation in what Hawkin called acquisition and transition (in children's intellectual development); the third, the opportunity to make original interpretations of children's behaviour; and the fourth the process of experimental abstraction in building experimental models. Special emphasis was laid by Steffe on the importance of *interdisciplinary work* in mathematics education research. This he described with respect to its conditions and possibilities in some detail on the basis of his personal very positive experiences. He indicated the most essential condition is that there be an implicitly understood and accepted hard core for any potential research program.

According to Brousseau, the *didactical contract* is a significant component of a didactical situation. With respect to a given task or phase within the teaching-learning process, it consists of an agreement between teachers' expectations and pupils' acceptances of the goals for their actions. Most frequently the contract is met tacitly and implicitly with some explicit elements in it. However, in the process of struggling with the problems and in their active acquisition of knowledge, the boundary conditions of the original contract for pupils may change, sometimes drastically. This may be because of fundamental difficulties the students have in coping with the problem, or because of errors they make, or because of various kinds of teacher intervention. Brousseau explained these phenomena by referring to some typical effects which he has identified in a metaphorical sense in order to develop a phenomenology of didactical contracts such as the *Topaze effect*, the *Jourdain effect*, and the *metacognitive shift*. The Topaze effect, historically exemplified in the first scene of a piece by Marcel Pagnol, is typical of a certain didactical situation in which some questions are explicitly put to the pupils. However, the teacher takes over the essential part of the work, thus preventing the pupil from actively acquiring the intended knowledge. By interfering, the teacher creates a new type of contract which necessarily has to be broken by the student in order to succeed with his own learning. Brousseau especially pointed out how these various effects were at work during the so-called new-math reform and how their insufficient understanding caused many of the failures attributed to this reform. Further case studies in this field and further elaboration of the fundamental concepts involved were considered as being most important for progress in research and practice in mathematics education.

Tom Cooney (USA) started out by saying that he could not imagine a more difficult topic to address than one involving the notion of theory on the one hand and teacher education on the other. As he viewed the situation, 'research on mathematics teacher education remains sparse. Our collective wisdom about teacher education may be considerable but it rarely finds its way to the pages of professional journals, particularly research journals. I conclude that our approach to teacher education is more akin to that of a practitioner than that of a researcher'. Cooney emphasised the conceptions teachers hold about mathematics and the teaching of mathematics, and the origins of these conceptions as a central aspect of the foundations of teacher education, as well as the present discussion. Results from his own case studies and evidence from other literature indicate that teachers make decisions about students and the curriculum in a rational way according to the conceptions they hold. Thus, Cooney came to the conclusion that we must come to a better understanding of the processes and components of teacher education. 'We need theory, good questions that can guide us in understanding what is achievable and what is not. ... An understanding of why teachers think and act as they do, an appreciation for their decisions and actions, insight into the origins of their beliefs are so very basic to the business of teacher education'.

Bent Christiansen (Denmark) gave a brief account of the history of the BACOMET project. It was established in 1979 when he, Geoffrey Howson and Michael Otte invited a group of mathematics educators to participate in co-operative efforts to identify components of the didactics of mathematics which — in the setting of teacher education — could be characterised as *fundamental, elementary and exemplary*. The project involves fifteen members from eight countries, and the work has been based on meetings of the full group each year since 1980 and upon related studies and research performed by the members at their home institutions. In the early years, high priority was given to an investigation of the relationships between the evolving process and the intended products. Christiansen described how a protocol for self-education (for educating the group by the group) then became established. A major product of the process will be a survey volume describing and analysing such basic components. The target audience will be the international group of teacher educators, but the organising principle will be the situation and needs of the teacher-to-be, although not in a direct, empirical sense. On the contrary, theoretical investigations of what is practice had to be carried out in order to ensure a deeper and more general value for the volume. Systematic attempts were made to use the major difficulties in the BACOMET working process as vehicles for constructive development. This principle was brought to bear on the following problem areas:

• to exploit the differences in knowledge and know-how within a group, and to proceed from the specificity of personal shared knowledge;
• to identify and develop knowledge which is appropriate for action;
• to exploit principles and hypotheses which are simultaneously the objects for analysis and development;
• to investigate the limitations and potentials of textual materials.

Knowledge about knowledge became a dominant perspective in the working process and in the development of the chapters of the BACOMET publication. Thus, it was considered important to distinguish between the two complementary categories of knowledge, specific information and overall awareness, as well as between different levels of knowledge. These distinctions are important for the teacher in his support of the student's construction of shared knowledge from personal knowledge, and of the student's access to objectified knowledge. Illustrations were given from the chapter on Task and Activity which, for Christiansen, supported the view that high priority is needed at all educational levels for the constitution of intended object for the student's activity.

Titles and Authors of Presentations made during the Congress Sessions

Session 1:
An introduction to the Theory of Mathematics Education (TME). *Hans–Georg Steiner* (FRG)
A view of didactics of mathematics which particularly reflects recent research activities in France. *Nicolas Balacheff* (France)
Towards one possible discipline for mathematics education. *John Mason* (UK)

Session 2:
Short presentations and discussion on the question: 'Is there a theoretical basis for mathematics education which links research, development and practice?'

Session 3:
Mathematical concepts in didactical situations as complex systems: the case of probability. *Heinz Steinbring* (FRG)

An educational scientist in mathematics education: observer, teacher and theoretician. *Leslie Steffe* (USA)
The role of didactical contract as a central component in analysing and reconstructing teaching-learning situations in the mathematics classroom. *Guy Brousseau* (France)

Session 4:
The contribution of theory to mathematics teacher education. *Thomas Cooney* (USA)

TOPIC AREA: TEACHING STATISTICS

Organiser: Lennart Räde (Sweden)

The aim of these four sessions was to discuss various aspects of why, when and how probability and statistics should be taught at the school level. The sessions were organised by the International Statistical Institute Task Force on International Conferences on Teaching Statistics whose members are: Lennart Räde (Sweden), Chairman, Peter Holmes (UK), Gottfried Noether (USA), John Oyelese (Nigeria).

The state of probability and statistics teaching at school level throughout the world has recently been surveyed in V. Barnett, *Teaching Statistics in Schools Throughout the World* (ISI, 1982). This survey shows that in most countries the extent to which probability and statistics should be introduced in the school curriculum is still under debate. In some countries, for example, England, New Zealand and Sweden, there are substantial courses in probability and statistics in some parts of the secondary school. With this report and discussions at international and national conferences as background, the following trends concerning the teaching of probability and statistics at the school level may be identified:

- more emphasis on statistics than on probability with special emphasis on descriptive statistics including the use of exploratory data analysis methods;
- emphasis on applications and model building;
- use of simulation both as a practical and a didactical tool;
  use of calculators and computers.
- use of project work.

These five trends were evident at the sessions on teaching statistics at the ICME 5 Conference. Reports from the organisers of each of the four sessions follow.

Session 1: Improving Statistical Literacy in the Schools
Convenors: Ann Watkins (USA), James Landwehr (USA), Jim Swift (Canada)

This session reported on efforts of a joint committee from the American Statistical Association and the National Council of Teachers of Mathematics to improve the teaching of statistics in the K–12 curriculum. The project involves developing new materials and training teachers in their use.

Progress towards a greater emphasis on statistics is being influenced by four major factors. Several recent national reports on education in the United

States have given strong support towards more emphasis on statistics. Graduation requirements in mathematics are increasing, and the study of statistics in an extra mathematics course is a reasonable option for many students. Microcomputers are becoming available in the schools, and the study of statistics and probability is enhanced by using computers as well as representing an appropriate application of them. Finally, the quality and productivity concerns of North American industry may cause schools to give greater emphasis to useful quantitative methods and ways of thinking, such as statistics.

Two current projects were described. First, a computer bulletin board and data library, to be known as the Nightingale Network, will be created. It will encourage and facilitate the exchange of interesting data between those active in teaching statistics. This project is the result of needs identified by 50 outstanding high school teachers interested in statistics and probability at a recent Woodrow Wilson Institute at Princeton University.

The major part of the session was taken up with describing the current activities of the second project, the Quantitative Literacy Project, which is administered by the American Statistical Association. Four flexible units have been drafted and are being field-tested in many types of classes in junior and senior high school. These units are titled: I — Exploring Data; II — Probability; III — Simulation; and IV — Information From Sample Surveys. The first and fourth units were described in some detail in the session.

The position taken by the unit on Exploring Data is that students can learn useful ways to graph, summarise and interpret data without first studying probability. The unit uses real data that is interesting and important to the student. Interpretation and learning how to think about data are stressed, not just the route application of some methods. The statistical techniques that are used include: line plot, stem-and-leaf plot, median, mean, quartiles and range, box plot, scatter plot, fitting straight lines to scatter plot, and smoothing scatter plot over time.

Unit IV, Information from Sample Surveys, is an introduction to confidence intervals. Students use simulation to construct sampling distributions for samples of a fixed size drawn from various binomial distributions. They learn which samples are 'likely' and which are 'unlikely'. Finally, a confidence interval for the percentage of successes in an unknown population from which a sample has been drawn is defined as the set of all binomial probabilities for which that sample is likely. Data from real surveys is used throughout.

Following the presentation, the audience raised questions and made comments about training teachers to use the booklets, including traditional topics such as histograms and pie charts, and about the level of rigour appropriate for secondary school students.

Session 2: Statistics in Non-Mathematics Courses for Students Aged 16–19
Convenor: Peter Holmes (UK)

This presentation was based on experience over the last three years with the Statistical Education Project 16–19, in Sheffield (UK). In this age group in England, there are three main school subject areas that include a substantial statistical component. These are geography, economics and the biological sciences (including psychology). Statistics has been included in these courses not as a result of pressure from statisticians but because the subjects themselves are becoming more numerical. This is a consequence of the desire to quantify concepts and measure effects which were previously only discussed qualitatively. It is this desire to use real data that motivates the student and makes the study of statistics important to the subject area.

The problem of teaching statistics to students in these courses is no small one. In England, there are approximately 90 000 students using statistics in such courses compared with about 10 000 who are doing a mathematics course which includes some statistics. As not all the 10 000 are part of the 90 000, there are many students who are having to learn statistics with no current mathematical background. The extent of the problem is emphasised by considering the syllabuses and questions asked. The type of questions asked and the level of understanding required are by no means trivial.

Even a cursory glance at the syllabuses shows that there are very different requirements in different subjects. It would not be possible to provide a single service course for all the different student requirements. Further problems are posed by the fact that much of the material is being taught by teachers who are not statistically qualified. The teacher with a traditional statistics background has much to offer but will not be completely competent since s/he will not usually know the subject area in which it is being applied.

The types of questions asked can be classified in some detail. The easiest type are those which may be classified as using data as a *wallpaper background*. This means that, although the statistics are given, they are not really used. They become a background against which the student makes his comments.

Students begin getting into statistics proper when they come to questions that require them to read tables of data. This is by no means as trivial as it may sound. It requires consideration of the source, accuracy of the data, consideration of frequencies and percentages, and searching for overall patterns.

Next there are cases where the student is required to interpret and draw graphs. These can vary greatly in level of difficulty. In general, it is harder to draw graphs than it is to interpret them, but even interpretation can be difficult. Interpretation can involve complex analysis of conditions that would lead to particular time series graphs, and even involve elementary ideas of inter-correlation between two time series (though such correlation is not calculated). Interactions may have to be considered. Some examples combine interpretation of tables and graphs together.

The next stage is those questions which require students to obtain information and consider experimental design. This includes the design of questionnaires and consideration of the connection between scientific design and statistical design. Elementary ideas of probability occur as in work on genetics. This can be followed by questions requiring estimation, hypothesis testing and inference, including the significance of Spearman's rank correlation coefficient.

From the above analysis of syllabuses and questions, a number of basic principles emerge for the design of relevant courses.

- The courses should be oriented more towards data than probability.
- Courses should include a good background in reading data and tables.
- Reading statistics is more important than doing statistics for these students.

Although doing statistics can be an aid to reading, it should not become an end in itself.

- The statistics arise in real contexts and are to be used, hence we should use real data and get the students involved in practical work.
- The usual introductory course in mathematical statistics is not appropriate. A different level of understanding is required.
- Teachers need a background in both statistics and in the subject area. This may be overcome by having joint responsibility for the course.

The STEP project has been working in this area for the last three years and is about to publish its material.

Session 3: Applications, Case Studies and Model Building in Teaching Statistics

Convenors: David Vere-Jones with assistance of Russell Dear, Lawrence Langston and R. Comwell (New Zealand)

Introducing the session, David Vere-Jones explained that the project had its genesis in the booklet, 'Statistics at Work', published by the New Zealand Statistical Association. The booklet consists of eleven studies, based on the work of local statisticians, and linked to topics in the Statistics section of the Applied Mathematics course for the final year of secondary school. It represents an attempt to move teaching away from the formal, mathematical aspects of the statistics course towards an understanding of its real-life applications. However, while many schools had purchased copies, he found that few teachers made use of it in the classroom. In asking why this should be so, it was suggested that the length of time needed to develop a case study could be a contributory factor along with mathematics teachers' unfamiliarity with this type of approach. He also briefly outlined the statistics program in New Zealand schools, pointing out that more than 50% of all students entering for the final examination took the Applied Mathematics course with the statistics section.

Dear invited the audience to take part in a simple simulation — estimating the number of tagged fish in a pond. His demonstration showed that such simulations could easily be undertaken within a 40 minute class period, and need not cause undue classroom disruption. He stressed the importance of finding forms of practical work that could be accommodated within these constraints. An alternative was project work which could be done outside the regular class periods but still form part of the assessment. Examples of such project work from a sixth form class (ages 16–17) were displayed for the group.

Langston expressed his belief that practical work in statistics should not start in the final years, but should be developed right through the secondary school program. It takes time for students to appreciate the nature and purpose of statistical work, particularly the inevitably tentative character of its conclusions. In real life, most statistical work was used to support a case or recommend a choice of action. He suggested that, in teaching, it was important to start with a problem, and then show how statistical tools could be used to resolve the problem. Examples of such problems that had been used with groups in the lower secondary grades were cited. In one study the class was asked to consider the question, 'Which are better at mathematics, boys or girls?' In another, they were asked to make a case for postponing the start of the school day after studying delays in train arrivals. In another, students were asked to recommend a policy on the number and location of overhead projectors needed for classroom use. All of these situations were studied by using descriptive statistics and with the aid of simple simulations.

Comwell described his efforts as a school inspector to encourage the use of practical work in statistics. He organised an inservice course during which the participants compiled a set of classroom experiments, which were published and distributed to schools in his region. Subsequently, he found that the teachers using them were mainly drawn from those who had participated in the course. He pointed out that recommendations to include more practical work in the New Zealand courses dated back to G.H. Jowett's pioneering efforts in the late 1960s, and suggested that significant changes in classroom practice would not occur until the practical work was explicitly assessed or supported by suitable questions in the final examinations.

In the discussion which followed, contributors described other ways of using case study material. One suggestion was to ask students to read through one study in books such as the 'Statistics by Example' series by Mosteller et al, and to prepare a report on the study. Examination questions could ask students to describe some real-life applications of particular statistical topics. Case studies need not be of great length, and could include a wide variety of model- building and practical aspects of statistics.

Session: Computer Animations: A powerful way of teaching statistical concepts
Convenor: Daniel Lunn (UK)

The session consisted of both discussion and demonstration of the use of modern technological aids in the teaching of basic concepts. Particular emphasis was given to model building, simulation, and applications, with the main focus of attention being the development of the right kind of intuition and understanding in the mind of the student.

Two kinds of animation were outlined and demonstrated, one kind being produced by a main frame computer and recorded, one frame at a time, onto videotape, the other kind being executed in real time on a microcomputer. The demonstration showed that the nature of the animations dictates the choice of method but, whichever way is adopted, the designer of such animations must first of all concentrate on interpreting the basic concept pictorially; particular thought must be given to the way in which the pictures move in real time.

For animations involving complex movements which must be executed smoothly and with careful timing, the method of production is to record directly from a main frame computer onto videotape. Each separate frame is drawn directly onto the tape by the computer to be viewed subsequently at 25 frames per second. Design of such animations was discussed and examples were shown in which probability distributions were explored.

Animations for which the timing is not so critical, smooth movement is not essential, and complex pictures (for example, three dimensional pictures drawn in perspective) are not required, can be produced adequately on a microcomputer with good graphics capability.

Once again, it is important to design the right kind of visual image. This problem was discussed using two illustrative examples, one involving computer simulation of a battle, and the other involving the teaching of the concept of confidence. In both examples, the design of the animations was discussed and then the finished products were shown.

In conclusion, it was pointed out that the animations shown were for demonstration purposes in the classroom. Such demonstrations require immediate follow-up work by the students and should be designed with this in mind. Furthermore, it is desirable to make the microcomputer animations student interactive wherever possible.

All of the animations shown are used in the television programs associated with a new Open University course on basic probability and statistics. The course, M245: Probability and Statistics, has sixteen such programs which are transmitted on main network television in the United Kingdom. Details are given in the appendix.

Appendix

M245: Probability and Statistics.
Course Team Chairman: Daniel Lunn
BBC Series Producer: David Saunders
Programs:
      Chance
      Probability
      A Probability Model for Rare Events
      The Genetics Game
      The Poisson Process
      Confidence
      Testing for Telepathy
      Conflict
      The Normal Distribution
      Regression
      Sampling
      The Central Limit Theorem
      Estimation
      Goodness of Fit
      Finding one's Bearings
      Decision Tree Analysis

TOPIC AREA: WOMEN AND MATHEMATICS

Organisers: Joanne Rossi Becker (USA) and Mary Barnes (Australia)

The four sessions on Women and Mathematics provided an opportunity to review the status of women and mathematics from an international perspective, particularly the current situation of girls in schools, with respect to participation and performance in mathematics classes. The Second International Mathematics Study enabled several contributors to compare experiences and results in their country with others also involved in the study. In general, SIMS provided an opportunity to examine the state-of-the-art on an extensive international basis.

The opportunity to share experiences with those involved in successful intervention programs and to again consider the factors that influence the still relatively low participation rates of girls and women in studying mathmatics enabled a wide discussion of the purposes and perceptions of mathematics. A brief summary of these discussions is reported here.

Women in Mathematics: An International Update

Discussions for the first two sessions focussed on reports on the SIMS data for six countries together with an overview of the isssues addressed in considering sex differences in performance and participation in the international study. The following short papers formed the basis for the discussion.

A conceptual framework for analysing sex differences on an international basis: *Erika Kuendiger* (Canada)

Gender and mathematics achievement: An examination of the Canada (Ontario) data from the Second International Mathematics Study: *Gila Hanna* (Canada)

IEA Second International Mathematics Study: England and Wales: *Lesley Atkin* (UK)

Sex differences and SIMS data from Scotland: *Pat Hiddleston* (Scotland)

Student gender, home support and achievement in mathematics: *Magdalena Mok* (Hong Kong)

Some gender differences from the IEA Survey in New Zealand: *Helen Wily* (New Zealand)

Sex differences in participation and performance in mathematics in Australia: *Gilah Leder* (Australia)

An international survey in 1982 showed that there are major differences between countries with respect to the availability of information about sex-related differences in achievement and participation in mathematics, and the amount and quality of research carried out to find explanations for differences. The Second IEA Mathematics Study (SIMS), which was conducted in more than 20 countries, will provide more information than previously available about gender differences. Moreover, for those countries which participated in the first and second studies, comparisons of achievement differences over 20 years will be possible.

Kuendiger, who is analysing the total SIMS data set for gender differences, provided a model for considering gender differences in achievement and course-taking in mathematics.

The figure below illustrates that model, which follows from a general belief that differences are caused by sex-role perceptions which influence achievement, expectations, and attributions of success and failure in mathematics. The arrows in the figure illustrate major directions of influence, starting with the two categories of general beliefs that are held in a community and personal learning experiences.

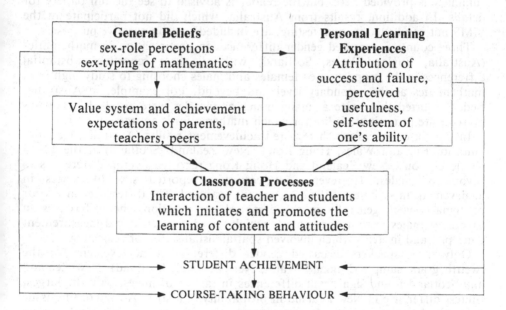

This model informs the analysis of the SIMS in which Kuendiger plans to investigate the following major questions:

- How much do countries differ in respect to sex-role perceptions and sex-typing of mathematics? Are these differences related to achievement differences and course-taking behaviour?
- Which of the motivational variables discriminate the best between males and females in a country? What is the relationship between these variables and achievement?
- Do male and female teachers in a country differ in their attribution of unsatisfactory progress in their classes? Do they have the same ideas about how to make their teaching more effective?
- What does the learning environment in single sex schools in a country look like? What impact do single sex schools have?

Although no data from the overall analysis were available, five countries were represented by women who had examined the data for their individual countries.

The SIMS tested samples of students in the equivalent of eighth and twelfth grades, with the latter samples including only students enrolled in college preparatory mathematics. At the eighth grade level, the major categories of the achievement test were arithmetic, algebra, geometry, probability and statistics, and measurement. At the twelfth grade level the major categories were algebra, probability and statistics, calculus, geometry, trigonometry and linear algebra. In addition, questionnaires were given to students and teachers to determine course-taking behaviour, sex-typing of mathematics, self- esteem, usefulness of mathematics to society, view of mathematics and perceived home support for students; and sex-typing of mathematics, view of mathematics, and teaching environment, for teachers.

Of the five reports on SIMS results, two focussed on achievement, one primarily on attitudes, and two on both achievement and attitudes. A summary is provided here, but the reader is advised to see the full papers for details. In addition, results from Australia, which did not participate in the SIMS but conducted its own testing, are included for comparative purposes.

Three countries reported gender differences in participation in mathematics (Australia, England/Wales, Scotland) with all three reporting substantial differences in the proportion of females and males choosing to study high level mathematics at the secondary level and beyond. For example, in Australia, Leder reported that twice as many males as females were enrolled in courses prerequisite to tertiary studies for which mathematics is an integral part.

In the six countries which reported achievement results (Australia, Canada: Ontario, England/Wales, Hong Kong, New Zealand, Scotland) at the 13–14 age levels, only New Zealand and Hong Kong reported overall differences in favour of males. However, four countries reported sex differences in achievement in subtests of the achievement test used. The differences in favour of females were generally in the areas of computation; the differences in favour of males were in the subtests which tested geometric and measurement concepts, and in items which involved spatial visualisation or reasoning.

Only three speakers discussed gender differences in achievement for the twelfth grade sample (England/Wales, New Zealand, Scotland). New Zealand and Scotland found significant differences in favour of males, with the largest subtest difference in New Zealand in probability/statistics. An interesting issue was raised concerning single sex versus coeducational schooling. Australia reported that girls attending single sex schools or single sex mathematics classes perform better than those in coeducational settings. However, this topic needs more study to determine if single sex classes per se contribute to higher performance.

More substantial gender differences were reported in attitudinal variables investigated in SIMS. These results generally support other research in the USA and Australia, but a dearth of research from non- Western countries makes if difficult to generalise. At both the eighth and twelfth grade levels, males considered mathematics more important and useful to them. Confidence was another variable which showed gender differences in SIMS and other research, with girls less confident of their abilities. Some intriguing results were reported by Mok for Hong Kong. The Chinese culture historically exerts a great deal of pressure on children to do well academically. Mok found that seventh grade girls and boys were receiving the same amount of parental pressure to do well in mathematics. However, high parental pressure was associated with decreased performance for both sexes, with the adverse effect more pronounced for boys. Thus, the parental pressure served to bring achievement differences closer together, and if pressure were decreased, an increase in sex-related differences in achievement favouring males could be expected.

This sampling of results from SIMS and comparable ones from Australia indicate some interesting patterns in gender differences. We must wait for the planned volume by Kuendiger (expected in 1985) to determine if these patterns occur in other participating countries.

Bringing About Change In Women's Participation In Mathematics.

Four reports on intervention programs to increase the participation of girls and women in mathematics provided the basis for discussion for the third session. The presentations provided an opportunity to briefly review research in the field as well as provide first hand reports of intervention models.
USA approaches to increasing the participation of women in mathematics.
     Joanne Rossi Becker (USA)
Beyond EQUALS — college math courses for adult women. *Ruth Afflack* (USA)
Women learning mathematics with computers. *Diane Resek* (USA)
Intervention programs in Australia — progress and problems. *Mary Barnes* (Australia)

Research in recent years on gender and mathematics has suggested that social and experiential factors play a major role in determining participation and achievement in mathematics. Variables which have been identified as major contributors to the observed sex differences include attitudes towards mathematics (confidence, anxiety), perception of the usefulness of mathematics, career awareness, stereotyping of mathematics as a male domain, achievement in mathematics, spatial visualisation, encouragement to study mathematics, and the presence of role models. A variety of programs aimed at increasing the involvement of women and girls in mathematics has flowed from this research, each focussing on one or more of the variables mentioned above. Development of such intervention activities began first in the United States. As a result, programs developed there have served as models for other countries such as Australia.

Becker presented an overview of USA approaches. Of those aimed directly at students, several have the primary aims of changing girls' career awareness, increasing their perceptions of the relevance of mathematics to their future lives and providing them with role models. These programs include conferences for schoolgirls, which provide 'hands-on' activities in areas of mathematics or science as well as opportunities to learn about careers in those areas, and a lectureship program which organises visits to schools by women working in mathematics-related fields.

Another group of programs focusses on the development of problem solving skills, by means of strategy games, puzzles and spatial activities. These stimulate interest and at the same time build confidence by providing students with experiences of success in mathematics. Programs of this type have been organised for students at all levels, from elementary school pupils to adult women reentering education. a friendly atmosphere and an emphasis on cooperation are an important part of all these activities.

Other intervention approaches have taken the form of inservice or preservice courses for teachers. These typically involve activities to raise teachers' awareness of the causes and consequences of mathematics avoidance, and to develop strategies to improve student attitudes towards, and understanding of, mathematics.

The greatest number of intervention programs in the United States has been directed towards college level students. Many of these involve psychological counselling to help students overcome anxiety about the subject. Others concentrate mainly on teaching mathematics.

Afflack described two courses designed to combat the avoidance of mathematics by adult women. 'Mathophobia' is a 15 hour intensive weekend workshop, designed to start women thinking mathematically and to encourage them to enrol in further mathematics courses. 'Math without Fear' is a one semester course offered to university students who have a weak background in mathematics, but who want to prepare to take further courses in algebra or statistics. Both have the primary aim of building the students' confidence in their ability to do mathematics. They provide a supportive, non-threatening environment and attempt to give women positive experiences of success in mathematics. Emphasis is on the use of concrete materials to promote an understanding of mathematical concepts, and on developing problem solving techniques, logical reasoning and spatial abilities. Risk-taking is constantly encouraged. Students learn that forming hypotheses and then testing their correctness is an important part of the mathematical process. For this reason guessing is encouraged, followed of course by checking the correctness of the guess, and then improving it. This process was illustrated with a novel approach to function machines and word problems.

Resek showed us how she uses computers to teach mathematics to women. She explained that mathematics education for women and girls needs to build on the interests and intuitions which most women possess, and at the same time develop other intuitions and processes which many women lack, but which are central to mathematics learning. She believes that, while computers are not essential for good mathematics teaching, they open up new opportunities for curriculum design and can be used to promote the learning of mathematics by women and girls. One example makes use of young women's natural curiosity about other people to stimulate interest in data analysis. Using a data base and data analysis package, students are encouraged to develop hypotheses and then use the computer to check them. In another example, students work in groups to play a strategy game and are then asked to 'teach' the computer their strategy. The computer can then play the game over and over, so that the success (or otherwise) of the strategy can be seen. This activity also encourages students to formulate hypotheses and then check and improve them. It stresses the importance of making precise mathematical statements so that the computer will interpret them correctly. Moreover, it builds on girls' interest in games and their preference for working cooperatively in groups. The only competition involved is against their own previous record. The final example showed how computer graphics could be used to develop students' intuitions about statistical concepts, again through a game played in groups.

Intervention programs in Australia have developed only during the last four or five years. Barnes reported that at first most of these drew on ideas and experience from the United States, and many had as their main focus changing girls' perceptions of career opportunities and emphasising the importance of mathematics in keeping open a wide range of career options. Many people are realising today, however, the need to complement such activities with others which change both classroom atmosphere and interaction patterns and the way in which mathematics is taught. One recent approach has been to encourage teachers to monitor and analyse classroom interaction patterns and modify unequal treatment of males and females. Another approach which is being tried in several places is the formation of girls-only classes within coeducational schools. Wherever this has been tried, girls have shown they appreciated the extra attention they received from their teacher, the absence of harassment or distraction from boys and the more peaceful, cooperative atmosphere in the classroom. Attitudes to mathematics improved in these classes.

Educators in Australia today recognise the need to modify the curriculum to meet the needs of girls by paying special attention to areas in which girls have been found to be weak, such as applications, measurement and geometry. They are encouraged to adopt teaching methods which suit girls' preferences such as the encouragement of cooperation and discussion, the use of themes which relate to their interests, and the provision of more opportunities to use concrete materials and to develop problem solving skills. To be effective, and to reach those who need them most, intervention programs need to be school-based, to involve the whole of the school community, teachers, students and parents, and to be an on-going process, not just a single event.

Women, Culture and Mathematics

The fourth session enabled the group to consider once again the influences that impinge on attitudes to and confidence in learning mathematics. The influence of cultural bias and perceptions of the nature of the subject were highlighted, together with some consideration of the place and role of mathematics in our culture, both in the past and the possible future. The three presentations were:

The relationship of confidence and other attitudes to mathematics study at the college level. *Roberta Mura* (Canada)

Attitudinal differences in perceived usefulness of mathematics. *Carole Lacampagne* (USA)

Women, culture and mathematics. *Nancy Shelley* (Australia)

In introducing this topic, Mura suggested that the under- representation of women in mathematics raises three types of questions:

- Why are there so few women in mathematics, and what can be done to increase their number?
- What is the 'mathematical experience' of those women who have chosen this field of study?
- Which characteristics of mathematics are the consequence of the fact that it has been built predominantly by (western, upper-class) males? What would a mathematics built by women look like?

Most research so far has concentrated on the first of these questions. Mura reported on two studies belonging to the second category which investigated sex differences in attitudes toward mathematics among university students who had chosen to specialise in mathematics. In a large scale study, students were asked to predict their final marks in their mathematics course. Both men and women tended to overestimate, but men more so than women. Even when level of confidence was controlled for, women expressed fewer intentions than men to enter doctoral programs. Those women who did intend to attempt a doctorate had lower expectation of success than men. Mura believes these results can be explained by cultural differences between men and women in attitudes to success and to risk-taking.

In a second study, students were interviewed about their relationship to mathematics and its environment. Many more women than men said they were attracted to mathematics because of its logical rigour, whereas men repeatedly mentioned the appeal of challenge. These attitudes afford a cultural interpretation: women may seek in the logic and rigour of mathematics an objectivity and fairness that is often denied to them elsewhere in society. Men, on the other hand, are culturally conditioned to link their personal value to their external achievements, and mathematical problems offer endless opportunities to reassure themselves of their value.

Mura emphasised that women's present minority status in mathematics is not a reason for regarding their attitudes as 'deviant' and men's attitudes as 'normal'. In case of conflict between women's attitudes and the mathematical environment, both are candidates for change. While both kinds of change might be necessary, she believes the latter solution is likely to be more beneficial for the development of mathematics itself.

Lacampagne discussed some of the prevailing myths about women's role in society which affect their learning of mathematics, and contrasted these with the realities of the situation. These myths are long enduring and hard to change. The prevailing myth is that woman's principal role in life is as nurturer and helpmate, whereas in reality women work outside the home for the majority of their working lives. But as a result of the myth, women do not prepare themselves for careers, especially in technical areas. Instead they cluster in jobs which fulfil the nurturing image, such as jobs in service areas and in professions such as teaching, nursing and social work.

Sex-related differences in teaching and learning mathematics and in attitudes towards mathematics cross national boundaries. Too few women are entering mathematics-related careers, and this is a loss not only to women but also to society. What can we do? We need to change women's attitudes to mathematics and convince them that it is useful. But we need also to consider what mathematics we should teach, at what level and pace, and how we should teach it. Girls may be discontinuing mathematics because they find it dull and rule-oriented, because too many isolated bits of mathematics are crammed down their throats at too fast a pace and with too little understanding. We do not at the moment teach mathematics as we do mathematics. Only when we have brought the myths and realities of our culture and the myths and realities of our mathematical subculture into perspective can we expect to find females and males equally involved in mathematics.

Shelley explained her view of the interlocking nature of women, culture and mathematics by selecting each of the three in turn and looking at the other two through the perspective of that one. In her words:

the total picture reveals that most women live in poverty, are oppressed, and see their children dying of malnutrition and disease. They are victims of violence and rape, work 2/3 of the world's working hours, earn 1/10 of the world's income and own 1/100 of the world's property. It also reveals that world-wide the dominant culture is that of militarism, with an arms race which threatens the annihilation of humankind and of the planet; with development of 'intelligence' networks for surveillance and control of ordinary citizens; with despoiling and pollution of the environment to the point of impending total breakdown of the eco-system. This culture of militarism is exercised, for the most part, by white men. It reveals that mathematics has high status in the world today, that qualifications in mathematics confer access to power, and that the greatest area of mathematical work is in problem-solving for the military and in laying the foundations for military science.

In the past, women have been excluded from mathematics and from power by such arguments as 'their brains are too small'... Today women, and people from other cultural backgrounds, are excluded by the bogus construct of 'ability'... Increasingly, it is acknowledged that the historically determined values of the white male warrior have been built into science and technology. It will be argued that this is also true of mathematics. What needs to be acknowledged is that granting primacy to logic has affected both mathematics and the structure of our thinking... It has affected our concept of truth, our notions of intelligence, many of our attitudes in society (particularly those relating to power), and our obsession with objectivity and prediction... We have failed to recognise the coercive power of logic... In supposing mathematics to be both culture-free and value-free, we have failed to acknowledge that it is culturally arrogant...

Other systems of mathematics are possible, which... offer the possibility of a different world. Women have the opportunity to develop a mathematics which... can provide the world with another way so that the planet lives and people the world over are fed and healthy.

Mura began by asking 'What might a mathematics built by women be like?' Shelley here proffers her answer to this question. Her view of the role of logic in mathematics contrasts with that of the women students in Mura's study, for whom the logical rigour of mathematics was its principal attraction. Further discussion of these issues seems warranted.

KANGAROO

PRESIDENTIAL ADDRESS:
MESURES ET DIMENSIONS

Jean–Pierre Kahane
Universite de Paris-Sud
France

Je commencerai par une petite histoire que je tiens de mon père et qui s'est passée, il y a plus de cinquante ans, à Paris, à la Faculté de Pharmacie. Au début de ce siècle, l'enseignement primaire en France était très ambitieux; on y enseignait les formules donnant les surfaces et les volumes des corps usuels, et tous les enfants étaient censés connaître les formules $S = 4\pi R^2$ et $V = \frac{4}{3}\pi R^3$ pour la surface et le volume d'une sphère (on ne disait pas encore 'une boule'). Dire qu'ils comprenaient ces formules, c'est autre chose. Ici commence l'histoire. En travaux pratiques de pharmacie, des étudiants ont à fabriquer des boulettes d'un certain produit. Mon père, comme moniteur, s'adresse à un groupe qui a modelé une boule trop grosse. 'Fabriquez donc des boules plus petites, disons, de diamètre deux fois plus petit.' Puis il y a l'idée d'une petite question: 'combien de petites boules pourrez-vous obtenir à partir de la grosse si vous réduisez le diamètre de moitié?' (figure 1). Hésitation des étudiants, réflexion, évaluation: 'Hum, peut-être quatre ou cinq!'. De la formule $V = \frac{4}{3}\pi R^3$ ils avaient peut-être retenu le coefficient $\frac{4}{3}\pi$, mais pas l'essentiel, à savoir qu'on multiplie V par 8 quand on multiplie R par 2.

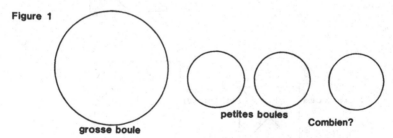

Figure 1

grosse boule   petites boules   Combien?

Comment varient les diverses mesures (les longueurs, les arcs, les volumes) quand on dilate un objet? Quelle dilatation faut-il opérer, par exemple, pour doubler la surface ou doubler le volume? Ces questions sont parmi les plus anciennes que nous rapporte la littérature. Je tenterai de montrer qu'elles ont des prolongements actuels intéressants, et qu'elles méritent de passere dans la conscience commune.

La plus ancienne leçon de mathématiques que nous connaissons est une illustration de la maïeutique, l'art de faire accoucher l'élève des connaissances qui sont en lui. Socrate s'adresse à l'esclave de Ménon, il lui pose une série de questions, et il lui fait découvrir comment doubler un carré. C'est Platon qui relate l'histoire, et voici un résumé du dialogue, sous forme de figure (figure

2). L'admirable dans Platon est qu'il n'y a pas de figure, mais qu'on peut les reconstituer sans difficulté à partir du texte. Je ne suis pas du tout convaincu que l'esclave de Ménon ait redécouvert quelque chose qui se trouvait déjà en lui, mais à coup sûr Socrate lui a enseigné une belle propriété de la diagonale du carré. Dans d'autres dialogues, il apparaît que Platon est fasciné par les propriétés arithmétiques de $\sqrt{n}$, dont le premier exemple est l'incommensurabilité de la diagonale et du coté, et aussi par le problème du doublement du cube. Dans sa théorie des polyèdres réguliers, les faces sont triangulées, et Platon introduit explicitement des échelles de dilatation permettant de doubler la surface du cube ou de tripler celles du tétraèdre, de l'octaèdre et de l'icosaèdre (figure 3): ainsi, à chaque élément — feu, air, eau, terre — correspond une forme — tétraèdre, octaèdre, icosaèdre, cube — mais plusieurs tailles. L'ensemble des cubes de Platon est invariant, semble-t-il, par le groupe des dilatations de rapports $2^{n/2}$, et l'ensemble des autres polyèdres par le groupe des dilatations de rapports $3^{n/2}$. C'est ainsi que, dans la chimie platonicienne, la formule feu + eau = air s'exprime comme la recomposition des petits triangles constituant les faces d'un tétraèdre et d'un icosaèdre sous la forme d'un octaèdre (figure 4).

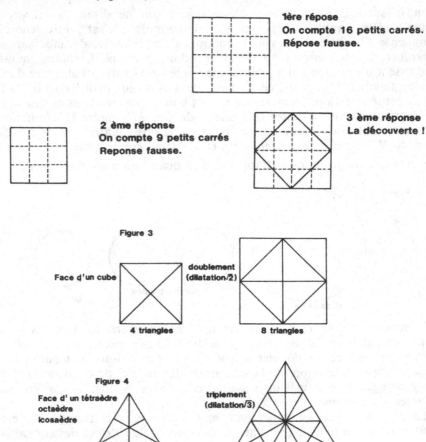

1ère répose
On compte 16 petits carrés.
Répose fausse.

2 ème réponse
On compte 9 petits carrés
Reponse fausse.

3 ème réponse
La découverte !

Figure 3

Face d'un cube

doublement
(dilatation√2)

4 triangles

8 triangles

Figure 4

Face d' un tétraèdre
octaèdre
icosaèdre

triplement
(dilatation√3)

6 triangles

18 triangles

Dans tous les cas, l'essentiel est ceci: si l'on dilate un objet dans le rapport λ, on multiplie sa mesure en dimension d par λ^d. Nour verrons tout à l'heure comment cela permet d'élargir considérablement la notion de dimension, et de parler de dimensions non entières. Mais avant de plonger dans les mathématiques du 20ème siècle, voici une très belle application de la notion de dimension, qui remonte à Galilée et peut être considérée comme la base de la biométrie. Pourquoi un éléphant n'a-t-il pas la même forme qu'une mouche? Tout simplement parce que si on dilatait une mouche jusqu'à la taille d'un éléphant, dans un rapport $\lambda \simeq 10^3$, son volume, donc sa masse, serait multiplié par $\lambda \simeq 10^9$, tandis que la surface porteuse des petites pattes serait multipliée par $\lambda^2 \simeq 10^6$; la pression par unité de surface serait multipliée par λ $\simeq 10^3$, donc la mouche agrandie à la taille d'un éléphant s'écraserait sur le sol. Dans la biométrie contemporaine, on s'intéresse depuis un siècle au volume du cerveau des mammifères en fonction de leur poids (Lapicque, etc...). La première approximation est la formule

$$v_c = Cp^{2/3}$$

où v_c est le volume moyen du cerveau pour une espèce, p le poids moyen de l'espèce, et C une constante indépendante de l'espèce. Cela signifie, grosso modo, que le volume du cerveau croît proportionnellement non à la taille ni au volume des animaux, mais à leur surface. En fait, les terminaisons nerveuses ne sont pas également réparties à la surface, la puissance 2/3 n'est pas tout à fait correcte, et C définit des classes d'espèces de mammifères, avec une place un peu privilégiée pour l'homme. Cela suggère déjà que les dimensions entières ne sont pas les seules à intervenir dans la nature. Nous y reviendrons.

Dans le développement des mathématiques du 20ème siècle, les notions de dimension et de mesure ont été élaborées avec des motivations et des applications bien différentes. Elles ne se sont rejointes qu'en 1919, avec la théorie de la mesure et de la dimension de Hausdorff, et elles prennent actuellement des aspects nouveaux. C'est cette histoire que je voudrais maintenant retracer.

La notion de dimension est à la source des travaux de Cantor sur la cardinalité. Tout au cours du 19e siècle, les mathématiciens avaient manipulé des variétés d-dimensionnelles, et non seulement des courbes (d = 1), des surfaces (d = 2) ou des volumes (d = 3). Localement, une variété d-dimensionnelle est l'image d'un cube I^d de dimension d (d = 1, 2, 3, 4, ..). Cantor tente d'élaborer l'idée intuitive qu'une variété de dimension d contient plus de points qu'une variété de dimension d' si d > d' A sa grande surprise ('je le vois, mais je ne le crois pas' écrivait-il à Dedekind), il démontre le contraire: il existe une correspondance bijective entre I^d et $I^{d'}$ quels que soient d et d'. On dit maintenant que toutes les variétés ont la même puissance, ou le même cardinalité: celle du continu. Dedekind, consulté par Cantor, confirme que la démonstration est correcte, mais il considère que le problème doit être posé d'autre façon: si $d \neq d'$, il n'existe pas, pense-t-il, de correspondance bijective et bicontinue entre I^d et $I^{d'}$. Cela devient, pour plusieurs décades, *le* problème de la dimension, finalement résolu par Brouwer. En effet, il n'existe pas de correspondance bijective et bicontinue entre I^d et $I^{d'}$, c'est-à-dire que I^d et $I^{d'}$ ne sont pas topologiquement équivalents. La dimension peut être définie comme un invariant topologique: toutes les variétés topologiquement équivalentes à I^d ont pour dimension d. Il existe ainsi une échelle de dimensions entiéres, telle que toutes les courbes sans points doubles aient pour dimension 1 , toutes les surfaces aient dimension 2 , etc... C'est la dimension topologique. Les articles et traités sur la dimension, jusque vers 1950, concernent principalement la dimension topologique.

Quant à la mesure, sa théorie moderne dérive de la thèse de Lebesgue. En trente ans (1903–1933) la théorie de la mesure renouvelle complètement l'analyse et les probabilités. L'outil essentiel est la nouvelle intégrale, l'intégrale de Lebesgue, et ses généralisations. En identifiant probabilité et mesure, l'axiomatique de Kolmogorov donne à la fois un cadre abstrait très général, et le domaine des applications les plus importantes dans la pratique. Ainsi, à première vue, et dans beaucoup de traités, la théorie moderne de la mesure apparaît comme une théorie abstraite, très puissante, fondement commun des théories de l'intégration, de l'analyse fonctionnelle, des probabilités.

Or, pour Lebesgue, l'aspect géométrique est également important. Lebesgue avait développé la théorie de la mesure sur la droite. Carathéodory l'avait étendue à d'autres variétés. Hausdorff en 1919 en donne une version géométrique très générale, qu'on peut résumer ainsi. Pour mesurer les parties d'un ensemble E , on dispose d'une infinité dénombrable de petites pièces e_k , de poids p_k. On suppose que chaque partie de E , soit A , peut être recouverte par une collection de pièces e_k de diamètres inférieurs à un $\epsilon > 0$ donné, et cela quel que soit ϵ . A chaque recouvrement (e_k) on associe la somme des poids p_k correspondants. Quand A et ϵ sont donnés, on prend la borne inférieure de ces sommes Σp_k correspondant aux recouvrements de A par des e_k de diamètres inférieurs à ϵ . Cette borne inférieure (qui peut être nulle, ou infinie) ne peut que croître quand ϵ décroit. Sa limite quand ϵ tend vers zéro est la mesure de Hausdorff de A, H(A) . Ainsi toute partie A de E a une mesure H(A) , qui peut être nulle, finie non nulle, ou infinie. Au sens de Lebesgue ou de Carathéodory, c'est ce qu'on appelle une mesure extérieure.

Pratiquement, on prend souvent pour e_k des intervalles, ou des disques, ou des boules, on choisit pour p_k une certaine fonction du diamètre des e_k

$$p_k = h(\text{diam } e_k),$$

et on choisit la collection des e_k assez riche (par exemple, la collection des boules à centres et à rayons rationnels). En particulier, si $h(t) = t^d$, H(A) est, à une constante multiplicative près, la mesure d-dimensionnelle de A . On le vérifie dans les cas usuels, et on le prend pour définition lorsque, par exemple, d n'est pas entier. Ayant ainsi défini la mesure d-dimensionnelle, que je note maintenant $m_d(A)$, il est facile de vérifier que

1) $m_d(A \cup B) = m_d(A) + m_d(B)$ si A et B sont à une distance positive (ou plus généralement si $m_d(A \cap B) = 0$ moyennant que A et B soient 'mesurables')

2) $m_d(A) = m_d(TA)$ quel que soit le déplacement T (on désigne par TA l'image de A par le déplacement T);

3) $m_d(\lambda A) = \lambda^d m_d(A)$ quel que soit $\lambda > 0$.

En conséquence, l'effet d'une dilatation de rapport λ est de multiplier la mesure d-dimensionnelle par λ^d, ce que nous avions reconnu comme une relation essentielle entre mesure et dimension.

On voit que, dans la théorie de Hausdorff, d n'est pas nécessairement un entier. On voit aussi que, si $m_d(A)$ est fini et non nul, la mesure δ-dimensionnelle de A est nulle si $\delta > d$, et infinie si $\delta < d$. Généralement, la borne inférieure des δ tels que $m_\delta(A) = 0$ est égale à la borne supérieure des δ

tels que $m_\delta(A) = \infty$. Cette borne commune est, par définition la dimension de Hausdorff de A . Il est facile de voir que cette dimension est toujours supérieure ou égale à la dimension topologique.

Voici tout de suite quelques illustrations de ces notions.

1. L'ensemble triadique de Cantor. C'est l'ensemble des points du segment [0,1] dont l'écriture, dans le système de numération à base 3 , peut se faire uniquement avec des 0 et des 2 . On l'obtient par une construction détaillée sur la figure 5.

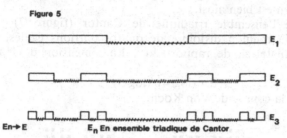

Figure 5

$E_n \rightarrow E$ E_n En ensemble triadique de Cantor

Admettons que l'ensemble triadique de Cantor, E , porte une mesure qui donne la même masse à deux portions égales, et soit m cette mesure. Comme E est la réunion disjointe de ⅓E et de ⅓E + ⅔, c'est à dire de la portion portée par [0,⅓] et de la portion portée par [⅔,1] on a

$$m(E) = m(\tfrac{1}{3}E) + m(\tfrac{1}{3}E + \tfrac{2}{3}) = 2m(\tfrac{1}{3}E).$$

Si c'est une mesure de Hausdorff finie et non nulle en dimension d, on a donc

$$1 = 2 \times (\tfrac{1}{3})^d$$
$$d = (\log 2)/(\log 3) = 0.631$$

c'est en effet la dimension de Hausdorff de l'ensemble E .

2. La courbe de Von Koch. Weierstrass avait donné en 1872 le premier exemple d'une fonction continue sans dérivée. Von Koch, en 1904, a voulu montrer qu'une construction géométrique simple conduit à une courbe qui n'admet de tangente en aucun point (figure 6). Nous indiquons deux manières de construire la courbe: 1) comme limite de lignes polygonales 2) comme intersection d'une suite décroissante d'ensembles plans. Cette courbe admet un paramètre naturel, donc chaque arc a une mesure qui est la différence des

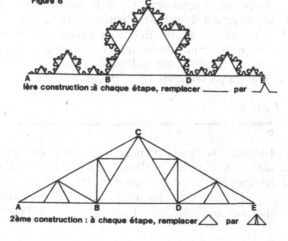

Figure 6

1ère construction : à chaque étape, remplacer ——— par ⌐_⌐

2ème construction : à chaque étape, remplacer △ par ⟁

valeurs du paramètre aux deux extrémités. Or, visiblement, les quatre arcs AB, BC, CD, DA sont égaux. Et l'arc AB est obtenu à partir de l'arc AE (la courbe entière) par dilatation de rapport ⅓ . Donc

$$m(\text{arc AE}) = 4m(⅓\,\text{arc AE}).$$

Si c'est une mesure de Hausdorff finie et non nulle en dimension d , on a donc

$$1 = 4 \times (⅓)^d$$
$$d = (\log 4)/(\log 3) = 1.262$$

et on vérifie qu'il en est bien ainsi.

3. Le carré de l'ensemble triadique de Cantor (figure 7). Là encore, l'ensemble E est la réunion disjointe de quatre portions égales, obtenues à partir de E par dilatation de rapport ⅓ . La dimension de Hausdorff est encore

$$d = (\log 4)/(\log 3)$$

la même que pour la courbe de Von Koch.

Figure 7

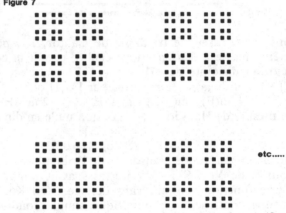

4. Les ensembles du type de Cantor à rapport de dissection ξ (0 < ξ < ½), et les courbes du type de Von Koch à rapport ξ . La construction est indiquée sur les figures 8 et 9. Les dimensions sont respectivement

$$d = (\log 2)/(\log 1/ξ)$$
$$d = (\log 4)/(\log 1/ξ)$$

Si ξ tend vers ½ , la première tend vers 1 , et la seconde tend vers 2 . Pour ξ = ½ , la première construction donne simplement la paramétrisation ordinaire du segment [0,1]. La seconde donne une paramétrisation intéressante d'un triangle isocèle, autrement dit, une courbe qui recouvre toute une surface: c'est une version, due à Cesaro et retrouvée indépendamment par Paul Lévy, de la fameuse courbe de Peano (qui recouvre tout un carré).

Figure 8

Il existe bien d'autres exemples d'ensembles dont on peut calculer la dimension de Hausdorff. A la suite de Hausdorff, c'est surtout Besicovitch et ses élèves qui ont développé les aspects géométriques de la théorie de la mesure et de la dimension de Hausdorff. Avant de revenir aux mathématiques, il est temps d'indiquer que ces notions sont bien adaptées à l'étude du monde

Figure 9₁

naturel. A propos du mouvement brownien, Jean Perrin écrivait déjà au début du siècle que 'c'est un cas où il est vraiment naturel de penser à ces fonctions continues sans dérivées que les mathématiciens ont imaginées, et que l'on regardait à tort comme de simples curiosités mathématiques, puisque l'expérience peut les suggérer' (voir fig. 10 le tracé donné par J. Perrin d'une trajectoire brownienne, approchée par une ligne polygonale joignant les points images d'une progression arithmétique). Et Jean Perrin ajoutait:

'On fera des réflexions analogues pour toutes les propriétés qui, à notre échelle, semblent régulièrement continues, telles que la vitesse, la pression, la température. Et nous les verrons devenir de plus en plus irrégulières, à mesure que nous augmenterons le grossissement de l'image toujours imparfaite que nous

nous faisons de l'Univers. La densité était nulle en tout point, sauf exceptions: plus généralement, la fonction qui représente la propriété physique étudiée (mettons que ce soit le potentiel électrique) formera dans toute matière un continuum présentant une infinité de points singuliers, et dont les mathématiciens nous permettront de poursuivre l'étude.'

Une bonne partie des travaux de Wiener sur le mouvement brownien allait sortir de cette intuition de Jean Perrin. Le mouvement brownien et les processus qui en sont dérivés constituent d'ailleurs un outil précieux pour obtenir des ensembles aléatoires de dimension arbitraire. On a bien étudié les propriétés des images d'ensembles donnés par de tels processus aléatoires, des graphes, des lignes de niveau. On dispose maintenant de véritables catalogues d'ensembles obtenus par des procédés aléatoires et dont on connaît la dimension.

Figure 10 La trajectoire du mouvement brownien selon Jean Perrin

d=2

Figure 11

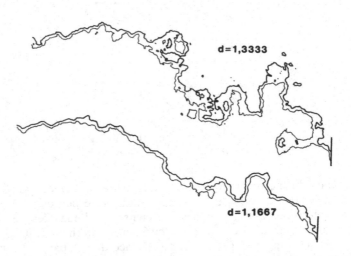

d=1,3333

d=1,1667

D'autre part, il apparaît que beaucoup d'objets naturels ont une forme ou une découpe qui rappelle les ensembles de Cantor ou les courbes de Von Koch, mais plus encore les objets aléatoires dont nous venons de parler. La forme d'une cote découpée telle que celle de la Bretagne en France, ou celle de l'archipel des Philippines, rappellent les ensembles de niveau de certaines fonctions aléatoires de deux variables réelles. Une foule d'exemples et de données se trouvent dans le livre de B. Mandelbrot sur les fractals, avec des illustrations de plus en plus belles au fur et à mesure des éditions (figure 11).

Ainsi, lorsqu'il s'agit d'une côte découpée, la notion de longueur n'a pas de signification. Suivant l'échelle, on obtient des longueurs différentes, et le modèle mathématique qui s'impose est celui d'une courbe non rectifiable. Pratiquement, si l'on désigne par $N(\epsilon)$ le nombre minimum de disques de diamètre ϵ permettant de recouvrir l'ensemble, on voit que $N(\epsilon)$ se comporte comme une puissance de $1/\epsilon$ quand ϵ tend vers 0, c'est-à-dire que le graphe de $\log N(\epsilon)$ en fonction de $\log(1/\epsilon)$ est à peu près linéaire. Si l'on désigne sa pente par d , c'est une bonne approximation de la dimension de Hausdorff du modèle. Il est donc raisonnable de parler de la dimension de la cote de Bretagne, ou de la dimension de la cote de Cornouailles et le fait que ces deux cotes sont découpées de manière analogue s'exprime par l'égalité approximative de leurs dimensions (figures 12, 13, 14).

Actuellement, des calculs de dimensions apparaissent dans beaucoup de disciplines expérimentales ou d'observation: le relief de la glace de la face immergée des icebergs, la répartition des nuages, la surface des substances favorisant la catalyse, les phénomènes de percolation sont des exemples où des dimensions sont effectivement considérées et calculées. Benoît Mandelbrot a attiré l'attention d'un large public scientifique sur le fait que la géométrie naturelle est bien souvent la géométrie des formes irrégulières, qu'il appelle

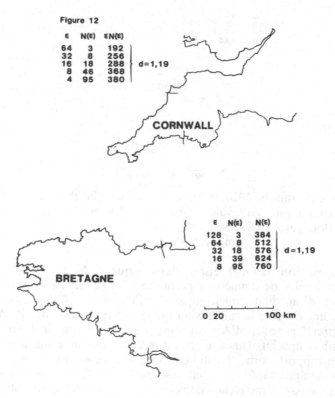

Figure 12

ε	N(ε)	εN(ε)	
64	3	192	
32	8	256	
16	18	288	d=1,19
8	46	368	
4	95	380	

CORNWALL

ε	N(ε)	N(ε)	
128	3	384	
64	8	512	
32	18	576	d=1,19
16	39	624	
8	95	760	

BRETAGNE

0 20   100 km

géométrie fractale, tandis que la géométrie de la droite, du cercle, des variétés régulières est le plus souvent celle des créations humaines.

En même temps que la dimension de Hausdorff commence à devenir familière aux scientifiques non-mathématiciens, les mathématiciens se préoccupent de plus en plus des aspects géométriques de la théorie de la mesure, des aspects métriques de la notion de la dimension.

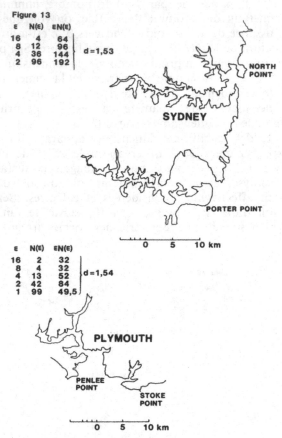

Figure 13

ϵ	$N(\epsilon)$	$\epsilon N(\epsilon)$	
16	4	64	
8	12	96	$d=1,53$
4	36	144	
2	96	192	

NORTH POINT

SYDNEY

PORTER POINT

0   5   10 km

ϵ	$N(\epsilon)$	$\epsilon N(\epsilon)$	
16	2	32	
8	4	32	
4	13	52	$d=1,54$
2	42	84	
1	99	49,5	

PLYMOUTH

PENLEE POINT

STOKE POINT

0   5   10 km

Au cours des années 1920–1950, c'est surtout la théorie du potentiel et l'analyse harmonique qui avaient provoqué des études de dimensions. En particulier, Bouligand avait considéré l'indice

$$\lim_{\epsilon \to 0} \log N(\epsilon)/\log (1/\epsilon)$$

et aussi sa version homogénéisée, dans laquelle $N(\epsilon)$ désigne le nombre minimum de boules de diamètre ϵ permettant de recouvrir l'intersection d'un cube fixe et d'un dilaté quelconque de l'ensemble. Ce second indice de Bouligand coïncide avec la 'dimension métrique' introduite par P. Assouad en 1977. Le premier souci d'Assouad est de définir un invariant pour les transformations lipschitziennes, c'est-à-dire celles qui ne modifient la distance que dans un rapport borné. La dimension métrique est un tel invariant, et de plus elle détermine (sauf quand elle est entière) s'il est possible ou non de plonger un espace métrique dans un espace euclidien $\mathfrak{R}^n$ par une

Figure 14

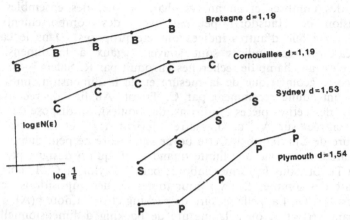

Bretagne d≈1,19

Cornouailles d=1,19

Sydney d=1,53

log $\varepsilon N(\varepsilon)$

Plymouth d≈1,54

log $\frac{1}{\varepsilon}$

transformation lipschitzienne: c'est possible si la dimension métrique est inférieure à n , impossible si elle est supérieure à n . A titre d'exemple, la droite réelle sur laquelle la distance est définie par

$$d(x,y) = |x - y|^\alpha \quad (0 < \alpha < 1)$$

a pour dimension métrique $d = 1/\alpha$. Le théorème de plongement dit donc que, pour $n > 1/\alpha$, il existe une courbe M(t) dans $\mathfrak{R}^n$ telle que le rapport $(\|M(t) - M(T')\|)/(|t - t'|^\alpha)$ soit compris entre deux nombres positifs indépendants de t et de t' . Quand $\alpha = (\log 3)/(\log 4)$, la courbe de Von Koch fait l'affaire. Quand $\frac{1}{2} < \alpha < 1$, une variante de la courbe de Von Koch répond à la question. Mais pour $\alpha < \frac{1}{2}$, la construction est moins facile.

Ces courbes d'Assouad Γ ont une belle application, fondée sur le théorème de prolongement de H. Whitney. En effet, si l'on considère le paramètre t comme fonction du point M , on a sur la courbe Γ

$$t(M) - t(M') \approx \|M(t) - M(T')\|^\alpha$$

le signe $\approx$ signifie que le rapport des deux membres est compris entre deux nombres positifs fixes. Le théorème de prolongement de Whitney nous dit alors — au moins dans les cas où d n'est pas entier — qu'il existe un prolongement de la fonction t(M) à tout $\mathfrak{R}^n$, soit $\theta(M)$, tel que la fonction $\theta( )$ prenne ses valeurs dans [0,1] et soit de classe C^d (c'est-à-dire différentiable jusqu'à l'ordre [d], et telle que les dérivées d'ordre [d] satisfont une condition de régularité de Hölder d'ordre d-[d]). Ainsi il existe une fonction de classe C^d, appliquant $\mathfrak{R}^n$ sur [0,1] et dont toutes les valeurs sont critiques (c'est-à-dire que leur préimage contient un point où le gradient est nul). C'est la réciproque d'un théorème de Sard (1942) disant que, pour une fonction de classe C^n définie sur $\mathfrak{R}^n$, l'ensemble des valeurs critiques est de mesure nulle.

En fait, pour $d < 2$, l'exemple est dû à H. Whitney, et c'est même l'origine du théorème de prolongement. Curieusement, l'exemple de Whitney a précédé le théorème de Sard. C'est un travail qui se trouve au carrefour de la théorie métrique et de la topologie (1935).

La théorie du potentiel avait suggéré d'autres définitions de la dimension. Ainsi, plusieurs notions de dimensions capacitaires ont été proposées par Polya et Szegö, et d'autres sont possibles. Toutes ces définitions équivalent, en fait, à celle de Hausdorff. C'est, essentiellement, l'un des résultats majeurs de la thèse de Frostman (1935), à compléter par des travaux suédois récents.

En théorie des nombres et en analyse harmonique, des ensembles ayant même dimension de Hausdorff peuvent avoir des comportements très différents, de sorte que d'autres indices ont été proposés. Dans le cas des ensembles aléatoires, ces indices sont souvent égaux à la dimension de Hausdorff — c'est un champ de recherches introduit par R. Salem — .

Dans la théorie géométrique de la mesure et de la dimension, une notion nouvelle a été introduite récemment par C. Tricot. Au lieu de recouvrir un ensemble A par des petites pièces e_k (disons, des boules), on dispose des boules disjointes b_k centrées sur A , de diamètre inférieur à ϵ , et on considère la borne supérieure de $\Sigma h(\mathrm{diam}\ b_k)$. Cette borne supérieure ne peut que décroître quand ϵ décroît. Elle a donc une limite quand $\epsilon \to 0$, qu'on désigne par P(A). La mesure de Tricot (sous la forme élaborée par J. Taylor et C. Tricot) est la borne inférieure des sommes $\Sigma P(A_n)$ pour toutes les décompositions de A en sous-ensembles A_n . On l'appelle mesure de packing et on la note p(A).

En choisissant $h(t) = t^d$, on a la mesure de packing d-dimensionnelle. Au lieu de la dimension de Hausdorff, on peut définir la dimension de packing, Dim A, comme la borne inférieure des d telles que la mesure de packing d-dimensionnelle soit nulle. On a toujours

$$\dim A \leqslant \mathrm{Dim}\ A.$$

Pour les produits cartésiens d'ensembles, on a les inégalités remarquables

$$\dim A + \dim B \leqslant \dim(A \times B) \leqslant \dim A + \dim B \leqslant \mathrm{Dim}(A \times B) \leqslant \mathrm{Dim}\ A + \mathrm{Dim}\ B.$$

Un sujet de recherches tout à fait actuel est l'étude des ensembles A pour lesquels $0 < m(A) = p(A) < \infty$, et de ceux pour lesquels $0 < m(A) \neq p(A) < \infty$. Les premiers, dans un certain sens, sont lisses, tandis que les seconds sont rugueux ou éclatés. Il se peut que la comparaison des mesures de Hausdorff et de Tricot donne une définition mathématique de la fractalité. Il se peut aussi que la fractalité reste une notion métamathématique, exprimant une attitude de l'esprit autant qu'une réalité objective. Actuellement, la meilleure et la seule définition d'un objet fractal, ou d'une fractale, est ce que Benoît Mandelbrot, l'inventeur du terme, reconnaît comme tel.

Si j'ai choisi de parler de ce sujet à Adélaïde, c'est pour une série de raisons que je veux énumérer en conclusion.

1. Mesures et dimensions traversent toute l'histoire des mathématiques, et c'est un sujet particulièrement vivant depuis quelques années.

2. L'aspect géométrique, secondaire pendant un demi-siècle, revient au premier plan. A la suite de Benoît Mandelbrot, de nombreux mathématiciens utilisent les ordinateurs pour visualiser des ensembles fractals (courbes ou arbres étranges, ensembles du type de Cantor, etc...). Des articles et des ouvrages sont en cours de parution sur la géométrie des fractals.

3. Cette géométrie exprime des phénomènes naturels importants, et qui s'imposent à l'attention, dans le monde physique comme dans le monde vivant.

4. Quoique très différente de la géométrie euclidienne, cette géométrie est également élémentaire, dans le sens que beaucoup de gens peuvent comprendre les résultats et s'attaquer aux problèmes. En particulier, c'est un champ de recherches possibles pour des enseignants qui n'ont pas la possibilité de s'attaquer à des sujets nécessitant beaucoup de connaissances ou de techniques mathématiques.

Au niveau même des enfants, la facilité que nous avons maintenant à tracer des figures et à manipuler de grands nombres me semble rendre les notions de mesures et dimensions plus faciles à assimiler qu'autrefois.

PUBLIC FORUM
1. Inquiry into School Teaching of Mathematics in England and Wales

Sir Wilfred Cockcroft

In 1978, as a result of apparently widespread and increasing dissatisfaction on the part of employers and others about the mathematical abilities of school leavers, Her Majesty's Government decided to set up a Committee of Enquiry to consider the teaching of mathematics in England and Wales. Education in Scotland is the responsibility of the Secretary of State for Scotland and is separate from that in England and Wales which comes under the Secretary of State for Education and Science and the Secretary of State for Wales.

I was asked to be Chairman and given a brief which reflected this dissatisfaction, requiring the Committee to consider the teaching of mathematics in primary and secondary schools in England and Wales with particular regard to the mathematics required in further and higher education, employment and adult life generally, and to make recommendations.

In other words, we were asked to concentrate on the outcomes of schooling.

But our brief reflected more than just dissatisfaction with outcomes. It implied that success in mathematics is judged by the world at large, through the eyes of government, to be a vitally important outcome of our educational system, whether the school leaver requires knowledge and understanding of the subject to move further up the educational system, or for employment, or simply as a 'life-skill'.

Why is this so? Why is this attitude not restricted to any one country? Many would, of course, immediately reply that it is because mathematics is 'useful', but in so doing usually imply that the way in which it is useful is somehow different from the usefulness of some other school subjects. Pressed further, I believe that those who react in this way would argue that it is useful because it provides a unique and particularly effective means of communication.

If that is so, then I must regretfully record that the results of a study carried out for the Committee of Enquiry show that, in England and Wales at least, there are many people whose mathematics lessons at school have not led them to see mathematics in this way. Far from seeing it as a means of communication, they would avoid using it if at all possible. Indeed, they see the subject as one in which you have to follow rules even though you do not understand them, in which it is necessary to use 'proper' methods to get 'exact' answers. Far too many people react with feelings of anxiety, helplessness, even guilt, when faced with the simplest of practical situations put in everyday terms using transport timetables, restaurant bills, *etc.*, and involving elementary mathematical concepts.

If we are to improve attitudes of this kind we must surely, above all, find a way to teach the subject so that we establish confidence in its use. Mathematics must be presented as a subject which can be used to solve practical and everyday problems by the largest majority of pupils.

The Committee of Enquiry did not believe that such an aim should be in conflict with presenting the subject as one in which enjoyment could be found, nor that it should inhibit the teaching of the most mathematically able of our pupils. We were not led to believe, from our own experience and from the wealth of evidence which we received, that mathematics is an easy subject to teach and to learn. We recognised that children learn mathematics at very different speeds and that in the nature of the subject, a move to one stage usually depends on an understanding of a previous stage.

We became convinced that mathematics should take account of all these facts if it is to build up confidence. In particular, it led us to advise that all mathematical syllabuses should be constructed from the bottom upwards, that 'watering down' that which was appropriate for the most able had produced a situation in which irrelevant material was being presented to the less able children, not being understood by them, building up a false mystique, and leading to a bewilderment and lack of trust in their ability to even start to consider using mathematics as a means of communication, let alone as a method of solving problems.

If any theme in the Committee's report has lessons to be considered beyond the confines of our two countries, I would recommend this, the building of confidence, by recognising appropriate aims for all abilities. We had much else to discuss and report upon, but I remain convinced that the vital question to ask is: if it is assumed vital for all pupils to have confidence to use mathematics as a means of communication, each to his or her level of ability, then are we not obliged to reconsider, and continue to reconsider, the methods we use to teach the subject and the support we give to those who have the difficult task of teaching it?

2. Policies for Education in the Mathematical Sciences in the United States of America.

Katherine Layton

Several national reports on precollege education have been issued in the United States in the past year and a half. Printed handouts are available at this forum on the following:

- Executive Summary of 'Educating Americans for the 21st Century' (the report of the National Sciences Board Commission on Precollege Education in Mathematics, Science and Technology);
- 'The Mathematical Sciences Curriculum K–12: What is Still Fundamental and What is Not' (the report of a conference held at the suggestion of the NSB Commission);
- 'New Goals for Mathematical Sciences Education' (the report of a conference sponsored by the Conference Board of the Mathematical Sciences).

America's educational systems, both formal and informal, are facing a major challenge. A large proportion of young Americans are not equipped to participate in our increasingly technological society. Far too many have emerged from the nation's elementary and secondary schools lacking an adequate background in mathematics, science and technology. This significantly limits their ability to acquire the training, skills and understandings that are needed today.

In response to a growing concern about these problems, the National Science Board established the Commission on Precollege Education in Mathematics, Science and Technology in April, 1982. The National Science Board is the policy making body of the National Science Foundation. The twenty members of the Commission represented a broad cross-section of interests and experiences in education. The Commission's task was to define a national agenda for improving mathematics, science and technology education in the United States. This task included the development of an action plan defining the appropriate roles and responsibilities of federal, state and local governments, of professional and scientific societies, and of the private sector in addressing this national problem.

The report of the Commission, 'Educating Americans For The 21st Century', was prepared after an intensive seventeen-month study. The work of the Commission involved specially conducted studies, public forums, and interviews and discussion with hundreds of educators, scientists, government officials, technologists, business leaders, industrialists and others concerned with elementary and secondary education.

The report defines educational excellence as educational offerings, teaching techniques and commitments that will not only enlarge the pool of students of highest potential, but also will encourage and enable all students to achieve at a level equal to their full capability.

Recommendations of the Commission include changes in standards of achievement, in the breadth of student participation, in the mechanisms for monitoring such achievement and participation levels, in the methods and quality of teaching, in the preparation and motivation of children, and in the content of courses.

To initiate these difficult changes an eight-point strategy is proposed:

- building a strong and lasting national commitment to quality mathematics education for all students;
- providing earlier and increased exposure to mathematics;
- providing a national system for measuring student achievement and participation;
- retraining current teachers, retraining excellent teachers and attracting new teachers of the highest quality and strongest commitment;
- improving the quality and usefulness of courses that are taught;
- establishing exemplary programs in every community to foster a new standard of academic excellence;
- utilising all available resources, including the new information technologies and informal education;
- establishing a procedure to determine the costs of required improvements and how to pay for them.

Underlying every Commission recommendation is one basic object: 'The improvement and support of elementary and secondary school systems throughout America so that, by the year 1995, they will provide all the nation's youth with a level of education in mathematics, science and technology, as measured by achievement scores and participation levels, that is not only the highest quality, but also reflects the particular and peculiar needs of our nation'.

Here is a brief summary of the major recommendations:

- The President should establish a National Education Council;
- States should set up Governors' Councils to stimulate change;
- School Boards should foster partnerships with business;
- Students must receive earlier and more instruction in mathematics;
- The school day, week, and/or year should be substantially lengthened;
- The Federal Government should encourage and partly finance exemplary programs;
- The mathematics teaching pool needs to be expanded with the help of industry and the military;
- The Federal Government needs to assure appropriate retraining is available for teachers;
- States should develop teacher training and retraining programs;
- States should adopt rigorous certification standards for mathematics teachers;
- Elementary and secondary school teachers should be computer literate;
- State and local governments should work to improve the teaching environment and provide ways for teachers to move up in salary and status without leaving the classroom;
- The federal government should finance and maintain a mechanism for

measuring student achievement and participation;
- Federal regulations should include a required period of educational programming on commercial television stations.

The Commission recommends both short term and long term actions. In the short term we propose to raise our standards of achievement and course requirements. In the long term, we must provide a new, coherent pattern of mathematics education for all children — a pattern that begins in kindergarten and continues in a coherent and consistent manner throughout elementary and high school. We must increase early experiences with mathematics. We should consider mathematics specialists in the elementary grades — teachers who like mathematics can pass on this enjoyment to the students.

In the short term, our school curriculum must include more time, higher expectations, and more rigorous standards of achievement in mathematics.

The time suggested for mathematics instruction is as follows:
- In the elementary school, the students should study mathematics 60 minutes a day;
- In grades 7–8, a full year of mathematics should be studied each year;
- For graduation from high school, a student should complete three years of mathematics, including algebra;
- For college entrance, a student should have studied mathematics for four years, including a second year of algebra and course work covering probability and statistics.

In the long term, curriculum changes are recommended. Analysis of current student performance in mathematics indicates the students are learning to be technicians not problem solvers. Opportunities should be provided for the application of arithmetic and mathematics in a variety of areas — in real life situations where analysis through mathematics is possible. Computers and calculators make it possible to present simulation as a problem-solving tool with many important applications.

A principal theme of kindergarten through eighth grade mathematics should be the development of number sense, including the effective use and understanding of numbers in applications as well as in other mathematical contexts. Instruction at this level should be designed to achieve the following outcomes:
- Understanding of arithmetic operations and knowledge when and where specific operations should be used;
- Development of a thorough understanding of and facility with one-digit number facts;
- Ability to use calculators and computers selectively to help develop concepts and to do many tedious computations. These machines should be introduced at the earliest grade practicable;
- Development of the ability to use informal mental arithmetic;
- Development of problem solving abilities;
- Understanding of elementary data analysis, elementary statistics and probability;
- Knowledge of place value, decimals, percents and scientific notation;
- Understanding of fractions as numbers, comparisons of fractions and conversion to decimals;
- Development of an intuitive geometric understanding and ability to use mensuration formulas for two- and three-dimensional figures;

- Ability to use some algebraic symbolism and techniques.

At the secondary school level the content, emphasis and approaches of courses in algebra, geometry, precalculus and trigonometry should be re-examined in the light of the new computer technology. Serious consideration should be given to the development of an integrated secondary school mathematical sciences curriculum. The current sequence which isolates geometry in a year-long course, rather than integrating aspects of geometry over several years with other mathematics, must be reconsidered. Some concepts of geometry are needed by all students. Components of the current curriculum can be streamlined, leaving room for important new topics. In addition to achieving many of the usual desired outcomes from algebra and geometry, instruction must now be designed to achieve the following additional outcomes:

- Knowledge of discrete mathematics (basic combinatorics, graph theory, discrete probability);
- Understanding of elementary statistics;
- Knowledge of computer science;
- Familiarity with the philosophical basis of calculus and understanding of elementary concepts of this subject.

The developments in computer science as well as computer technology suggest new approaches to the teaching of all mathematics in which emphasis should be on:

- Algorithmic thinking as an essential part of problem-solving;
- Student data-gathering and exploration of mathematical ideas in order to facilitate learning mathematics by discovery.

The new technology may lessen the importance of some of the traditional mathematics curriculum. For example, symbolic manipulations will allow students to do symbolic algebra at a far more sophisticated level that they can be expected to do with pencil and paper.

As a result of this and other reports, there has been activity at the federal, state and local levels. Our Congress substantially increased appropriations for science education programs under the direction of the National Science Foundation. These funds are being used for Presidential Awards for Excellence in Mathematics Teaching, for research on and development of instructional material in mathematics for students and teachers, for programs to provide incentives for teachers to stay in mathematics teaching and to improve their mathematics capabilities, for support for local professional development programs, and for expanding the dissemination of information about successful projects. At the state level, forty states have strengthened requirements for high school graduation and thirty-six states have initiated actions to bring about curriculum reforms. At the local level, graduation requirements in mathematics have been increased, teacher retraining programs have been initiated, and curriculum revision is taking place.

I conclude by suggesting the three things that are most important for the United States. We need to provide a basic understanding of mathematics for every child. We must provide a consistent and coherent pattern of mathematics education that begins before kindergarten and continues through high school. We must have national and local leadership, commitment and resources. It is an exciting time to be a high school mathematics teacher.

3. Changes in Social Aspects in Japan and Their Effects on the Reform of Mathematics Education: Conjecture and Expectation

Tadasu Kawaguchi

Recent actions towards reform of the educational system in Japan

I suppose that educational reforms of this kind are taking place worldwide, even though the degree of change will differ from country to country. In Japan, the present situation is so serious that opinions from various groups all suggest that the entire structure of our educational system be thoroughly reviewed in order that it might cope with current problems.

As a consequence, the Japanese Government has introduced into the Diet (Parliament) a Bill to set up a Deliberative Council under the direct control of the Prime Minister. This Council will be responsible for developing a plan to reform our educational system. The Bill is expected to be passed by the Diet in August.

As soon as the Bill is passed, the Council, consisting of twenty-five regular members together with nominated specialists and consultants, will begin work. It is anticipated that it will take a long time for the Council to produce a concrete proposal for reform since the issues it has to consider are such important issues affecting the nature of our education system. The Council will investigate:

- a six-year secondary school system, unifying the two systems of lower and upper secondary, each of which is now for three years of schooling;
- a new organisation of the school system by various combinations of schooling for kindergarten, elementary and secondary grades, for example, suggesting the unification of some of the upper grades of kindergarten and the lower grades of elementary school into an early childhood grouping;
- a plan to improve the University entrance examination system;
- problems of normalisation to cope with the frequent occurrence of violence on lower and upper secondary school campuses.

All of these problems are related directly to the organisation of the school curricula, and those concerned with proposals to reform the school curriculum will not be able to proceed far until the Council has made its recommendations for the whole education system. The motivation of the Japanese Government in setting up the Council is in response to the Government's belief that the great changes affecting our society have produced a serious crisis in education, and that this crisis must be overcome as a matter of urgency.

Social changes associated with Japan's economic development

Problems associated with the explosive increase in student population

In Japan, compulsory education was extended from six years to nine years in 1974, and the attendance rate for students in these schools is 99.99%. Furthermore, 94% of those who complete compulsory schooling go on to upper secondary schooling, and the attendance rate for those who go on to university or junior college reached almost 40% in 1976.

Consequently, the unique curriculum enforced by the Government cannot cope with the wide differences in aptitudes among students. In mathematics teaching especially, the present situation is such that, even by using group teaching according to ability, these difficulties cannot be easily overcome. This has become one of the influences causing problems of various types during the latter part of lower secondary and through to upper secondary, where self-consciousness grows stronger.

The increasing tendency to attach great importance to nominal academic records and the intensification of competition in entrance examinations

The universities and junior colleges in Japan are not necessarily few in number when compared with the number of secondary school graduates. For example, the total number of graduates in 1983 was 1 519 424 while there were 989 such institutions.

Number of Institutions (1984)

	National	Public	Private	Total
Universities	95	34	328	457
Junior Colleges	36	51	445	532
Total	131	85	773	989

And so, the average number of student admissions per institution may be estimated as

$$1\ 519\ 424\ /\ 989\ =\ 1540$$

which is acceptable in terms of intake capacities of the institutions.

But in Japan, there is a nation-wide tendency to think that the name of a graduate's university symbolises his or her social status and, moreover, is also a measure of one's worth in economic terms. Therefore, most candidates naturally want to go to the so-called prestige universities. This forces competition for entry to become so serious that there are chains of prestige kindergartens, elementary schools, lower secondary and upper secondary schools, all aiming towards the prestige universities.

The prosperity of education outside the regular system and decline of reliance on school education

One notable aspect of the Japanese national disposition is its strong sense of equality. Hence, from the point of view of the national education administration, whose basic policy is democratisation, it is desirable that all of the students in public schools be educated with the same educational content. Attempts towards curriculum organisation according to differences of personality or aptitude among students have, in most cases, not been accepted by the people. This can be seen from the following example.

Scores in CFAT* (1983)

Subject	Number of examinees	Mean score	Highest score	Lowest score
Mathematics I	342 762	69.69	100	0
General Mathematics	17	34.44	86	9

* Common First Achievement Test

The number of candidates who selected Mathematics I as an examination subject in the CFAT conducted in the national universities in 1983 exceeded 340 000, but those in secondary schools who selected General Mathematics, which is the subject for comparatively non-gifted students, numbered only 17, as shown in the table above. This tendency has remained unchanged since 1979 when this examination system commenced.

If the philosophy of equality of opportunity is to continue as a basic tenet in public school education as democratic national policy, then the standard of education may be lowered inevitably.

On the other hand, those who want to win elite social status in a free, competitive society must pass the severe entrance examinations, a factor which is surely a source for injecting dynamic energies into Japanese society, a society with extremely poor resources. Consequently, even the general citizenry who lay stress on democratic equality come to feel deeply uneasy about the scholastic development of their own sons and daughters under the policy of equality on which public school education is based, and many dare to seek education outside the regular system in order to promote their scholastic development.

Thus upper secondary school students go to 'Yobiko' and elementary and lower secondary students go to 'Juku' in order to prepare for the entrance examinations for the prestige schools which are their respective targets. Furthermore, an unusual recent phenomena has been noted in the establishment of 'Juku' for young children to prepare them for the entrance examination for prestigious kindergartens.

Nationwide Survey on the 'Juku' (Ministry of Education, 1977)

1. Number of 'Juku'   50 000
2. Number of users: (elementary and lower
   secondary students)   3 100 000 (20.2% of total)
3. Number of 3rd grade students of lower   38% of those who wish
   secondary among (2) above   to attend upper sec.
                                                           (45% in cities)

It has been estimated that almost all upper secondary students go to 'Yobiko', though no authoritative statistics are available.

The reform of the system of entrance examinations

For the entrance examinations from lower secondary to upper secondary, the following reforms are now in force but are not necessarily successful.

• In large cities, freedom of choice of desired public school is restricted by

grouping schools on the basis of candidates' place of residence.
- The selection of students is based on data obtained from students' school records and from results achieved in the entrance examination. This system encounters problems because it is out of favour with gifted students.

All applicants for entrance to every national or public university are required to take the Common First Achievement Test (CFAT) composed of 7 subjects in 5 courses. This test is carried out simultaneously throughout the country by the University Entrance Examination Centre (UEEC). The final decision to admit a student is left to each university after considering the results obtained in a second examination administered by the university itself (if necessary, combined with the results of CFAT obtained by UEEC). But this system is now severely criticised for its bad influence on secondary school education, especially in the case of mathematics education. Because there are more then 300 000 candidates for CFAT, objective type items are used with responses recorded on marked sensed cards for easy computer marking and rating.

Harmful effects associated with the development of mass evaluation techniques by the education industry
Standard scores, calculated from CFAT examinees' marks and the results of large-scale trial entrance examinations conducted by the education industry, are collated and used to construct a mark distribution which is used to grade students and to rank schools. This determines the future prospects of students, including their allocation to courses. Systems like this are bound to produce inequities and the deficiencies of this system are now appearing.

Frequent occurrence of misconduct or violence by students
Owing to factors such as variance of social surroundings accompanied by material wealth, imbalance between physical, emotional and mental development, maladjustment to the uniformity of education, immaturity of adaptability to a free society, or destruction of the home caused by the mental or economic breakdown of parents, misconduct or acts of violence by students, especially in lower and upper secondary schools, occur so frequently that this has become a big problem in Japanese society.

The Actual State of Mathematics Education and its Assessment under such Social Surroundings in Japan

A view gained from the achievement results in the IEA Mathematics Study
Japan is a member of the International Association for the Evaluation of Educational Achievement (IEA), and twice has joined in the IEA Mathematics Studies, the first in 1964, and the second in 1980–81. The National Reports for the second study were published in three volumes during 1981–83 by the National Institute for Educational Research (NIER).

Our forecast was that the achievement level of Japanese students in the second study would be markedly lower than in the first study under the social conditions mentioned above. But the results were contrary to our expectations. There were no signs of lower achievement as far as the results were concerned, and some improvement could even be acknowledged in the calculation area.

It was pleasing for our mathematics teachers to hear that Japanese students achieved such results in the IEA Study. Furthermore, it was happy tidings to hear formally that the calculating ability of our students was not lower, as this result coincided with our own national survey.

Anxiety about the apparent scholarship of students
But in spite of these results, we feel great anxiety and doubt as to whether or not the scholarship of Japanese students is truly genuine, for the following reasons:
• Student achievement in the area of higher order thinking processes was found to be lower in the second study than in the first even though achievement in the area of calculation had increased.
• Japanese students master the skills for solving a group of fragmentary problems very quickly, as they are preparing for severe entrance examinations and receive special training from the experts in 'Yobiko' or 'Juku' who are very skillful in teaching them tactics for these entrance examinations.
• Moreover, the CFAT for entrance to the national public universities is an objective test similar to that used for the IEA Study. Therefore it could be surmised that the system could operate to raise the scores of students in the IEA tests because they are very familiar with those kinds of items.
• It is thought that students in such educational surroundings are deeply influenced in thinking of mathematics as fixed and static, and are very weak in thinking flexibly or creatively.

Sense of crisis, with respect to education, held by the economic world

It is the economic and financial worlds which have begun to feel a sense of crisis, reacting sensitively to this kind of educational situation, and many organisations in those worlds have presented fairly drastic suggestions for the reform of education.

The dominating opinion is that the three problem points of 'closure', 'uniformity', and 'non-internationalism' are the barriers which hinder Japan's rapid progress to the 21st century. Thus it is important to break these down and cultivate individualistic 'creativity' and energetic 'diversity' in the system.

Important viewpoints regarding the reform of mathematics education in Japan

Although the reform of mathematics education in Japan should be included as part of the general reform of the educational system which is being energetically pursued by the Japanese Government, the following radical ideas are necessary for the reform of mathematics education if it is to meet the needs of our society in the 21st century:
• Development of a curriculum having a flexible basis to enable it to cope with the varieties of individual characteristics and abilities of students;
• Development of a curriculum having the construction and contents to make it an effective stimulus for creative activities by students;
• Study and implementation of teaching methods which emphasise the development of useful student learnings rather than providing only fragmentary knowledge and formal skills;
• Study and development of a curriculum which encompasses the whole range

of mathematical activity as well as content. The traditional curriculum has a tendency to be concerned only with the arrrangement of content.

For the realisation of such a reform of mathematics, education curriculum construction based on the idea of 'Matheracy', as proposed at the Tokyo Conference of October 1983, could well provide a sound starting point.

[Proceedings of ICMI-JSME Regional Conference on Mathematical Education, Tokyo, 1983, pp. 3-9]

4. Response by the Minister in South Australia for Education and for Technology

Hon. Lynn Arnold

Thank you for the opportunity to respond to the presentations made in this Public Forum. Can I say that as a person whose mathematics history is moderately mediocre, I was feeling some diffidence this week in preparing myself for to-day's session, and I spent some time just trying to delve into a mathematics climate by thinking mathematics and hoping that might help me. The only problem was that in doing some very general reading I came across a little episode that apparently happened some time in the nineteenth century when the French Academy, for reasons totally unknown to me, went to speak to the ten year old son of a Sicilian shepherd and ask him a problem. They said to him, 'What satisfies the condition that its cube plus five times its square is equal to forty-two times itself increased by forty?' Apparently the lad answered in sixty seconds, and that did nothing at all for my self confidence, knowing I would come into an audience where all of you are now working that out and will very shortly have the answer. I then had a moderate boost to my confidence because I attempted it, and I do not know if I got it out in sixty seconds, but quite to my surprise, I did get it out. Then coming along feeling somewhat refreshed by that, feeling I could at least cope in front of nearly 2000 mathematicians, Sir Wilfred Cockcroft dismayed me again in the first part of his contribution this morning when he said, 'People use idiosyncratic methodologies to get the answer.' and I realised that the method I had used had been most idiosyncratic, and I do not intend to share it with you now.

I certainly have found the contributions we have heard this morning very interesting, and I have found them very interesting in my joint roles as a State Minister for Education and a State Minister for Technology. I believe that there are many issues that have to be seriously answered by educators, by the community, and by the Government at whatever level it may be. Those question have been posed this morning, and I want to make a few comments on some of the issues that have been raised.

I found one of the recurring issues was that of methodologies. I mentioned it just a moment ago in terms of Sir Wilfred's comment on how people have learnt methodologies at school, and then, for a variety of reasons, have abandoned them. I suggest that raises a very big question about the examination of methodologies that we do in fact use. Of course there have

Note: After a career as a teacher and also as a student leader, Mr. Arnold entered the House of Assembly in South Australia for the seat of Salisbury in 1979. In November 1982 he became Minister for Education and for Technology.

been significant changes internationally in the methodologies we use in the teaching of mathematics, which is one reason why parents can no longer help their children with their homework, and it certainly is an area that should not be taken as fixed.

Similarly, the question that was raised by Sir Wilfred concerning the curriculum, and the angle from which to approach it, needs to be addressed, whether you approach it from the desire to create an elite of mathematicians and work downwards to provide for that group of students, or alternately work from the bottom up. The legitimate answer to that problem probably is that there have to be elements of both involved. We here in South Australia have been very conscious of that and have been attempting to widen the curriculum design by means of changing the authorising or credentialling authority that is responsible for senior secondary education in this State, known as the Senior Secondary Assessment Board. That has been an attempt to provide, through the structure of the education system, more appropriate responses that might give a better range of curriculum choices in, for example, mathematics to all students, not just those designed to go on to tertiary education.

I was interested in another comment Sir Wilfred made about the element of practicality that is essential. I suppose it has always been a problem for those on the fringe of mathematics, like myself, that we have constantly to come to grips with how mathematics is integral to everything about life, and that what we have learnt about mathematics is indeed important and practical. Most recently I have been impressed to see work that has been done within our education systems in Australia and by bodies outside our education system to ensure that students studying mathematics are aware of its practical application at all stages. It is happening, as I said, within the system, but one example that came to mind was the Institution of Engineers of Australia who are now supporting the production of mathematics books designed for use on excursions to factories and doing all sorts of mathematical calculations as they pass through various parts of the factory.

Then Sir Wilfred raised the matter of assessment, which was also raised by Professor Kawaguchi and Katherine Layton. I certainly have to say that I agree with the comments made. One issue I think we also have to face is that very often when we are trying to say, 'What are we actually doing by assessment?', that is, trying to determine the extent to which an individual has achieved his or her capabilities in an area and what level of proficiency that individual has reached, we also come across the question of standards. I do not want to undermine the issue of standards, they are critically important, but we must make sure that when we are debating assessment, we are debating it on proper educational terms, conscious of the need for the maintenance and development of standards while at the same time conscious of the individuality question and the need to make some measure of assessment of each individual's proficiency or performance or capabilities in that area. I have also noted, Sir Wilfred, your comments about Government expenditure required, and I have an instruction from the Treasurer just to say 'I note the comments'.

During Professor Kawaguchi's address I was most interested in the summary he gave of the significant changes that have taken place in Japanese society and the way in which the education system must respond to those changes.

One of the things that has always happened in the Australian context is that there is a tendency to say society is changing, therefore education must solve the problems. It is certainly true that unless education is prepared to be part of the act of meeting change, then we will be in diabolical trouble. But it is not sufficient for us to say it is education that will solve our problems, because education left alone may not be able to. Then, of course, education is blamed.

I think we need to take the wider view, and I was interested in the comments Professor Kawaguchi and Ms. Layton made in the same area, of the way in which the educational contribution integrates with the other areas of change. I noted the comments the Professor made about the lower and senior secondary system in Japan and the consideration that has been given to developing or changing the system with a view to possibly one form of secondary education. That probably was noted with great interest by Australians in the audience because we have had similar debates going almost the other way, with the sugestion of the development of senior secondary colleges, and I think that is an issue that will be debated more fully as the week goes on. The explosive increases in student population is clear in the Japanese circumstance from the numbers and percentages quoted by the Professor, but it is also clear from other information we have had about the Japanese education system. We certainly have had growth rates in participation in Australia, but they have not been anywhere near as satisfactory. It also raises the important issue, that if we are able to improve participation in education, we have to make sure that a few other questions are raised, in other words, that there is a purpose to that participation. I have said on other occasions that we must make sure that the increased participation in the education system is not just the result of economic push from the community, but rather the result of educational pull, that there is some legitimate reason why students are in fact there and that we can in fact address the needs of each individual student. We in this country are engaged in what we call participation and equity programs. I hope that kind of program will help us meet the increasing numbers, and that these numbers are explosive in this country too, with education programs that will excite and not disillusion.

I was interested to note the vast difference in the numbers of those who study for the two levels of the CFAP. I noted the ranges of scores were different with one range from 0 to 100 and the other 10 to 86. I would be interested to know what the median and mean scores were in each case to consider the distribution, because it seems important to me to analyse whether or not the two different kinds of mathematics subjects are meeting a purpose.

The general problem of students doing subjects that may not be the most appropriate for them is one that we have to tackle in this country, and I think we are attempting to. I raise the question whether or not we should spend more time talking about sub-units or semester units or the like, rather than discrete year-long units, so that a student does not in fact do twelve discrete year-long courses in mathematics over his or her schooling, but rather does a basket of units, some of a more sophisticated nature and some of a less sophisticated nature. Being smaller units enables you to get away from the problem of status units as opposed to non-status units.

The report Professor Kawaguchi gave of the evaluation of mathematics in Japan the figures were very interesting, and I noticed the differences between 14 and 18 year olds. I also noted that over the period 1964 to 1981 there had been significant improvements in the calculation skills of the 18 year olds, 19.1% was recorded. I wonder what changes have taken place to lead to that result. I know you asked us to be careful in studying that figure and not to place too much reliance upon it, but I wonder what studies have been done within Japan on changes in methodology, curriculum or the private tuition system, or indeed, just luck that might have resulted in that significant improvement in the calculation skills of 18 year olds. I would also be interested to know the role of the abacus in the Japanese education system where it still has an important part to play I understand, in the development of mathematical skills, particularly calculation skills. The question there of why things have been changing I think also needs attention.

Coming to Katherine Layton's presentation, first of all my ears pricked up immediately when she mentioned challenges to formal and informal education systems. I have a strong personal interest in that issue. I think it is critically important that we recognise that education is coming from a number of sources, from formal as well as informal ones, and educational planners, be they educators or governments, need to prepare for both of those. The fact that you further linked mathematics with science and technology also took my interest, particularly coming back to the practicality point that was raised earlier. Here in Australia we have two task forces, one in this state and one national, looking into the question of education and technology, and I am hoping that in both cases they will address issues such as the role of mathematics education.

Can I also make mention and refer back to a point that was raised by all of the speakers, and that is the reasons why people do or do not do mathematics, the barricades that we set up for ourselves that preclude some of us from doing mathematics. I am most interested to see that the Institution of Engineers and ACER in this country are joining together to survey students and past students to find out why they did or did not do mathematics. I think that will be a most useful exercise and I think all of us should pay very close attention to the results. It has been a point of concern to me that when we have analysed some of the groups who are not doing mathematics, and let us take certain particular subgroups, girls for example, one of the major barriers that seems to have been there is very often an internal one, the barrier of the individual saying, 'I cannot do mathematics, therefore I won't do mathematics.' and they are their own worst enemy. They have a lot of barriers to overcome once they have overcome their own, but it has often been their own that has deterred them in terms of participation.

I have been interested to hear the comments on the work in the executive summary, the questions of commitment to quality mathematics, the exposure right across the curriculum from a very early stage. Assessment, and the question of assessment that I referred to earlier is one on which all of us would appreciate more debate as well as the question of the retraining, retaining and recruiting of teachers, the latter a very difficult point, certainly here in Australia and in other parts of the world. My fellow Ministerial colleagues around Australia and I have had to tackle this issue. The answer simply is that we cannot find enough mathematics teachers to meet our needs, and we are having to make internal responses within our education systems in an endeavour to retrain as we can, or prepare for retraining as we can, as well as hoping the tertiary system can provide us as early as possible with more teachers with these particular skills. I think there will be many mathematics teachers in this country who will be most interested to hear the views of others from overseas as to how your respective systems are responding to what I believe is an international problem and not just one localised to Australia.

The matter of curriculum overhaul again is very important, and comes back to the methodology question. The issue of exemplary programs that was raised I find interesting and I would be interested to hear more about how that system will operate. You also raised the question of the utilisation of all resources, new information technologies and again, informal education. Clearly it is important that we make sure that we adequately use new technologies as best we can. I suppose it is one of the feelings of many educators that we had a relatively new educational technology in the early seventies with the video recorder, and, in an educational sense, I have to say I think we fluffed it. I think that we missed maximising the use of that technology for educational purposes as far too many children spend far too much time sitting down watching video tapes of dubious value that are not integrated into the curriculum that they are studying. Let us make sure that in bringing new information technologies that we do not make the same mistake again, that we are concerned that its integration is substantial and educational.

Katherine Layton also made some mention about teachers and their status within the teaching system and the need to keep people of calibre within the classroom. You would have noticed the response from the audience. This has been a matter of considerable debate in South Australia over the last ten years and it has been a fascinating educational debate that will, at some stage, have to lead to some conclusion, because it is important that we do not lose people of calibre from the classroom.

In conclusion, I think it is true that people do have major mental barriers to mathematics, and while it is true that all of us use mathematical calculations of one sort or another every day of our lives, I suspect that part of the problem is that most people when they go through a set of procedures do not count them as mathematics, do not consider them as mathematics at all and therefore still believe in mathematics as something distant which they cannot relate to. They can handle the ten numbers and the plus, minus, division, multiplication buttons on a calculator, but all those obscure buttons at the top and at the side they leave well and truly alone. Now, what they are doing, of course, is defining mathematics as those other buttons, and what they are doing as not mathematics at all. That is a conceptual hurdle that we need to overcome. To overcome the fact that many people regard mathematics as a kind of mystic mumbo-jumbo for an elite, as Thomas Brown said in the seventeenth century, a secret magic, is a major problem. And many regard it as of dubious relevance. (We all know the problems we are given to solve — 'My brother's sister is five foot six, who is married to my uncle who is six foot, therefore what time do I get up in the morning?'). I never quite understood how to go about those sorts of problems, and I think that is part of the conceptual hurdle we need to overcome.

Secondly, in terms of mathematics participation, one of the critical issues that comes through is that everybody is going to have to be involved in mathematics education, therefore our curriculum must address everybody and not leave any forgotten group out on the assumption that they cannot or should not do mathematics. Everybody should be able to do mathematics and feel some satisfaction in doing it.

Finally, in talking about mathematics, and in fairness to the other areas of my education portfolio, I must bring in a little English literature to represent one other major area and use it as an example about part of the problem that may exist. It is a far stretched example, I admit. Elizabeth Barrett Browning wrote: 'How do I love thee? Let me count the ways.' She did not say: 'Let me *list* the ways.' or 'Let me *itemise* the ways.'; she did choose a mathematical term and say: 'Let me *count* the ways.' But then, of course, like many of us, she forgot to go on really to answer the full mathematical question that she had asked in the first place. She did not say: 'How many ways do I love thee?'; she said: 'How do I love thee?', but she did not provide mathematical analysis or formula. Maybe her education did not give her those skills.

THE EFFECTS OF TECHNOLOGY ON THE MATHEMATICS CURRICULUM

H.O. Pollak
Bell Communications Research, Inc.
USA

Mathematics is one of a very small number of subjects which is, in most countries, taught to all students every year throughout many years of schooling. Why does society give us all this time — when it will do the same for almost no other subject? A variety of reasons is given for this unexpected phenomenon, and we wish to begin by mentioniing several of them. Mathematics is an essential part of human culture which the educational system is designed to transmit. This reason fits well into the traditional liberal education of Western Europe, as designed in the 17th and 18th centuries, with an emphasis on classics, literature, natural philosophy and mathematics. Mathematics is the best way to teach youngsters how to think; this purpose fits well into the ideal of encouraging students to go as far in their education as their talents and motivation will permit. Mathematics is beautiful; a personal aesthetic experience of mathematics which we fervently hope will become real to more than a small percentage of our children, associates well with progressive education, discovery learning, and individualised instruction.

I have a strong personal affinity for each of the above reasons for giving lots of time to the learning of mathematics. However, I believe that each of them is secondary to the most fundamental reason why we place so much emphasis on mathematics, and that is its *usefulness*. Every occupation which our students may choose to pursue, and much of their everyday lives, are full of the opportunity, and the need, to apply mathematics. That, probably more than any other reason, is why we teach so much of it.

Let us now examine the question of the need to change the curriculum from each of these points of view. If we teach mathematics for its general cultural value, then it will probably change quite slowly, and only when the longevity of a new direction has been established beyond question. If we teach mathematics for its impartation of reasoning power, then we will change only if a new branch bestows demonstrably superior ratiocination to a previous one. If we teach mathematics for its aesthetic experience, then a new topic had better be more beautiful than the old one it replaces. At this stage of the argument, we are confronted with conservative forces indeed!

Let us now pursue the usefulness of mathematics in more detail, and how this might affect the susceptibility of the curriculum to change. We will return to the other reasons for teaching mathematics at the end of this presentation. If usefulness is indeed a fundamental reason for teaching as much mathematics as we do, then we must exhibit and exercise and emphasise this usefulness at every opportunity. Furthermore, as what is useful for society changes, we must change what we teach and how we teach it. But why should what is useful to society change? Because of technological change! I can do no better than to quote from the report by the Conference Board of the Mathematical Sciences of its meeting in September, 1982, entitled *The Mathematical Sciences Curriculum K-12: What is still Fundamental and What is Not*: "We concluded that the widespread availability of calculators and computers and the increasing reliance of our economy on information processing and transfer are significantly changing the ways in which mathematics is used in our society. To meet these changes we must alter the K-12 curriculum by increasing emphases on topics which are fundamental for these new modes of thought." Our task then is to examine, in some detail, how technology relates to the teaching of mathematics. We shall find that the effects are unexpectedly numerous and far-reaching.

The first and most readily apparent effect of technology on the teaching of mathematics is the use of technology in teaching existing mathematics — in helping to overcome the innumerable pedagogic difficulties with which we are so familiar, in helping to motivate students, in helping the teacher to do a better job. We can use the microcomputer to provide practice for the student with a new technique, to tutor the student at a place where the background is weak (after the computer has found where that is!), to show some new applications of the current subject matter, to diagnose a persistent pattern of error, to try out special cases in a situation in which the mathematical pattern is not clear, or to manage a series of individualised tests. Many of these are wonderful opportunities and worthy of detailed discussion, but they do not imply change in the content of the curriculum as such, and we shall not pursue them further at this time.

A second effect of technology on the teaching of mathematics is that it makes certain subjects possible to teach — subjects which we have always wanted to include in the curriculum, but which we were simply unable to handle pedagogically. A good example of this kind is data analysis. We wish to have youngsters taking data, learning what the data are trying to tell them, and then drawing informal conclusions therefrom. In the course of this, they will need to summarise the data, graph them, transform them, and be able to compare a set of data with another set, or with some predetermined distribution. The pedagogic difficulty has always been that if all students in a class take their own data, it will take a very long time for the class to agree on any single computation done with these data. Numbers from the real world are dirty and messy, and by the time a class has found any answers, they have all forgotten what the question was. Notice that even college-level textbooks on statistics are full of artificial sets of data, carefully designed to make the arithmetic come out easy, and contain very few opportunities for students to take their own data! With a hand-held calculator, the computational problem disappears. The calculator doesn't care whether the numbers are clean or dirty. The pictorial representation of the data, and the trial of various transformations, become easy on a microprocessor. All of this can be — and has been — done in the schools, even with kids in the upper elementary grades.

As a broader consequence of this line of thought, the whole nature of science in the elementary school, and the relationship between science and mathematics, can now be reconsidered. Elementary science has always been descriptive and non-quantitative, not because this is the way we believe science really is or would prefer to teach it, but because we couldn't handle the numbers that come from real experiments. This argument has now lost its validity and, as a result, we can seriously plan to do real science even in the elementary school. Consequently, we can also use science as motivation and application of mathematics, and some long-desired unity can appear in our total curriculum.

A third effect of technology on the teaching of mathematics is that it makes certain subjects necessary. Their importance to concurrent and future work in technology simply requires them to be there. In this category you find, for example, discrete mathematics, subjects like combinatorics and graphs and logic. These are part of "mathematics for computer science", the tools that have to be available for the student to understand how you do things on a computer, and why. Since elementary computer science is an educational goal whose validity can be argued to equal that of elementary calculus, it follows that the mathematics necessary for computer science can compete with the pre-calculus sequence. This helps to make curriculum planning for the secondary school a much more challenging task than it has been before.

To generalise from some of the previous remarks, a fourth effect of technology is that the overall priorities in school mathematics have changed because of the technology. The mathematics which is important for all students simply isn't the same as it was. Certain topics like estimation are even more important than they used to be; we have mentioned others above. Drill in the elementary operations is less important than before, and it is possible to argue that long division might all but disappear. Less familiar is the thought that much symbolic manipulation is easily done on the microprocessor, so that this aspect of the traditional algebra courses could well be deemphasised.

Consider, for example, a traditional kind of algebra exercise:
"Simplify"
$$1/x + 1/y - 2/z$$
The "answer" expected in the usual course is:
$$(yz + xz - 2xy)/(xyz)$$
Are you crazy? Do you seriously wish to maintain that the second expression is simpler than the first? Just look at the number of chicken scratches it took to write them! Seriously, in what sense, if any, is the second expression "simpler" than the first? An answer is that the first expression has three indicated divisions, while the second has only one. If you imagine a value system in which addition and subtraction cost nothing, multiplication costs ϵ, division costs 1, and division by messy numbers costs 100, then the second expression is indeed the cheapest. And what is this value system which I have described? It is the value system of paper-and-pencil arithmetic! But on a hand-held calculator, division is no more expensive than addition, and you'd be crazy to transform the expression in the way we have indicated. The lesson is that symbolic manipulation is not only easy to do on a microprocessor, but that what is important has changed, and it is necessary to rethink the whole content of algebra as well as the pedagogy.

Another example of the change of priorities in school mathematics is the greatly increased emphasis on algorithms. The subject of how you actually do something in mathematics as efficiently as possible has become much more interesting than it used to be. Think of all the ways you *might* find a least common multiple. Which one do you want to carry out in practice? How do you actually want to solve simultaneous equations, or find square roots, or construct shortest networks in some given situation? How do you most efficiently evaluate a polynomial, or find a given power of a matrix? Much traditional mathematics takes on a different coloration in this new environment.

The discussion of algorithms foreshadows a fifth effect of technology on mathematics education, and that is the change in mathematics itself. There is an excellent beginning of a discussion of the effect of technology on mathematics in the recent ICMI discussion document, *The Influence of Computers and Informatics on Mathematics and its Teaching*. That document points out that methods of writing and of calculation have always affected mathematics itself, and that we can expect many new points of view on traditional material. The nature of proof in mathematics is changing, as we find ourselves examining on the machine a collection of special cases too large for humans to handle by conventional means. My distinguished colleague, Phil Davis, has examined most carefully the varying degrees of confidence and certainty which we express in mathematics, and how these have been changed by the technology. But most of all, perhaps, the computer is encouraging us to practise, unashamedly and in broad daylight, certain customs in which we indulged only in the privacy of our offices, and which we never admitted to students: *Experimentation*! To a degree which never appears in the courses we teach, mathematics is an experimental science. We show our classes carefully crafted sequences of clever theorems, and thereby give them a picture of mathematics as beauty without the associated hard work. In fact, we all know that the 10% inspiration comes with 90% perspiration, but we don't say this in public. The computer, I believe, has become the main vehicle for the experimental side of mathematics. We should now be honest with the students, and let them participate fully in the joy as well as the hard work of finding out what's going on in a given situation, and then proving it after they've found out.

Effect number six is in the nature of the incoming student, and it is a problem of very serious concern in the United States. We have been struggling for decades with inequality of access, and with the effect on education of the divergence of home background. The technology has the potential of making this problem worse; a terminal in the home is one more way in which one home can better prepare a youngster for education than another. A school district which buys microprocessors and the time for its teachers to learn to use them wisely will give its students an advantage over a school district which does not. Furthermore, the home effect and the school effect may be correlated: a school district consisting of homes likely to have computers is perhaps itself more likely to have computers! The hope is that this kind of a problem, now that it has been widely recognised, is also on its way to being solved. Responsible government action has done much, and will do more, to provide to all students equal access to technology.

Effect number seven is perhaps a transient problem, but the time constant is sufficiently long — a generation — that we must deal with it. This change concerns the pedagogy appropriate for the topics we have been talking about

— the topics connected with the computer technology. We are currently in a situation without precedent in the history of education — for the very first time, as far as I know, we are trying to teach a subject about which some, and perhaps most, of the students in the class know more than the teacher! Most teachers know very little about microcomputers and, if the truth be told, are rather afraid of them. The students have been brought into, and brought up in, a world in which, for them, this technology has always been there. They are at ease with it and utterly unafraid.

So what? Why does this present any pedagogic problem? The difficulty is that most teaching of mathematics (and of all other subjects) is authoritarian. The teacher expects to act as the fountain of wisdom, to be the boss and behave accordingly. A group of students who know more than the teacher about the subject at hand is certainly not consistent with this form of pedagogy. Now we are not claiming that an authoritarian style of teaching is the best way to present mathematics; far from it. Open-ended, discovery teaching has been recommended for decades. The experimental evidence is, however, that most teachers (the figure quoted is 80%) are very authoritarian in their instincts. For teaching a subject in which the teacher knows more than any of the students, this may not be optimal, but it is certainly workable. But how can you be authoritarian in a subject in which the students know more than you do? The pedagogy must change to a fully participatory pattern in which the teacher acts as moderator of the discussion and not as the source of all knowledge. Right now, for many teachers, there is no other way; in 20 years, when the teachers will themselves have been brought up with the technology, this problem will disappear. By that time, however, we hope that discovery teaching will have established itself.

It is worth noting that the technology in its best pedagogic use itself encourages discovery learning. Students have the opportunity to experiment and find out for themselves. The computer can be programmed to diagnose patterns of student learning and gaps in this learning, can summarise this information for the teacher's benefit, and can supply hints to the student if this is desired. This brings us to our final major point in our discussion of the effects of information technology on mathematics education. As I look around the world at a variety of research and prototype development and experimentation, I see two different instinctive tendencies in the work — and, of course, many points on the line segment joining these two. The computer technology is being used, on the one hand, to help the teacher become a *better* teacher, to remove some of the tedious and routine components of the job, and to enable the teacher to spend more time at a higher cognitive level. The computer technology is being used, on the other hand, to *replace* the teacher, to give up on the teacher's ability to handle certain subjects and approaches, and simply to circumvent the teacher. "This program will teach the students x and y and z" — with no word about the teacher. A current shortage of well-trained mathematics teachers cannot help but encourage this tendency. I am perhaps most heartened by the work at the Shell Centre for Mathematical Education in Nottingham which has shown perhaps better than any other research centre how the technology can make a mediocre teacher good, and a good teacher even better.

This whole discussion, you may recall, began with the assertion that the primary reason Society gives us so much time to teach mathematics is its usefulness. Thus, we claimed, when what is useful changes, mathematics educaton must change, and we have been examining the consequences of this point of view. But how is technology related to the other traditional reasons for teaching mathematics? What about culture, and learning to think, and beauty" It is interesting how relevant technology is to these other points of view as well. There has been a recent realisation in the United States that technology and applications of mathematics have become a "new liberal art". The Slcan Foundation has sponsored much effort at the college, *i.e.* tertiary, level to experiment with materials and teaching to add fundamental understanding and appreciation of technology to the experience of *all* well-educated students. The program has been called "the new liberal art", and recognises the place which the technology now has in our total culture.

With regard to mathematics teaching youngsters to think, there may, in fact, be no better vehicle than computer programming. With all of its famed precision of thought and standard of proof, compared to computer programming, mathematics is a pretty sloppy racket. We are forever pushing symbols around without understanding, we fail to cover lots of special cases, omit many details in our proofs, and when all else fails, leave it as "an exercise for the reader". When you write a program, you can't do those things! Most computer languages are totally unforgiving; make a single mistake, and the program won't run. Can you think of a better example of totally rigorous thinking?

Finally, the beauty of mathematics. The appreciation of this is intensely personal, it goes with personal discovery of unexpected structure, of seeing things that were a disorganised jumble of ideas fall into place. The feeling of beauty is, in my opinion, associated with true discovery learning. When used at its best, as we saw in our discussion of the relation between the computer and the teacher, the computer is a very great help in bringing this kind of experience to the student. It can help to guide each individual student to the "aha!" of discovery, guided by the inductive knowledge of the student's pattern of thought and individual strengths and instincts. If we bring the mathematics student to such experience, we can ask for nothing more.

THE NATURE OF PROOF

Philip Davis (USA)

I should like to talk about certain changes that are coming in what might be called the 'philosophy of mathematics'. Part of these changes is due to technology and part simply due to a perception that the classical philosophies do not provide an adequate description of how mathematics is done by those who do it.

This is a mathematised world. I am sure you will agree. It is hardly news, but what I should like to emphasise in that statement is the suffix, the 'ised', a 'mathemat*ised*' world, and there I mean to imply that we are doing the mathematising. It is not coming down from above. The ghost, Pythagoras, is not whispering something to us, that we ought to be doing this. We are doing it ourselves and that, essentially, in one sentence, is my philosophical message, and this is at some odds with previous philosophies. But what I have to do now is twofold. First of all, to expand this a little bit, and secondly, to say why this is relevant to a congress on mathematics education.

I think that a new formulation of the philosophy of mathematics is taking place which, for the first time, gives proper account of the teaching process. I think former philosophies ignored this.

To go back just a moment to the idea of the mathematised world. I am sure you feel it; I feel it.

Very recently, my wife and I took a trip just before coming here in the State of Oregon, which is a fairly large place, and when you get out on the road you like to call up ahead, perhaps a couple of hundred miles, and make an advanced booking. Well, first of all, I had to find the number of the motel two hundred miles down the road and that took eleven digits. Then, I had to pay for the telephone call, so naturally, being an advanced plastician, I wanted to pay for it with plastic so I called in, punched in my telephone credit card number: that was fourteen digits. Then, of course, the motel did not know me, so they wanted some sort of an assurance that I would show up. I could not give them that assurance but they were delighted to have my credit card number which was sixteen digits and then the expiration date 9–86 was an additional three digits, so it added up to forty-four digits. It seems to me that we are floating in a sea of digits. This is mathematisation at the very elementary level. Of course, you can say, this is all unnecessary really; that in the future, my voice print will have built into it my credit number and all of that kind of stuff. This is related to one of the points I would like to make: that the proper function of applied mathematics is always to get rid of itself.

I suppose it is true to say the world is becoming mathematised and, therefore, the young ought to get with it and learn mathematics and this will open up options.

But there is another way of putting it which may be a little more conservative: I like to play the conservative role. Strangely, we don't say the world will become mathematised only to the extent that we will go along with it. The world will become mathematised only to the extent that we agree with the process. At the moment, we agree with it a hundred percent but whether this is going to be so in the future is a matter for history to say, and history has exhibited many strange turns in its thousands of years.

In view of this mathematised world, I think it is an interesting abstract problem for us to ask ourselves, 'What are the intellectual prerequisites for that mathematisation?' Attempting to say what the intellectual prerequisites are for this or that may seem pompous, but I was led to answering this question because I found an answer to a similar question for technology in a famous series on the philosophy of science from Rydell Publishers in Holland. In an article on technology by a fellow by the name of Rapp, it was suggested:

'The intellectual prerequisites for technologisation (I have that correct) are as follows:

Valuation of work;

Efficient management;

Rational thought; (I don't know what that is, but the author does);

Objectification of nature. (This I understand because objectification of nature means we conceive of nature in a non-animistic way. There are no spirits that are resident in this lamp over here);

A mechanistic view of nature;

Experimental investigations;

Creation of mathematical models.'

Everybody is fond of lists these days; I'll make my own list shortly.

If you have these eight or nine points, then society is ready to technologise. Well, what are the corresponding things in my list for mathematisation? I think that you could have drawn a similar list for yourself so I'd like to suggest that the following are the points; perhaps I have forgotten a few.

First of all, you must have an ability to and a willingness to symbolise, to abstract and to generalise primary experiences of counting and of spatial movement. You must have a sharpened sense of space and quantity and possibly of time. You must have an ability and willingness to dichotomise sharply. You must say yes or no. You must say zero or one. You must say true or false. In the mathematical world you must say that, although life is not that way, not in my experience anyway. Unless you have that dichotomisation, you have no mathematics. You must have an ability to discern causal chains, if A then B, and you must be willing to concatenate those, to put them into chains and to reason about those chains, and here is where proof comes in. You must extract from the real, an abstract surrogate. Then you must be willing to accept the formal manipulations of the abstract surrogate as an adequate expression for the behaviour of the real. This is modelling. You must have an ability and a desire to play with symbols even in the absence of concrete reference for the symbols. In this way, you create a world of objects of the imagination.

Now, I can say nothing about proof as it exists in, let us say, the elementary grades. I have no experience with that. All I can say is that proof is that wonderful element that seems to reside only in mathematics. Proof is what made me a professional mathematician and probably made you one also. It is something which hits young people at a very early age when they perceive that, in proof, there is something that is vitally different from that which exists in any other place in the intellectual world.

Now, proof, when I was a student, resided solely in geometry. Algebra had no proof. I do not know why. That's the way it was in those days. In fact, you could have defined algebra by that very qualification: it was the mathematical subject in which there were no proofs. So, you might say that geometry, in the older sense of geometry with the proofs and the problems that required proof, made a whole generation of mathematicians. But proof, as an exclusive ingredient in teaching, is also something which can kill mathematics for lots of good people so that we have to understand what its proper role is in pedagogy, and we should also understand what its role is within the philosophy of mathematics.

Now, proof serves many purposes in my view. I suppose that it is conceived of as validation. Validation of a statement or of a complex situation is where the truth of mathematics is said to reside, and I use the word 'said' advisedly. Proof also is discovery because proof can lead to discovery of new mathematics. Proof, of course, involves formalisation. Formalisation is certainly a very important ingredient in modern mathematics. Proof is also ritual ... we do it in courses for ritualisation purposes. Proof is also a debating forum. Proof is where errors are eliminated. I can give you a half a dozen more reasons, or purposes, that proof serves. But I would like to tell you a story about the value of proof as a debating forum.

A few months ago, I met a lady who is a very fine teacher of mathematics and I made this statement to her; that proof was a debating forum, and she said she never thought of it that way; that proof was an argument. She said that she was brought up as a young girl to wear gloves and hats, and young girls in those days were never expected to argue about anything, and therefore, how could she conceive that mathematical proof was really an argument in the sense of a fight in which the whole thing is thrashed out before the thing, in fact, is set down. But, this is proof in the ideal sense.

We might also like to have some opinions about how proof is regarded by our students when we are able to present proofs to them. As I say, I have no personal experience about this so I went to one of the leading experts on problem-solving and proof-devising, a man by the name of Alan Schoenfeld, who is participating in this conference in a parallel session. He is an expert in cognitive science, and Schoenfeld reports to me the following. According to his students he has tested, proof is a classroom ritual that confirms what someone else tells you is the case. Secondly, proof is absolutely irrelevant to understanding and to discovery and, thirdly, and this came as a shocker to me, he said that students, having proved that something is true, then go on to make conjectures that violate what they have just proved. This is far from the ideal of the way the mathematical theoreticians conceive of the nature of truth.

To some, proof is the name of the mathematical game and they insist on it even though it may kill the understanding. To me, proof is just one of the many games that mathematics exhibits and the teaching of mathematics requires a certain balance. Of course, a lie to the idea of proof and to questions of what is appropriate to teach, is the question of what mathematics 'is' really. What is it? I suppose that you have presented your students with your favourite definitions and the definitions of what mathematics is are time dependent. It depends on what hour of the day and whom you ask. It may be the science of quantity and space or perhaps it is the science of deductive structures or perhaps it is the science of abstractable patterns. Well, as you define your subject so you will teach it, but there are other routes. In any case, deduction and proof lead to a philosophy of mathematics. Any philosophy of mathematics must account for proof in some sense.

Why, as teachers, should we be concerned with the philosophy of mathematics? Well, it is necessary to have a philosophy of mathematics in order to lend sense to the symbols. The symbols are formal abstract things, and just written down on a sheet of paper or on a computer screen or wherever, they have no sense. In order to relate these symbols to something called an exterior reality, you must formulate some sort of a philosophy. Most mathematicians do not do this consciously. They simply inherit the current philosophical stance of the mathematical culture and they accept it for the most part. Many people do not accept the classical stances and attempt to re-adjust or reformulate, to re-argue these things in a number of different ways. In the last one hundred years, the philosophy of mathematics has largely been devoted to one question and that question is, 'What makes mathematics true?' This is by no means the only interesting question that you can raise of a philosophical nature about mathematics, but it has monopolised the discussion of the last hundred years to a remarkable extent. I think the reason for this has been the longevity of Bertrand Russell. Bertrand Russell, not quite singlehanded but with Whitehead and Frege and other people, initiated modern mathematical logic and modern philosophy in mathematics. He was a very beautiful writer, a very exclusive writer, and he was an interesting character in many ways who lived well into his nineties: the influence of Russell has been pervasive. His students, of course, carried on in this tradition.

So let me answer the question, 'What makes mathematics true?' Well, there are probably as many answers to this question as there are individuals seated here. Let me give you a few of them.

Mathematics is true because it is God-given. It is true because man has constructed it. (Sort of the opposite you see.) It is true because it is nothing but logic and what is logical must be true, or mathematics is true because it is tautological, or it is true because it is like a fabric that has been knit from its axioms as a sweater is knit from a length of yarn. It is true because it has been constructed. It is true because it is beautiful. It is true because it is coherent. It is true because it is useful. It is true because it reflects accurately of itself the phenomena of the real world. It is true because it has been elicited in such a way that it reflects accurately the phenomena of the real world. These two things sound alike but not to a philosopher. Mathematics is true by agreement. It is true because we want it to be true, and whenever an offending instance is found, the mathematical community rises up and extricates that instance and then re-arranges its thinking. If you want to see humorous instances of this, take a look at Lakatos' book on proofs and refutations. Mathematics is true because, like all knowledge, it is based on tacit understanding. It is true because it is an accurate expression of a primal, intuitive knowledge. Well, pick one.

Is the philosophy of mathematics relevant to doing mathematics? Probably some who do mathematics say no. Is it relevant to logic? Probably those that do logic say no. Is is relevant to teaching? Well, I suppose it is in a certain way because, I suggest, what one expresses about this subject reflects the current mainstream of philosophical thought, but that is not my point here. My point is that it is relevant to teaching because there is a way of accommodating teaching into a reformulation of the philosophy of mathematics. This is my main point.

What is, or has been, the traditional philosophical stance? The stance is, or has been, that mathematics has foundations and the business of the philosophy of mathematics is to find those foundations. That is where the truth comes in and that the foundations of mathematics are to be delineated in terms of mathematical logic. This position, in my view, is inadequate and many people have attacked it. By way of contrast but within this particular view, there have been developed a number of divergent schools of mathematical philosophy that go by the names of Platonism, logicism, intuitionism, and so on. But they all adhere to that paradigm of those three points. They are all ultimately interested in that one question.

A formulation which comes from a young philosopher of science suggests that there are two kinds of mathematical theories that are possible. The first kind is a 'private' theory and the second kind is a 'public' theory. The traditional philosophies of mathematics support private theories — Platonism, logicism, formalism, intuitionism — all of those that are traditional in the past and support private theories of mathematics. However, I perceive that mathematics is a social practice and it must be supported by a community and not by isolated individuals. Therefore, in my view, a proper theory of mathematics must be a public theory of mathematics. Within a private theory of mathematics, mathematics is infallible. It is eternal. It is atemporal, that is to say, time plays no role in its statement whatever. It is unlimited in memory and complexity. It deals with mental beings without physical contexts, and the fundamental assumption is that the ideal mathematician is really potentially capable of being isolated from any other ideal mathematician. In effect, there is but one ideal mathematics and, in effect, there need be only one ideal mathematician. That's all you need — one. A private theory of mathematics reduces all epistomological problems to private or internal processes, or reduces them to private relations between ideal mathematics and something called mathematical reality. In a public theory, on the other hand, we cannot settle the question of, what does a mathematician know and how does he come to know it, by looking inside the mathematicians's mind. You do it by finding out the facts of the entire community and of that mathematician's position within the entire mathematical and scientific community.

Now, the relevance to teaching is as follows. Private theories such as Platonism, formalism or any of the others that I mentioned, need not mention teaching at all. They provide no account of teaching. Public theories account for teaching. Teaching has, therefore, in such a public theory, an important role in the philosophy of mathematics. The mathematical community perpetuates itself by initiating new members into that community. The procedures for initiation are largely determined by our teaching practices. The teaching practices are represented by the course structures we set up. This is the manner in which the teaching goes forward. Of course, it's true that most mathematicians earn their living by teaching mathematics and so, obviously, within some sort of a philosophy, one ought to account for the existence of this phenomena that we call teaching. Now, the activities for teaching are not just, let us say, if you are one of those people who assert that mathematics is the game of proof, the activities of teaching are not just the presentation of theorems and proofs and so on, proof formulation and checking, but the functions of teaching are organisation and understanding, and exploration and exercise, formulation and counter-example, formulation and error correction, and error detection, and the whole slew of things that you and I do all the time.

What are the fundamental things that we should be teaching? Well, the teaching community answers these questions as it has to take into consideration such things as the previous lecturer spoke of. But could you answer this on the basis of those, let us say, if you think for example, that we ought to be teaching calculus and linear algebra to, say, our college freshmen, or geometry, or probability, or computers, or micro or whatever? Do you think that you can arrive at this answer on the basis of the five or six intellectual prerequisites for mathematisation? You cannot do this. You must answer it on the basis of the total mathematical experience. Now, teaching can account for discovery. Private theories — and this is one of their shallow points, one of their great weaknesses — private theories of mathematics cannot account for discovery. It is that mysterious thing. It is obtained by a mysterious intuition or insight or something or the philosophers ignore it entirely. Teaching can account for the concepts, the value concepts, that we place on what we teach, why we practise mathematics as we do. If mathematics were totally mechanisable, as some people claim, then it doesn't need teachers. It needs only computing machines. How many? Well, ideally, it needs only one computing machine. This view, in my opinion, is false but it is an aspect of the private theories of the philosophy of mathematics.

Now, this talk was announced as one in which some words about proof, deduction and so on were to be said. I said a few words but let me say a few more words and then wind up. There is a view of proof or a view of mathematics which I disagree with and which I think is a myth, which says that mathematics is potentially, totally formalisable, and therefore, one can say, in advance, what a proof is, how it should work, etc. There is also an idea that the proofs that are given in books and in classrooms are totally adequate, or if they are not adequate, can be made adequate just by a little bit of cosmetology, let me say. What happens when you eavesdrop on a typical college lecture in advanced mathematics? Imagine that you have broken in in the middle of a proof. Now, ideally, since proof passes from assumption to conclusion by tiny logical chains of reasoning, you should be hearing the presentation of those small, logical transformations which are supposed to lead inexorably from assumption to conclusion. But you hear very strange things. You hear such noises as, 'it is easy to show that', or 'by an obvious generalisation' or 'by a long, but elementary computation which I leave to the reader or to the student', 'you can verify so and so'. Now, these phrases are not proof. These phrases are rhetoric in the service of proof and, recently, one young mathematician made quite an hiliarious list or taxonomy of this type of rhetoric that goes on in our proof- processing.

• 'Proof by example' is where the author will present the case and suggest

that it contains most of the ideas.
- 'Proof by intimidation' is where you say something is trivial.
- 'Proof by eminent authority' is 'I saw Jones in the lift and he suggested to me that such and such is the case'.
- 'Proof by cumbersome notation': best done with access to at least four different alphabets and special symbols.

There are twenty-four different categories, taxonomical classes that this individual has distinguished.

Now, of course, you can raise the objection that this is all fun and so on. You can raise the objection that behind these handwavings, behind these desk-poundings, behind any one of these twenty-four, behind our appeals to intuition, to pictures, to lack of counter evidence, to meta arguments, to the results of papers which have not yet appeared, behind all of this, there really is something which you might say is an absolute in the very Platonic sense, that watertight, inexorable list of transformations which take you from hypothesis to conclusion. But this is not the case. This is an illusion. This is a myth. It doesn't exist. It probably cannot exist. In the real world of mathematics, in my view, a paper does several things. It testifies that the author has convinced himself and his friends that certain results are true and it presents part of the evidence on which this conviction is based. My job here is not to undermine the proof process but simply to place it in a realistic way. It presents part, not all, because of certain routine calculations which, you see, are deemed unworthy of print and the reader is expected to reproduce them. But the point is that, in order to evaluate a mathematical paper, you must be part of a certain mathematical subculture. Once you are there, then you will know whether a paper has proved what it claims to prove or it has not.

Now, this is quite a serious matter because it means that, for example, if I am a specialist in numerical analysis, as I claim to be, whenever I take up a paper on, let us say, geometry, let us say, something in multi- dimensional geometry, I find I cannot follow it. It isn't for lack of space due to the fact that the editor wants to cut the length of the papers down, it is simply that the author of this paper in another mathematical subculture has all sorts of things in his head and mind and experience that he shares with the other members of that culture to which I am not privy. So, what does it mean for a mathematician to have convinced himself that certain results are true?

Well, the Chairman has reminded me that it is now 9 o'clock and I have a practice of stopping in the middle of a sentence, even in the middle of a question, so I will stop. Somewhere here I have a wrapup and so if I can find my wrapping-up paper, I'll go to that. In summary, while proof is important in the game of mathematics, it is not the whole game. A public philosophy of mathematics restores or places proof in a realistic position with respect to the total mathematical experience.

* This is an edited version of the transcription of the address.

DEBATE: THE MICROCOMPUTER: MIRACLE OR MENACE IN MATHEMATICS EDUCATION

Debaters: Hugh Burkhardt (UK) and Philip Davis (USA)

Chair: Henry Pollak (USA):

We have seen throughout this Congress the influence of modern technology on mathematics teaching. Many interesting talks have been given that have involved this topic. I think it has come up in every one of our formal sessions and in many of the other discussions that have taken place at this Congress. As was expected, there has been much disagreement about the place of technology in mathematics education, and so it was decided to bring these varying points of view to a focus at our final plenary technology session this morning.

The debate which is about to take place is entitled 'The Microcomputer — Miracle or Menace in Mathematics Education', which is a memorable mouthful indeed. I have asked each of the two debaters this morning to overstate his case a bit, to omit the careful qualifiers for which researchers in mathematics are so well known. The order of events will be first a presentation by Professor Hugh Burkhardt and then a presentation by Professor Phil Davis. Hugh will then ask Phil some nasty questions and Phil will ask Hugh some nasty questions. Maybe I will ask both of them some nasty questions and I hope there will be questions from the floor. Finally there will be a summary by Professor Davis ending with a summary by Professor Burkhardt.

So I should like to begin by introducing Professor Hugh Burkhardt from the Shell Centre for Mathematics Education at Nottingham University. A physicist and applied mathematician, he has been interested also in a great variety of areas of mathematics education. I have particularly asked Hugh to abandon the quiet self-effacing English understatement for which he is so famous and to go on the attack.

Burkhardt (UK):

Thank you Henry. It is not true that the motto of the Shell Centre is never knowingly undersold. You'll hear some of our guiding principles leaking into my talk this morning. It is a forbidding task, not only because my opponent is who he is but because, in this quasi-religious environment, I know I stand here as the representative of a religious movement. I will let some of them down and for that I beg forgiveness in advance. I should also add a similar prayer for forgiveness to the translators. My indiscipline on occasions like this is well known, although at least we have started on time. What I say will be about people, because mathematics education is centrally about people and their ideas and the things that they have managed to achieve. Not terribly impressive, taken all in all, I would say, but I hope at least it's proving great fun for all of us to tackle these enormous tasks that the field presents.

The future roles in the mathematics curriculum of microcomputers, calculators and their successors cannot be reliably foretold; not by Professor Davis or myself this morning nor by anyone else. The history of prediction in mathematics education should discourage us from believing our ideas. The pattern of modes of use, their successes and failures will only emerge from experiment planned or otherwise in classrooms, and in homes all over the world. However, the outcome will be affected, at least in the medium term, by the direction in which each of us decides to work over the next few years, and by the imagination and the care with which the ideas we choose to pursue are developed. The aim of this debate is to identify, and to illuminate some of the issues which need to be borne in mind.

In outlining the enormous educational potential of these devices which has already begun to emerge, I shall be very selective. I shall talk only of those places where it has been shown that the micro (I shall use that as the generic term) can help to produce a clear qualitative enhancement of the curriculum. I will not talk about 10% more on standardised tests or that sort of thing, although these need not be neglible achievements; I'm looking for clear qualitative educational enhancement. This operates typically through technology stimulating and supporting in ordinary classrooms learning activities of value that are rarely found in normal circumstances.

My emphasis on ordinary classroooms and ordinary teachers is an important one. There is a tendency, born of egalitarian wishful thinking, to assume that whatever one teacher can do, any teacher could do — with the right attitude and a bit of inservice training. In practice that does not prove to be true; this is unfortunate, but not surprising. In any skilled activity the range of performance is wide and mathematics teaching is no exception. Many of the changes that we advocate for the curriculum place extra demands on teachers. The achievements of the exceptional few are no guide to what can be expected from a larger target group of teachers in their normal working environment; that can only be determined by trying things out with a representative group under realistic conditions and by observing what happens.

In mathematical education, generally speaking, guesses don't work. It is important to recognise that what may be the best method for one target group of teachers may not be appropriate for another. Car drivers with exceptional skills who are in a hurry will go around corners will all four wheels sliding and the steering turned towards the outside of the curve — this is not necessarily to be recommended to all of us. Equally, one of my favourite metaphors, the Fosbury Flop effect in mathematical education, I propose to demonstrate to you. The Fosbury Flop is the most efficient way of clearing a high obstruction. Provided you are up to it, what you do if there is a reasonable height is that you run towards it and, in a curving path, you fling yourself backwards over the top and land on the back of your neck. This is not necessarily to be recommended to middle-aged men like me as a way of getting over obstacles.

To return to mathematical education, what do I mean by valuable, elusive learning activities? There is a general agreement, I believe, supported by some evidence, that a balanced diet in many mathematics classrooms should contain various elements. They were rather nicely articulated in the British context in the Cockroft Report (of which you would have heard a bit) and its famous paragraph 243. What they said was that we need exposition by the teacher (that is, explanation with illustrative examples). We need discussion of the mathematics between pupil and teacher and between pupil and pupil. We need practial work and we need consolidation and practice in appropriate skills. We need problem solving, including the application of mathematics to realistic situations from every day life. And we need open investigations in which the asking of the questions as well as the answering of them is for the student.

I think there would be little dispute within this hall on the importance of this balanced diet. In practice, all round the world in the vast majority of classrooms, only two of thcsc arc found. I am not offering prizes for guessing which they are. The standard method of teaching used by every mathematics teacher involves explanation, with illustrative examples, followed by many, many exercises which are close imitations of what the students have just been shown. This style of teaching, quite appropriate for many purposes, is accessible largely because it is single track, and the track is defined and controlled by the teacher. And the coaching when the pupils are in difficulties amounts to a reiteration of that standard track. Unfortunately, the problems which mathematics can help us tackle do not often present themselves in the standardised predigested forms, so flexibility and some ingenuity are needed in order to deploy our mathematical and other skills. It is the role of the four missing activities, discussion and practical work, problem solving and investigation, to develop these things; all children know far more mathematics than they can use.

I shall go on now to describe how and why the micro is an extra-ordinary supportive aide to teachers and children, allowing typical teachers to handle learning activities in their classrooms that are, at present, the preserve of the exceptional few. By this I mean certainly less than one percent and probably less than point zero one percent in all the countries from which I have what I would regard as sensible data.

You will gather as I have not yet said anything about technology, that I do not represent that theme in the conference; rather, the users of their enterprise — the problem solving theme — and many other crucial aspects, particularly, the professional life of teachers. If there is one single role of the micro that I value most highly at this point, it is as an inservice device showing teachers important possibilities for enhancing their teaching in the ways I have described.

I shall have difficulty in conveying to you in a few words these remarkable new classroom phenomena: indeed, it is not possible. I am no orator as Phil Davis is, but a teacher very much aware of limited powers and of the enhanced effectiveness that technology offers. In being reduced to the socratic debate format we have agreed not to use any technology; it leaves me standing here feeling quite naked, more or less.

I hope all of you have taken the chance to explore what was on offer in the various project displays and in the technology theme. Those of you, for example, who saw the theme summary presentation, were, I hope, entranced by the way the classroom dynamics was brought out; how, for instance, in the simulation of working with calculators, those in the audience were the pupils. A word of warning is necessary: it is not possible to see the classroom dynamics in the program. Those of you who have looked quickly at a screen and said, as so many do, that software is junk or it's marvellous: don't do it! You have to see what happens with the people who are the major part of the situation we're interested in — the students and the teachers.

I shall mainly concentrate in this opening statement on demonstrated successes subject to the severe criteria I've mentioned. But I shall have to say something about the many other exciting possibilities which are not yet proven in realistic curriculum terms. Again, may I say to you, I think the sensible position is to encourage those with the energy and imagination to explore these things and for the rest of us to suspend judgement until we can exploit their creative work. Instant judgements are a major plague. Many people who are happy to condemn or praise on superficial inspection, do us no service — it's like trying to review a film from the still pictures outside the cinema.

One final comment with reference to developing countries and then I'll begin to get down to work. All this talk about microcomputers may seem largely irrelevant in an environment where teachers, books, pencils and paper are even in short supply. That is a reasonable view. However, an alternative view expressed by some knowledgeable people and particularly eloquently by Ubi D'Ambrosio, is that microelectronic devices offer to developing countries the cheapest, most ready access to the ability to handle technology, which is crucial to any wish they might have (for better or for worse) to move into the industrially developed style of life. The rapidly falling cost of these technologies also means that, provided they make an important qualitative contribution, they may relatively soon seem to be affordable.

In encouraging you to look at the developments, I'd like to just remind you of the developments in Computer-Assisted Learning (CAL for short). I'd like to remind you of its precursor PAL: it couldn't want a friendlier user image than that. PAL (Paper Assisted Learning) has been with us for a long time. It started several thousand years ago when some creative people in Egypt found an unorthodox use for the Papyrus reed. It threatened established media: it had clear advantages of permanence over writing in the sand, and of ease and portability over engraving tablets of stone. It was, I think, immediately recognised that it would be enormously useful for official documents. But it was quite some time before there was paper for all books, not just religious books, but books about every conceivable thing; diaries, paper backs, kleenex, toilet paper, and so on were then gradually developed.

It will take some time before we make that progress with silicon as well. 1985 is the year, I am told, when it will be cheaper to store information on silicon than paper, but the equivalent of some of these uses that I have mentioned will surely be a long time in coming.

There is one other lesson here that I would like to commend to you. For reasons I don't understand, there are people who say, 'The real use of the micro is .....;' they haven't, I think, even begun to arrive. We don't have to decide whether we are book persons or toilet paper persons: if we did there is little doubt as to which we would have to choose. But the fact is that each of these uses is independently available, independently developed; so will it be with computers.

Now to some of the successes that are already happening on a wide scale and that amount to a qualitative enhancement of the general curriculum. First, a small proportion of children, presently a few percent, have become computer nuts, that is, computer enthusiasts. They are there when the school opens in the morning — the caretaker throws them out in the evening; they come in on Saturdays. What they are doing is activity project work, programming the micro in all sorts of ingenious ways for all sorts of ingenious purposes. Those of us who have watched them at work, I think can have little doubt that the sort of mental processes involved are worthwhile and exciting; they certainly amount to an active form of learning, which is rare elsewhere. When compared with the norm of classroom learning activities, this is surely a massive qualitative enhancement of the curriculum though it is interesting to note that it takes place outside the normal curriculum. How far this can be extended, what the proportion of potential computer nuts is, is an interesting question for the near future.

Secondly, in many places a small proportion of teachers, presently a few percent at most, are giving all their students a far richer, more varied diet of mathematical experience than they would receive without the support of the micro. This is occurring in a lot of ways, but many of them are linked by the micro world concept, a phrase coined, I think, by Seymour Papert in the context of LOGO. There is now a great variety of micro worlds, systems to explore and explain and exploit. They vary from the simple to the very rich and complex. Some of them simulate and support the whole range of mathematical activity: calculation, exploration, pattern spotting, conjecture, guessing rules and checking them, explanation, justification and proof, generalisation and extension in a variety of directions. These activities, as I have said before, are exceptionally rarely found in mathematics classrooms.

I would like to say a word about the position of LOGO in this spectrum. It was developed with exceptionally bright kids, by exceptionally bright teachers and works superbly for them. It is an interesting observational question as to how far in its widespread uses it is manageable by the range of kids who are exposed to it or how far they are thrown back into what are in fact, immutable activities. That sort of question has to be asked about every micro world, by its developers and, of course, others. Equally, the calculator enables children to explore mathematics more fluently and to learn to understand it and do it better.

Now, from what can happen in studies conducted with typical teachers and typical children, but not yet carried through on the large scale, we know of a number of modes of use which are a qualitative improvement. I shall mention a few of them. More children will get through to algebra with programming as a transitional activity. This was established, I think, particularly by the Minnesota group, David Johnson, Tom Kieren, Larry Hatfield (some of them are here), fifteen years ago. It is not yet implementable on a wide scale because it needs a one hundred micro-computer school. And even the favoured countries are still mostly an order of magnitude behind this; but when we have the provision it will play an important part.

Both the calculator and the micro make a wider range of realistic applied mathematics possible. Realistic numbers and data can be handled without swamping the kids with the cognitive load of hand processing. Computer games are a feature of the educational scene as well as life outside; games have a lot to offer in promoting strategic thinking. Interestingly, there is no reason why you need a micro to get mathematical games in the classroom but the coming of the micro has acted as the catalyst enabling more teachers to support this.

I shall finish the record of success by saying a few words more describing one application which has proved particularly rich: the use by the teacher in the classroom of a single micro programmed to be a teaching assistant. I do so for various reasons: it is less familiar to most of you, but I know it well because we have worked in that area and it brings out some general points about the overwhelming importance of the people (teacher and pupils) and the dynamics of their interaction. It is particularly relevant to schools as we know them because it seeks to enhance the performance of a teacher working with a group of children in a classroom in the normal way and doesn't require you to restructure your whole mode of operation. It also only requires one microcomputer per class rather than one per child. This mode of use has been shown to have remarkable effects in leading typical teachers in a quite unforced and natural way to broaden their teaching style to include open activities.

It is just worth saying how this happens. The micro is perceived by the children as an independent personality — it is astonishing; it's an illlusion but it's invaluable. It takes over for a time a substantial part of the teacher's load of explaining, managing and task setting (the things that every maths teacher does) and releases them for the less directive activities, that are essential to support problem solving activity. This sort of role shifting has proved very rich.

I would like to have talked to you about teacher lust, a most important effect to which we might perhaps return later, but I must stop. Among the exciting possibilities for the future that have still to be proven beyond the prototype stage are the algebraic manipulation packages, muMATH and so on; data-handling programs for real situations and modelling programs to describe them; shape processes related to the enormously sophisticated computer-generated design facilities now available. (Geometry is a little like virtue. We all are in favour of it, but there's not a lot of it about and some of us hope that we will be able to get through life without actually having to take it on board.) I should like to have said some words about how we may work to make this progress but I shall stop. Thank you.

Pollak:

Thank you very much Professor Burkhardt. Professor Burkhardt has, as requested, painted a chauvinistic picture and Professor Davis will now counter with the conservative position. Phil Davis is an old friend and a colleague of mine for over 35 years; in our youth we wrote six research papers together. Professor Davis is now Professor in the Applied Mathematics Division at Brown University.

Davis (USA):

Thank you very much. I suppose we all have our favourite anti-computer story and this is mine. Shortly before coming here to Adelaide, my wife and I had lunch together in the University cafeteria and, as it worked out, we selected the same items. We went through the check-out and there was a lady there punching the stuff into the check-out to figure out our bill and the bill came out to $2.85. And I said to her 'How can two identical lunches add up to $2.85? This confused her. I said, 'If you double any integer you get an even integer'; this confused her even more. She said, 'I don't know what you're talking about. I'll do it over again.' So she did it over again and came out with $2.85. Since this was thirty-five cents less than what it should have been, I decided to pay up and run.

Well, this, in a nut-shell, is what I have to say about the dangers of computation. But I would like to do now is to express these same ideas in more complicated words. I respect the wisdom of the organising committee and I'm delighted to have the opportunity of taking the negative here, but I venture that the committee could have found a more vigorous opponent. After all, those of you who know me realise that I have been in the computation business since before the digital computer was around and that, professionally, I have spent many years developing numerical analysis and teaching numerical analysis. Only a few months ago I was teaching a course in advanced matrix theory using a very nice package called 'Mathlab' and I have every intention of doing this kind of thing in the future. I was one of the first persons in the States to have run a course in what I call 'computer calculus' and I went on the lecture circuit advocating this kind of thing.

Now, in view of this background, how really can I defend this side? I think perhaps we could have switched a little bit and it might have come out the same. But I suppose that the answer really lies in a considerable underlay of the what I call 'doubt' that I have about the ultimate achievements of the computer path that our technological society is embarked upon. Let's call these doubts 'constructive doubts'. I hope they are constructive. Nonetheless, I would like to tell you, I would like to admit, that this morning I speak from the profundity of a split personality.

Well, there's a saying that you can't fight something with nothing. The something here is the splendid achievements — the splendid achievements of Dr Burkhardt and his group and many of us have gone up there and seen his shop and the very nice things that are there and one gets, as he suggests, the promise of the future along those lines. That's the something, now what's the nothing? The nothing, I suppose, is certain philosophical arguments that I would like to make. Certain perhaps metaphysical arguments, perhaps historical arguments based on not very much maybe, just the inner feeling that one has. Professor Burkhardt is talking about the here and the now: what he has to show you can be applied, let's say, over the next few years. My time scale is different. I'm trying to project forward and while it is true that no man is a prophet and I am not a prophet, I think it is a mistake not to try and prophesy a little bit.

There is no doubt that computers are compelling certain rearrangements in our lives and ideas; we often hear that it's a revolution. Well, I don't know what kind of revolution. Maybe it's like the revolution that led mankind from the feudal period in history to the capitalist period in history; I don't know. But it's pretty clear that the computer is probably one of the most fateful forces in modern life; and I'd like to have you concentrate on the word 'fate'. So there's no doubt that Dr Burkhardt speaks as a prophet when he says that no one knows what is at the end of this tremendous development that we have embarked upon; whether new prophets will appear or whether we shall arrive at a state of mechanical patrician embellished with convulsive self-importance. Maybe this will prevail.

As I said, I was perhaps not the optimum person to get to put the case for the negative this morning, so I would like to bring in, by way of support for my case, several individuals who are not present and allow them to do part of my work for me. One of the individuals is very sophisticated in the mathematical line, very sophisticated indeed, and the other individuals, as a group, really quite unsophisticated, beginning at the start of their careers as teachers of elementary mathematics. Let me start with the sophisticate.

I think that the best person that I could think of in the whole world to present the case for the negative is Professor Clifford Truesdell of John Hopkins University at Baltimore. Now Professor Truesdell is a distinguished applied mathematician and he is a profound scholar of the History of Science. He has written and is about to publish (or it has perhaps appeared, I'm not certain, but in any case he has been kind enough to lend me an advance copy) an essay and the title of his essay is 'The Computer Will be the Ruin of Science and a Threat to Mankind'.

Now let me give you a precis of Truesdell's points. And now I think it's important to understand the relationship between the beginning and the end here. I mean this point was made by several of the previous speakers. Mathematical research depends entirely on education and the direction it will take will be, in part, conditioned by the mathematical experience of school children so it is not unrealistic to talk in this session of say, 'maths at the research level'.

Now Truesdell is no fool, perhaps he is a bit over-dramatic, but he is not a fool. He grasps, for example, that space flight is an impossiblity without a computer. But he also points out that space flight was impossible without the equations of motion which themselves are the end products of perhaps four millenia of scientific thought. He says there is nothing yet derived from the computer that resembles say, Newton's Laws, and the flashy examples (that's my term) that you may have been shown recently all firmly lodged within this structure, have been labouriously built up over four millenia. Truesdell thinks that calculation without the classical standards of reference is dangerous and he asserts that in many of the current fields these standards do not exist or do not yet exist; for example, in biology or in the physics of particles. Truesdell asks, 'How do you verify an answer?' Truesdell points out that the calculation is nothing but a model of a model; he points out that the results of calculation often have no basis in human thought; he asserts that the computer may delay the solution of problems by what you might call classical mathematical means.

He notes that it can deliver any kind of results, good, bad or indifferent; he points out that it causes people to be indifferent to these categories as I noted in the beginning in my restaurant story. He asserts that the computer as a research tool promotes factual fraud: it replaces ends by means; it substitutes specific results for general understanding.

And finally, as a wrap-up of Truesdell's position, and I think this is the most important one — one of the important ones — he asserts that while the classical path promotes inductive and deductive models, computing promotes floating models. That is to say, a floating model is one that treats phenomena individually and with no subsumption under any general theory or from organised knowledge gained from experience. Well, this is Truesdell, so much for him. I'd better say something a little bit on my own line.

After all, the organising committee, when I promised to come here, promised a dinner of fried chicken if I would get up and state my own case. So I have to do this. Let me move, just a bit, towards the teaching function. Truesdell, after all, is directed more towards the research function. Now I have teaching experience only at the college level, so to fill myself in a little bit with the teaching at the lower levels, I was fortunate enough a few months ago to be able to give a course to a number of prospective secondary mathematics teachers, a course in what I call 'Nature in Mathematics' and I told them that this debate in Adelaide was coming up and I wondered whether they could help me formulate some points out of their experience. These are prospective teachers but they also would have some experience of classes working with the computer. So the points I am about to make are due in part from, or were arrived at with the help of, this class of a few months ago.

I should like to argue that excessive computerisation can result in the following things: first of all, the loss of a fundamental base of memorised or internalised knowledge, hence an excessive dependence on the computer and the subsequent (what they call) 'rotting of the mind'. This point was made to me very dramatically by one of the leading professors at Massachusetts Institute of Technology where I visited a few months ago. He said he was teaching a course in Advanced Calculus and of course they have any micro you want being taken into the class and these kids in Advanced Calculus are doing eight times seven on a micro computer. I shall argue that excessive computerisation will lead to a loss of human-teacher-researcher as a role model in the mathematical process. (Put it this way, do you want your grandchildren to be little R2D2's?) I predict that there will be decreased social interaction both in the scientific and in the humanistic areas of education and research. I should guess that what is called immediate intellectual gratification, that the quick response that the computer gives, leads to unrealistic expectations and to a sort of laid-back philosophy of scientific life. But you know no struggle, no evolution.

One of my friends here in Adelaide put it very nicely just last night. We went out to dinner with Professor Jerry Kautsky of Flinders University and he said to me: 'the only computer package that you can learn anything from is a bad package'. Now that's an interesting paradox — excessive formalism in which symbols are separated from their underlying meaning and become the sole arbiters of their own meaning. I think also that a false sense of how mathematics is created might be engendered. I think that, for example, we have had assertions about the perfect theorem prover or theorem discoverer that will find solutions from bare facts using undirected research. I think this is rubbish!

Finally, I think that excessive computerisation can result in the replacement of the total mathematical experience with its elements of creating and discovering and checking and arguing — particularly arguing. As I said a couple of nights ago, a proof is an argument really. All of this will be replaced by the idea of one universal, a temporal computer. Well, these are the points I have thought through and I would guess that this would bring to conclusion the first part of my presentation.

Pollak:

The next portion of the debate this morning is for the two gentlemen to debate with each other for a moment. Hugh, it is your turn to ask Phil an appropriate question that you might like to put to him.

Burkhardt:

Thank you. Let me put two questions to you. I think that perhaps we have some trouble finding a really sharp disputation in that the horror story that you have set out to paint is one that none of us would be happy with. Some aspects of it I think are more serious perhaps than others, but I would like to ask you the question as to how you see us maintaining the necessary vigilance to avoid the dangers that you see? The question about what you have said, like all speculations, must be how likely it is to occur and how we can prevent its negative aspects without losing the positive ones?

Davis:

My answer to that is very unexciting. It is simply to practise moderation. Make haste slowly as the saying goes. One of the things that distresses me about the whole computer world is the amount of bunk that is connected with it. You spoke of religious movements a moment ago — that the computer world is a religious movement. I believe that that's true. I know, for example, that in my university the Department of Computer Science is run by an Ayatollah.

But of course it is also a commercial movement. As we all know, there is an awful lot of money that's put at risk in this thing and one hears horror stories of people being replaced — well, not being replaced, but that the first priority is to the hardware and to the availability of the hardware. Some central government agency or foundation has made this tremendous hardware available to us and therefore whether we like to or not we must go along with it and computerise, *etc, etc*. Now this distresses me very much and I think that one ought to develop a certain resistance to this overwhelming wave of commercial and religious pressures that has developed.

Pollak:

Phil, would you like to ask him a question?

Davis:

Yes. There is one point in which we come in conflict and perhaps you could elaborate your position just a bit more. At the very end I said that one needs the whole human mathematical experience to work out the various components of mathematics like abstraction and generalisation, etc, etc. You also said that this was occurring within the computerised environment. I wonder whether you would elaborate on that a little bit.

Burkhardt:

Yes, well we are now just talking about a sort of experimental facts. It's perfectly true, of course, that examples do exist of the less attractive things that you described of people calculating and believing their numbers — not questioning what is going on. Serious examples occur in education and outside. For instance, corporations, the military, governments can get so hooked on their models, usually computer-based models, that they are not in a position really to understand in qualitative terms what they are saying.

Having acknowledged the negative side, it seems to me that this is basically a matter of having adequate questions of design that put things you want first at the top of the priority list. I think that it is not hard to visualise that mathematical tasks can be stimulated by computers which do take pupils of all ages and abilities through the full spectrum of mathematical activity. Probably LOGO is the best known of these but it's range is by no means the widest. Very simple challenges to explore in microworlds can, will and are being offered to pupils who do go through all these stages. There is one other thing I think which has shown up in all the serious studies of computer use and which is an interesting counterpoint to your worry about personal isolation. The one thing that has come out of all studies that I know of is that discussion between the participants is sharpened and increased in quality and quantity when the computer is used as a stimulus in this way. Certainly we, in our design methods, would want to take on board all your concerns.

Can I ask my second question? My second question concerns this interesting matter of what we might lose from the things we are gaining? I'd like to mention two examples from my own experience. One is the ballpoint pen. When I was at school I was forbidden — we were all forbidden — to use the ballpoint pen because it was well known it ruined everything from your handwriting to your moral code. In my case, the first predication was correct — I will not comment further. But, on the whole, this concern is not there.

The second example (I talked about PAL and CAL) is SAL. Saga Assisted Learning has dropped quite out of use. I suppose I belong to the last generation that learnt extensive quantities of poetry and prose by heart. It doesn't happen much in school now. I think it is quite suggestible (as you suggested about various aspects in the mathematical domain) that not learning things by heart is an impovishment but how do we tell and what would you do about it?

Davis:

I recall that I was of the generation that resisted learning poetry by heart. Then a strange thing occurred in that, as I got older, I managed to learn some poetry and that I found it became a common basis, common link between individuals and so I would certainly recommend that one puts in the memorisation of poetry together with other things like the sine squared plus the cosine squared is equal to 1. One ought to know a few of these things really. But you raised the larger question of, if one computerises is there a loss? What is the loss? I told you that many years ago, it must be fifteen or seventeen years ago, I tried in my university to put in a computer calculus course. Well, the Department of Mathematics objected greatly to this. It was over their dead body that I would put in a computer calculus course. Well, I did it over their dead body or something. And this was the question they asked me. You see, you have a certain amount of time to spend; part of it has to do with computer aspects. What are you going to lose? I said there would be a loss inevitably and I didn't know what the loss would be. It would probably be up to myself. It had to do with my own interaction between the computer and the class.

Well, I tell you what the loss was and it was the same seventeen years ago and the same in this course that I gave a few months ago in advanced matrix theory. The loss is the loss of proof — proof material. I find myself reducing proof material practically to zero. Not quite to zero but almost to zero. Well, I found that I had plenty to talk about. Overwhelmingly enough to talk about without proving anything and that this path was a lively and substantial diet for the students that were in my course. So it was proof in my case but it might be some other element in some other case — element meaning these traditional elements and it is something one worries about. I don't worry about it excessively but it is a bit of a worry.

Can I ask you one more question? I should like to address a question to Professor Burkhardt. This question may seem a bit irreverent, heretical even in this audience. You mentioned the fact that computerisation, with some of the delightful languages that we have around and so on, will lead more children to get through algebra than has formerly been possible. Why should we want more children to get through algebra?

Burkhardt:

That's a very interesting question. I would be the first to dump large sections of the mathematics curriculum. Not only logarithms and easy targets like that but the functional operations on fractions and so on, the ability to handle directed numbers — I think these really should be fighting for their space. I can only say those who speak algebra seem to derive great benefit from it. It is a value judgement we each of us have to make. Certainly the number who are subjected to algebra, without ever learning to speak it , is one of the many sad things about what goes on in the ten million maths classrooms around the world. but I don't think it is difficult to defend fluent speaking of algebra as a skill worth possessing. I could elaborate if I needed to but I don't think I do.

Pollak:

By the way, let me point out a circulatory in Hugh Burkhardt's original presentation and argument which might be interesting. He reminded us that people originally wrote on sand. Then they wrote on stone, then on papyrus, then on paper and now they're writing on silicon which we all know is sand.

I should like, I guess, to ask one question of both the gentlemen here. Professor Burkhardt has particularly given examples and has recommended the use of the technology to make a teacher better — to allow higher cognitive levels to go on in the classroom. In many cases that we see, however, the opposite is happening and the teacher is not becoming better, the teacher is being circumvented. Now, is that danger so great that we should perhaps not follow this direction to the extent that you've talked about?

Burkhardt:

That seems to put me on the spot. It's perfectly true that not particularly exciting things are happening with computers in classrooms. I don't think that they are less exciting than the things that are happening without computers in classrooms. Whether your drill occurs on a piece of paper or on a screen seems relatively unimportant. I should have mentioned in my talk some of the failures, and one of the failures is in the original idea of Computer Assisted Instruction. The problem of writing a program that will diagnose improperly learned skills and concepts and remediate them has been tackled by very able people for fifteen to twenty years now and their delivery date gets further and further into the future.

There are other failures too. Paper shirts after all are not in wide demand nor paper sheets and the silicon sand area will no doubt have its failures too. As to Henry's underlying question as to what we do about the unexciting things, I think vigilance and communication of the possibilities that are on offer is the most important aspect of it.

My view is that if curriculum change is going to fly on the large scale, it has to make life easier for teachers, more fun for teachers. It has to address a problem that they know they've got and it has to have some measure of public support. Most of the things we advocate fail on all four counts. They make life harder, they are threatening and therefore no fun. They set about giving teachers new objectives and therefore new problems that they didn't know they had and if the public cares it is possibly just to say : 'Why aren't you teaching my child any decent mathematics?' Now one of the characteristics of some of these hopeful technology-lead changes is that they do get perhaps even two and a half points out of four on, what I would call, 'the take-off scale'.

Pollak:

Thank you. I think our time is almost up at this point and I shall ask each of you to summarise very quickly. First Professor Davis.

Davis:

The computer offers us the prospect of modest assistance with our task, both teaching and research. It also offers the possibility of considerable derailment of the scientific process into a condition of mechanical stagnation. The watchword for the next generation must be to approach computerisation with constructive doubt. In his plenary lecture a few days ago, Professor Potts pointed to the possibility of ten year old children punching out solutions to non-linear difference equations. In my view, this is a short range position. The long range view, the view of constructive doubt, asks what is the scientific loss, the loss of meaning, that accompanies this act.

Burkhardt:

I think constructive doubt is always the right attitude to take. Man is a tool maker and a tool user — the things that distinguish us from the apes is our ability to use these resources. The fact that you can do very unpleasant and uncomfortable things to yourself with an axe should always be borne in mind. I can't resist telling you what teacher lust is. It is not something from page three of the 'Sun' — a great commonality between Australia and the United Kingdom. It is the inability to resist explaining. The computer seems to be an aid in shifting teachers from this position. Coming back we must not confuse our hopes with established facts. Education has suffered, still suffers, from a belief that wise men know the answer and that to cast doubt on the currently fashionable view is to undermine it and thus sedition. We will only make rapid progress if we take a systematic approach at once, hopeful and sceptical, developing and testing our ideas in realistic circumstances. If the Shell centre is to have a motto I would hope it would be 'Fail Fast and Fail Often'.

Pollak:

I should like to thank very much Professor Burkhardt and Professor Davis for a very interesting session. One of the things that must also be curbed is moderator lust to speak. Thank you very much.

SPECIALLY INVITED PRESENTATIONS

A Review of Mathematics Education in French Schools

Title: A survey of the increasing part of information in the French educational system
Presenter: Daniel Gras
Title: A scheme for teacher training in the use of new technologies in education
Presenter: Anne-Marie Bardi
Title: A year's work of the National Commission for mathematics teaching and learning
Presenter: Robert Amalberti

Some Aspects of Mathematics Education in the USSR

Title: On teaching algebra in Soviet schools
Presenter: S.M. Nikolski
Title: Requirements for training teachers of mathematics in the USSR
Presenter: N.I. Shkil
Title: The improvement of mathematics instruction and Soviet school reform
Presenter: V.M. Monakov
Title: Requirements for mathematical training in mathematics syllabuses in the USSR
Presenter: S.A. Teljakovskii

Commonwealth Association for Science Technology and Mathematics Education (CASTME)

Title: The CASTME Caribbean Project: Training technicians by radio
Presenter: Avi Bajpai (UK)

The Cockcroft Report: Mathematics Counts

Title: A review of the teaching of mathematics in England and Wales
Presenters: Freddie Mann (Department of Education and Science, UK),
    James Ridgway (University of Leicester, UK),
    Bent Christiansen (Denmark)
Respondent: Sir Wilfred Cockcroft (Department of Education and Science)

Films in Mathematics

Title: Art and Mathematics
Presenter: Michele Emmer (Italy)
    A film was shown followed by discussion of the relationships between art

and mathematics demonstrated through the medium of film.
Title: Mathematics Through Film
Presenter: Thomas Banchoff (USA) represented by Michele Emmer
A film was shown followed by discussion on the use of film for the representation of mathematical concepts and ideas.

Mathematics Education Research Group of Australia (MERGA)

Title: Review of Mathematics Education Research in Australia
Presenters: Dudley Blane, John Conroy, Graham Jones, Gilah Leder

Australian Academies Series

Three of the Australian Academies were invited to make presentation on matters of international concern in mathematics education, namely new applications of mathematics in disciplines within the ambit of each Academy, and the mathematical needs of students entering tertiary study in those disciplines.

The Australian Academy of Science
Title: New directions or trends in the application of mathematics to research in the sciences
Presenter: Gavin Brown (University of New South Wales)
Title: Mathematics necessary for tertiary study in the sciences
Presenter: Ian McCarthy (Flinders University)

Academy of the Social Sciences in Australia
Title: Mathematics in Social Science: Teaching
Presenters: Economics: Eric Sowey (University of New South Wales)
Education: John Keeves (Australian Council for Educational Research)
Physchology: Phillip Sutcliffe (University of Sydney)
Sociology: Robert Cushing (Australian National University)
Title: Mathematics in Social Science: Research
Presenters: Economics: Maurice McManus (University of New South Wales)
Education: Barry McGaw (Murdoch University)
Sociology: George Cooney (Macquarie University)
Psychology: John Keats (University of Newcastle)

Australian Academy of Technological Sciences
Title: Mathematics Education in Preparation for Technology
Presenters: L. Davies (University of New South Wales)
John Bennett (University of Sydney)
Le Roy Henderson (University of Sydney)

Hanna Neumann Lectures

Hanna Neumann (1914-1971) was one of the foundation vice-presidents of the Australian Association of Mathematics Teachers (AAMT); she was also the foundation Professor of Pure Mathematics at the Australian National University and a Fellow of the Australian Academy of Science. She was a distinguished mathematician with an active interest in the learning and teaching of mathematics at all levels. After her untimely death, her colleagues and friends from around the world contributed to a fund which is used to support a lecture at the biennial meetings of AAMT. As ICME 5 incorporated the biennial meeting of AAMT in 1984, it was decided to sponsor a named series of lectures by Australian mathematicians on mathematics relevant to the teaching of mathematics in schools and universities.

Normal Numbers. William Moran (University of Adelaide)

Optimisation without calculus. Bob Bryce (Australian National University)

Two Faces of Biomathematics. Michael Deakin (Monash University)

The Fascination of the Elementary. Paul Scott (University of Adelaide)

Mathematics and Weaving. Cheryl Praeger (University of Western Australia)

Projects

Projects: Displays

The following projects were displayed at ICME 5:

Argentina: Integrated Mathematics Project

Australia: Reality in Mathematics Education Project (RIME)

Bulgaria: Integrated Mathematics for 6–12 year olds

Israel: Activities from the Rehovot Program

Israel: Using Drama to Stimulate Mathematics Learning in Elementary Schools

Italy: Mathematics in The Real World

Japan: Mathematics Education With Modern Technology

Netherlands: OW and OC: Developmental Research in Schools (on Curriculum and Computers)

United Kingdom: Chelsea Centre for Mathematics Education

 Developing Ideas in Mathematics Education (DIME)

 Developments in Mathematics Education in England, Wales and Scotland

 Secondary Mathematics Individualised Learning Experiment (SMILE)

 Shell Centre for Mathematics Education, Nottingham University

 Open University Centre for Mathematics Education

 Practical Mathematics in Schools

 School Mathematics Project (SMP) 11–16

 Foundation Level Mathematics in Scotland

 Statistical Education Project (16–19)

United States of America: Improving Statistical Literacy in Schools

 Challenge of the Unknown, a film series on mathematics and problem solving

 EQUALS: Programs to Promote Paticipation of Women and Minorities in Mathematics

Projects: Presentations

Invited Project presentations were:

Dynamic approaches to the teaching of mathematics. Geoffrey Giles (Scotland)

Activities from the Rehovot Program (Israel). Rina Hershkowitz (Israel)

Challenge of the Unknown: a film series on mathematics and problem solving.

James Crimmins (USA)

Improving statistical literacy in the schools. Ann Watkins (USA) (Note that this session is part of the Teaching of Statistics series)

'Multiples and Divisibility' — A television drama about a dreary topic (elementary school mathematics). Nisa Movshovitz–Hadar (Israel)

Art and Mathematics: an interdisciplinary project. Michele Emmer (Italy)

Mathematics education in a society with modern technology. Hiroshi Fujita (Japan)

Integrated Mathematics education for pupils aged 6–12. Blagovest Sendov (Bulgaria)

Developing Spatial skills in 2D and 3D. Geoffrey Giles (Scotland)

The School Mathematics Project SMP 11–16. John Hersee (UK), John Ling (UK)

Statistics in non-mathematics courses for students aged 16–19. Peter Holmes (UK)

The concepts in secondary mathematics and science (CSMS) and The strategies and errors in secondary mathematics projects (SESM). Kath Hart (UK)

Practical mathematics comes into the school: mathematics at work for young people (12–16 years). Connie Knox (UK)

Developments in courses and inservice programs at the Open University. Joy Davis (UK), John Mason (UK). David Pimm (UK)

Roles of the System for and in the Japanese Project. Fumiyuki Terada (Japan), Ikuko Kaminaga (Japan)

Mathematics in the real world: how to motivate students (13– 18 years). Emma Castelnuovo (Italy)

SMP 11–16 Workshop. John Ling (UK)

More math EQUALS more options. Sherry Fraser (USA)

Integrated Mathematics in Argentina. Jorge Bosch (Argentina)

SMP — Individualised mathematics for the lower secondary. Nigel Langdon (UK)

Open University films and video. Jack Koumi (UK), John Jaworski (UK)

Short Communications

Oral Communications

Over 150 contributors were accepted to present a short talk on a topic of their choice, during the Congress program. These sessions ran on a tight schedule with a 15 minute time limit observed by presenters. Other participants were invited to attend for all or part of a given session. Abstracts for the talks were published separately.

Poster Sessions

Approximately 200 posters were on display at three venues throughout the Congress. Abstracts of these posters were published separately.

A condition of presentation was that each contributor agreed to be available near the poster on at least two of the four programmed 'Poster Session' times.

The time that an author wished to be available was indicated alongside each poster; other times were arranged specifically, either directly (using Congress participants' accommodation lists) or by leaving a message alongside the poster.

It provided a rare opportunity to make contact with a colleague with similar interests or experience which is an important post-congress benefit to all concerned.

Other Displays

Throughout the Congress, there were permanent displays in, and around, the venue.

1. Australian Association of Mathematics Teachers: Branch Displays

This was an extensive display of materials from around Australia, and was in two main sections:

Student-made materials; original work prepared for mathematical talent quests in the states of Victoria and Western Australia, together with student generated materials from these States and also South Australia and Newcastle, New South Wales.

The work of the Branches of the Australian Association of Mathematics Teachers from the States of South Australia and Western Australia, New South Wales and Queensland.

2. Computers and their Applications

This was an exhibition of recent books and periodicals, arranged and presented by the British Council. Over 300 books, pamphlets and journals are contained in the display which covers a wide variety of applications as well as technical aspects and social issues.

3. Women in Mathematics

A special display examined some historical aspects and social issues concerned with this subject. The exhibition was organised by the Mathematical Association of South Australia.

4. Special Displays

A special display of posters on journals from around the world involving problem solving for talented school students. (Arranged by Peter O'Halloran (Australia)).

A specially prepared display including video, on the Australian Mathematics Competition. This competition is the largest per capita annual competition in the world.

Displays from government Education Departments from the Northern Territory, Queensland, South Australia, Victoria and Western Australia.

A display of original material from a creative contest in South Africa amongst teachers. The prize? Funding to ICME 5 of course!

5. A Continuous Workshop

A group of Australian teachers, educators and students organised by Alistair Mcintosh (elementary), Steve Murray (Lower Secondary) and Bob Perry (Early Childhood) prepared a popular, activity-based, 'drop-in' centre for anyone in the Congress who wanted a quiet, central place in which to sit, try out stimulating material, look at a variety of printed materials and displays, meet colleagues, or simply have a cup of coffee!

Trade Exhibition

The following Companies agreed to participate in a Trade Exhibition:
Addison-Wesley Publishing Company, North Ryde, New South Wales
Blackie/Chambers, Glasgow, Scotland
Brian Bowman & Associates, Perth, Western Australia
Cambridge University Press, Middle Park, Victoria
Curriculum Development Centre, Dickson, Australian Captital Territory
D.A. Book (Australia) Pty. Ltd., Mitcham, Victoria
ERA Publications, Flinders Park, South Australia
Harcourt Brace Jovanovich Group Australia Pty. Ltd., North Ryde, New
    South Wales
Hargreen Publishing Co., North Melbourne, Victoria
Heinemann Education Australia, Richmond, Victoria
Holt-Saunders Pty. Ltd., Artarmon, New South Wales
Latitude Media & Marketing Pty. Ltd., Glen Waverley, Victoria
Longman Cheshire Pty. Ltd., Melbourne, Victoria
MacMillan Company of Australia Ltd., Hackney, South Australia
Northgate Computing Company Ltd.

Sponsors

The organising committees for ICME 5 gratefully acknowledge the support and assistance of the following organisations:
    Adelaide Convention and Visitors' Bureau
    Australian Academy of Science
    Australian Association of Mathematics Teachers
    Australian Government
    Australian Mathematical Society
    Australian National University
    British Council
    Canberra Mathematical Association
    Commonwealth Foundation
    Gateway Inn
    Grosvenor Hotel
    Hanna Neumann Trust
    Host institutions of all Australian Co-ordinating Organisers
    International Commission on Mathematical Instruction
    International Mathematical Union
    James Hardy & Sons
    Mathematical Association of New South Wales
    Mathematical Association of South Australia
    Mathematical Association of Victoria
    Mathematical Association of Western Australia
    New England Mathematical Association
    Northern Rivers Mathematical Association

Orlando Wines
Pembroke School
Penfolds Wines
QANTAS Airways
Rank Xerox
Satisfac Credit Union
B. Seppelt & Sons
South Australian College of Advanced Education
South Australian Government
South Australian Institute of Technology
Trans Australia Airlines (TAA)
Tyrrells Wines
UNESCO
University of Adelaide
University of Queensland
University of Sydney
Westpac Banking Corporation
Wolf Blass Wines

Related Conferences

Several conferences on special aspects of mathematics or mathematics education were held in Australia immediately preceding or following ICME 5. Many ICME participants took the opportunity to include one or more of these conferences in their itinerary.

August 13–17: Twelfth Australasian Conference on Combinatorial Mathematics and Computing, Perth.

August 16–19: PME 8 International Study Group for the Psychology of Mathematics Education, Sydney.

August 18–23: Conference for Administrators of the Mathematics Curriculum, Melbourne.

August 21, 22: Research Counciul for Diagnostic and Prescriptive Mathematics Pre-ICME 5 Conference, Melbourne.

August 23, 24: Pre-ICME 5 Conference of the International Study Group on the History and Pedagogy of Mathematics, Adelaide.

August 31: Theory of Mathematics Education Conference, Adelaide.

August 30–September 2: Conference on Computer Assisted Learning in the Tertiary Environment (CALITE), Brisbane.

September 2–5: Second Australian Computers in Education Conference, Sydney.

THE WORK OF ICMI

Geoffrey Howson

The two sets of initials ICMI (International Commission on Mathematical Instruction) and ICME (International Congress on Mathematical Education) are so alike as to cause considerable confusion, made all the worse by the fact that the only parts of the Commission's work with which most people are familiar are the four-yearly Congresses. I should like then to say a little about ICMI's other work.

First, ICMI is now a sub-commission of the International Mathematical Union. It is the IMU which pays our administrative expenses and the General Assembly of the IMU which elects ICMI's officers and executive committee. The Commission consists of National Representatives from some 55 different countries, not all of which belong to the IMU. These National Representatives were able to meet earlier this week at the General Assembly of ICMI. Many also met at an informal meeting held last year at the Warsaw International Congress of Mathematicians.

For some years now, ICMI has helped mount regional meetings on mathematics education throughout the world. In October 1983 a meeting was held in Tokyo in cooperation with the Japan Society for Mathematics Education. This was attended by educators from almost 30 different countries and studied the topic, 'School Mathematics in and for Changing Societies'. In May this year the South-East Asia Mathematical Society with ICMI mounted a conference in Thailand with the impact of the computer as its main theme. Preliminary plans are in hand for further regional meetings. An international symposium on 'What should be the goals and content of general mathematics education?' was held in Warsaw in August 1983 to coincide with the ICM.

Other conferences and workshops are regularly held by two official study groups of ICMI. The Psychology of Mathematics Education Group has met annually for some time: in recent years in Grenoble, Antwerp, Jerusalem and Sydney, and has published Proceedings of these meetings. The International Study Group on the Relations between the History and Pedagogy of Mathematics has mounted several workshops and now has its own newsletter. Further study groups are in the process of formation. International Mathematical Olympiads have been held successfully for many years. Now ICMI plays a part in the IMOs, for it is responsible for appointing the IMO Site Committee which determines the host country for the year. ICMI also operates nationally through its various sub-commissions. To give one example, the Belgian sub-commission recently mounted a very successful international seminar on the teaching of geometry, a report of which appeared in *ICMI Bulletin No. 13*.

Information on such activities is provided in the *ICMI Bulletin* which is published twice a year. This is sent to all National Representatives and is also distributed widely by UNESCO, whose financial support is essential to ICMI's well-being. The ICMI Secretariat cannot supply copies of the *Bulletin* and those who wish to receive it should contact their National Representative.. Information about ICMI can also be found (in English) in a special section of the Swiss periodical *L'Enseignement Mathematique*. ICMI has recently decided to mount four special studies. The first of these new iniatives, on the impact of computers and informatics on mathematics and its teaching, is now well under way. A discussion document has been published in *L'Enseignement Mathematique* and copies have been sent to all National Representatives. An international symposium at which questions raised in the document will be discussed in depth will be held at Strasbourg in March 1985. Proceedings will then be published and follow-up meetings arranged at the 1986 International Congress of Mathematicians at Berkeley, USA. Talks are proceeding with the Committee for the Teaching of Science on a joint study into 'Mathematics as a Service Subject'. It is hoped to have a first meeting of the planning committee later this year. Other studies are planned on 'Cognition', a look at the contribution of psychological research and theories on this particular topic, and on 'School Mathematics in the 1990s', an attempt to identify and illuminate key issues and principal problem areas.

To return to the International Congresses, the present Congress has been a great success. On behalf of the Executive Committee of ICMI and of its General Assembly, I wish to thank all of those, both from Australia and overseas, who have contributed to Congress planning.

Finally, to look ahead four years, the Executive Committee has accepted the invitation of the Janos Bolyai Institute to hold ICME 6 in Hungary. We look forward to visiting this country with its great mathematical traditions, and I hope that many of the participants in this Congress will be able to be reunited there. In the meantime, you are invited to take an interest in and to contribute to ICMI's other activities.

LIST OF DELEGATES WHO ATTENDED ICME 5, ADELAIDE, AUSTRALIA
AUGUST 1984

Abe, Professor Koichi (Osaka University of Education) JAPAN: Abellanas, Professor Pedro
SPAIN: Abramov, Dr Blanche (Pace University, Pleasantville) USA: Adda, Dr Josette (Universite
Paris 7) FRANCE: Afflack, Professor Ruth (C.S.U.L.B.) USA: Ahmed, Mr Afzal (West Sussex
Inst. of Higher Education) ENGLAND: Ahmed, Mrs Sheila (Worthing Sixth Form College)
ENGLAND: Aihara, Mr Akira (The Association of Mathematical Instruction) JAPAN: Aitken,
Brother Robert (Marist Brothers') NSW AUSTRALIA: Akers, Ms Joan (San Diego County Office
of Education) USA: Akhurst, Mr William T. (Narrabeen High School) NSW AUSTRALIA:
Akita, Mr Toshifumi (The Association of Mathematical Instruction) JAPAN: Alder, Professor
Henry L. (University of California Davis) USA: Aldridge, Ms Sharne Maria (S.I.E.) NSW
AUSTRALIA: Alexander, Dr Donald (Green River Community College) USA: Alexander, Mr
Noel Stewart (Ashburton College) NEW ZEALAND: Alexander, Mrs Sonna (Leahill Elementary)
USA: Alle, Mr Gunter (Grundschule Atzenbach) W.GERMANY: Allen, Mr Andrew (The Scots
College) NSW AUSTRALIA: Allen, Mr Bob (Hurstbridge Primary School) VIC AUSTRALIA:
Alshaikh, Dr Issam Mohamed Ali (Mosul University) IRAQ: Altamuro, Mr Vincent (Community
School District 10) USA: Amaiwa, Professor Shizuko (Iwaki Junior College) JAPAN: Amalberti,
Mr Robert FRANCE: Amin, Ms Aziza Abdel Azin (Ministry of Education) EGYPT: Amitsur,
Professor Shimshon Avraham (Hebrew University) ISRAEL: Anand, Dr Kailash K. (Concordia
University) CANADA: Anderberg, Professor Bengt (Hogskolan for lararutbildning) SWEDEN:
Anderson, Mr John Campbell (Miramar North School Wellington) NEW ZEALAND: Anderson,
Mr Kevin C. ACT AUSTRALIA: Anderson, Dr Margaret C. (University of Maine) USA:
Anderson, Mr Peter Charles (Armidale C.A.E.) NSW AUSTRALIA: Andrew, Mr David John (The
Heights School) SA AUSTRALIA: Andrews, Mr Anthony John SA AUSTRALIA: Andrews, Ms
Charlene (Oakland Unified School District) USA: Angus, Mr Alan Grant (Melbourne C.A.E.) VIC
AUSTRALIA: Angus, Mr Alexander Rob (Paisley High School) VIC AUSTRALIA: Annice, Dr
Clem NT AUSTRALIA: Anshen, Mrs Helen USA: Aoki, Mr Kazuyoshi (Chuo University)
JAPAN: Appleton, Mrs Susanne M. (Para Vista High School) SA AUSTRALIA: Arifin, Dr
Achmad (Institut Teknologi Bandung) INDONESIA: Arkinstall, Dr John Robert (S.A.I.T.) SA
AUSTRALIA: Armstrong, Mrs Alma Patricia (C.C.E.G.G.S.) ACT AUSTRALIA: Armstrong, Mr
Raymond John (M.C.E.G.S.) SA AUSTRALIA: Aroonsrisophon, Ms Somchai (Kasetsart
University) THAILAND: Asada, Mrs Teruko JAPAN: Asp, Mr Gary (Melbourne College of
Advanced Education) VIC AUSTRALIA: Atkin, Miss Jean Lesley (Dept of Education and
Science) ENGLAND: Atkins, Mr Warren (C.C.A.E.) ACT AUSTRALIA: Aub, Dr Martin Richard
(University of West Indies) JAMAICA: Auckland, Mr Lloyd Douglas (University of Waterloo)
CANADA: Audehm, Mr Manfred Otto VIC AUSTRALIA: Avery, Mr Glen Michael (Loxton
High) SA AUSTRALIA: Avery, Mr John Robert NSW AUSTRALIA: Avital, Professor Shmuel
(Technion I.I.T.) ISRAEL: Ayliffe, Mr John Maxwell (Morialta High School) SA AUSTRALIA:
Aylward, Mrs Margaret (Catholic Education Office) VIC AUSTRALIA: Babolian, Dr Esmail
(University for Teacher Education) IRAN: Bailey, Mr Maurice Ross (Department of Education,
Hamilton East) NEW ZEALAND: Bailey, Mrs Valerie Joan (Fairfield College) NEW ZEALAND:
Baird, Mr Bill (National Aboriginal Education Commission) AUSTRALIA: Bajpai, Professor Avi
C. (University of Technology Loughborough) UK: Bakker, Mrs Sally (Canberra College of
Advanced Education) ACT AUSTRALIA: Balacheff, Dr Nicolas (I.M.A.G.) FRANCE: Ball, Mr
Geoffrey Roy (University of Sydney) NSW AUSTRALIA: Ballard, Ms Anne (Annesley College)
SA AUSTRALIA: Bana, Dr Jack (W.A.C.A.E.) WA AUSTRALIA: Banat, Mr Wasfi Y. (Unrwa/
Unesco Agency) LEBANON: Banks, Mr Ian (Yarra Valley School) VIC AUSTRALIA: Barber,
Mr Donald John (Norlane High School) VIC AUSTRALIA: Barcham, Mr Peter J. (Gisborne High
School) VIC AUSTRALIA: Barker, Reverand Dean Ellis (Weston Creek High School) ACT
AUSTRALIA: Barnes, Ms Mary Stuart (University of Sydney) NSW AUSTRALIA: Barnes, Mr
Ronald Francis (Department of Education, Brisbane) QLD AUSTRALIA: Barnett, Mr Grant
Edmund NEW ZEALAND: Barnsley, Mr Rowan Hamilton (Professional Services Branch) NT
AUSTRALIA: Barnwanijakul, Miss Suma (Kasetsart University) THAILAND: Baron, Mr G.L.
(Daped Informatique) FRANCE: Barraclough, Mr Michael (Hawthorn Institute of Education)
VIC AUSTRALIA: Barry, Mr Colin John (S.A.C.A.E.) SA AUSTRALIA: Barry, Mr William
(Catholic College of Education) NSW AUSTRALIA: Bartlett, Mr Russell (Para Vista High School)
SA AUSTRALIA: Bartlett, Mr Walter David (Hobart Technical College) TAS AUSTRALIA:
Barton, Miss Bridget (Thomas Nelson Australia) NSW AUSTRALIA: Barton, Mr William David
(Onslow College) NEW ZEALAND: Bastow, Mr Barry W. (Education Department of W.A.) WA
AUSTRALIA: Bauer, Dr Carole (Triton College) USA: Baxter, Mr Jeffrey P. (S.A.C.A.E.) SA

AUSTRALIA: Beatty, Mr Don USA: Bebbe Njoh, Dr Etienne (Centre National d'Education)
CAMEROON: Becker, Professor Gerhard (Universitaet Bremen) W.GERMANY: Becker,
Professor Jerry (Southern Illinois University) USA: Becker, Dr Joanne Rossi (San Jose State
University) USA: Beckett, Miss Debra Elizabeth (Christchurch Girls' High School) NEW
ZEALAND: Beeby, Mr Terence John (Victoria College) VIC AUSTRALIA: Begg, Mr Andrew
John Cameron (Auckland Metropolitan College) NEW ZEALAND: Behr, Professor Merlyn
(Northern Illinois University) USA: Bekken, Dr Otto B. (A.D.H.) NORWAY: Bell, Dr Alan
(Shell Centre of Mathematical Education) ENGLAND: Bell, Ms Eileen Mervyn (Casula High
School) NSW AUSTRALIA: Bell, Mr Garry (Northern Rivers C.A.E.) NSW AUSTRALIA:
Bender, Professor Peter (Gesamthochschule Kassel) GERMANY: Bennett, Professor John
Makepeace (University of Sydney) NSW AUSTRALIA: Benson, Mrs Rosemary (Correspondence
School) NEW ZEALAND: Bentz, Professor Hans-Joachim (University of Osnabrueck)
W.GERMANY: Berdonneau, Dr Catherine (C.M.I.R.H.) FRANCE: Beresford, Ms Janice
Lorraine (Perth College) WA AUSTRALIA: Bergeron, Dr Jacques C. (Universite de Montreal)
CANADA: Bergman, Dr Marc (IREM - Faculti de Luminy) FRANCE: Bergstrom, Mr Lars
Carl-Olle (Polhemsgymnasiet) SWEDEN: Berlin, Dr Donna F. (Ohio State University) USA:
Bernard, Dr Donald H. (University of Florida) USA: Bernard, Mr Eric David (St Andrews
Cathedral School) VIC AUSTRALIA: Berrill, Dr Renee Anne (Newcastle University) ENGLAND:
Berry, Dr John W (University of Manitoba) CANADA: Berthelot, Mr Rene (Ecole Normale)
FRANCE: Berzsenyi, Dr George (Lamar University) USA: Bessot, Mr Didier (IREM de Basse-
Normandie) FRANCE: Bezuszka, Professor Stanley (Boston College) USA: Bibby, Dr Richard
Martin (The University of NSW) NSW AUSTRALIA: Bickersteth, Mrs Cecily Craig (S.C.E.G.G.S.)
NSW AUSTRALIA: Bidin, Dr Mokhtar (Universiti Kebangsaan) MALAYSIA: Biggs, Dr Edith
ENGLAND: Bijan-Zadeh, Dr Mohammed Hassan (University for Teacher Education) IRAN:
Billington, Mr Colin R.R. (Mitchell College of Advanced Education) NSW AUSTRALIA: Bishop,
Ms Glenys Ruth (University of Adelaide) SA AUSTRALIA: Bissessor, Mr Sunjeeth (Division of
Indian Education) SOUTH AFRICA: Bitter, Dr Gary (Arizona State University) USA: Bjorck,
Professor Goran (Univ. of Stockholm) SWEDEN: Bjork, Mr Lars-Eric (Sunnerboskolan)
SWEDEN: Bjorkqvist, Dr Ole (Faculty of Education, Abo Akademi Kyrkoesplanaden)
FINLAND: Blakers, Professor Albert Laurence (University of W.A.) WA AUSTRALIA: Blanc,
Mr Michel (Ecole Normale) FRANCE: Blane, Dr Dudley (Monash University) VIC AUSTRALIA:
Blanksby, Dr Peter Ernest (University of Adelaide) SA AUSTRALIA: Blazey, Mr Graham John
(Cranbrook School) NSW AUSTRALIA: Bloom, Dr Lynette Myra (W.A. C.A.E.) WA
AUSTRALIA: Bloom, Dr Walter Russell (Murdoch University) WA AUSTRALIA: Bobka, Miss
Susan (Whitney Young Junior High School) USA: Bolletta, Mr Raimondo (Cede-Centro Europeo
Educazione) ITALY: Bolton, Mr Stephen (Bendigo C.A.E.) VIC AUSTRALIA: Booker, Mr
George (Brisbane C.A.E.) QLD AUSTRALIA: Booth, Ms Ada (University of Santa Clara) USA:
Booth, Dr Drora NSW AUSTRALIA: Booth, Mr Kenneth William (Safety Bay High School) WA
AUSTRALIA: Booth, Dr Lesley Rochelle (Chelsea College C.S.M.E.) ENGLAND: Booth, Dr
Raymond Sydney (Flinders University) SA AUSTRALIA: Booth, Dr Thomas E. (Los Alamos
National Laboratory) USA: Borland, Miss Robin (Upper Yarra High Technical School) VIC
AUSTRALIA: Borovcnik, Dr Manfred AUSTRIA: Borowczyk, Mr (IREM) FRANCE: Bortei-
Doku, Dr Sam Okla (University of Cape Coast) GHANA: Bosch, Prof. Jorge Eduardo
(Universidad Caece) ARGENTINA: Bouchier, Mr Edward Allan (Phillip Institute of Technology)
VIC AUSTRALIA: Boudoukos, Miss Olympia VIC AUSTRALIA: Boutte, Mr Daniel (Universite
Louis Pasteur) FRANCE: Bowden, Mr Kevin Cyril (Mitchell Park High School) SA AUSTRALIA:
Bradburn, Dr Mary ENGLAND: Braden, Mr Lawrence Stewart (Iolani School) USA: Bradford,
Mr Anthony Francis (Stuart High School) SA AUSTRALIA: Bramley, Ms Gwenda (Canberra
C.A.E) ACT AUSTRALIA: Brazzatti, Mr John SA AUSTRALIA: Bremner, Mrs Alexandra
Patritia (St Mary's Anglican Girls School) WA AUSTRALIA: Bremner, Mr Frank James
(Plympton High School) SA AUSTRALIA: Brennan, Mr Edward Robert (Brisbane State High
School) QLD AUSTRALIA: Briedis, Mr Edgar John SA AUSTRALIA: Briggs, Dr John Thomas
Francis (Brisbane C.A.E.) QLD AUSTRALIA: Bright, Dr George W. (The University of Calgary)
CANADA: Briginshaw, Dr Anthony John (The City University) ENGLAND: Brine, Mr Graham
(Snowtown Area School) SA AUSTRALIA: Brink, Mrs Sue Dianne (Presbyterian Ladies' College)
VIC AUSTRALIA: Brinkworth, Mr Peter Charles (S.A.C.A.E.) SA AUSTRALIA: Britt, Mr
Murray Sarelius (Auckland Teachers College) NEW ZEALAND: Broadribb, Mr Kevin (Golden
Press Pty Ltd) NSW AUSTRALIA: Brock, Mr John Stuart (Camberwell Grammar School) VIC
AUSTRALIA: Brockmann, Dr Ellen (Totowa Public Schools) USA: Brolin, Dr Hans (Uppsala
University) SWEDEN: Bromilow, Dr Mick (C.C.A.E.) AUSTRALIA: Brooks, Mr Bevin John
(West Lakes High School) SA AUSTRALIA: Brooks, Dr Malcolm (C.C.A.E.) ACT AUSTRALIA:
Broomes, Dr Desmond (University of the West Indies) BARBADOS: Brousseau, Mr Guy (IREM)
FRANCE: Brown, Mr Alexander David WA AUSTRALIA: Brown, Ms Alice Morgan (Baltimore
City Public School) USA: Brown, Professor Gavin (University of NSW) NSW AUSTRALIA:
Brown, Mr Michael James (Victoria College) VIC AUSTRALIA: Brown, Dr Peter Joseph
(Canberra C.A.E.) ACT AUSTRALIA: Brown, Dr Sue (University Houston-Clear Lake) USA:
Brownbill, Mrs Molly (Radford College) ACT AUSTRALIA: Browne, Mr Paul Leonard (Trinity
Grammar School) NSW AUSTRALIA: Brownlee, Miss Mary (S.A.C.A.E.) SA AUSTRALIA:

Bryant, Mr Gary (Millen Primary School) WA AUSTRALIA: Bryce, Dr Robert ACT
AUSTRALIA: Buchanan, Mr Duncan John (St Patricks College) TAS AUSTRALIA:
Buchberger, Professor Bruno (Universitat Linz) AUSTRIA: Buck, Mr David Alan (W.A. C.A.E.)
WA AUSTRALIA: Buckland, Mr Harold Christopher NEW ZEALAND: Buckland, Mrs Mary
NEW ZEALAND: Bullen, Professor Peter (The University of B.C.) CANADA: Bullock, Mr Ian
James (Marion High School) SA AUSTRALIA: Bungey, Ms Diana Marion (M.A.S.) SA
AUSTRALIA: Burditt, Dr Chris (Napa Valley College) CALIFORNIA: Burgess, Mr Robin
(Kapunda High School) SA AUSTRALIA: Burkhardt, Professor Hugh (University of Nottingham)
ENGLAND: Burnett, Dr J. Dale (Queen's University) CANADA: Burns, Professor John Carlyle
(Royal Military College) ACT AUSTRALIA: Burrow, Dr John (Melbourne CAE) VIC
AUSTRALIA: Burrows, Mr Darryl (Brisbane C.A.E.) QLD AUSTRALIA: Burton, Dr Leone
(Avery Hill College) ENGLAND: Butler, Mr Douglas E. (The M.E.I.) ENGLAND: Butler, Mrs
Helen May SA AUSTRALIA: Butler, Mr Rory Hugh (Waiopehu College) NEW ZEALAND:
Button, Mr Trevor Martin (Nuriootpa High School) SA AUSTRALIA: Byrne, Dr Angelina H.J.
(Melbourne College of Advanced Education) VIC AUSTRALIA: Byrne, Mr Neal Fraser
(Melbourne College of Advanced Education) VIC AUSTRALIA: Byrt, Mr Edwin Andrew
(Victoria College) VIC AUSTRALIA: Callaghan, Mr John (Richmond River High School) NSW
AUSTRALIA: Campbell, Dr Patricia (University of Maryland) USA: Campbell, Mr Ross (Wesley
College) VIC AUSTRALIA: Campigli, Miss Robyn (Mt Barker Primary School) SA AUSTRALIA:
Cannizzaro, Mrs Bolletta Lucilla (Istituto G.Castelnuovo) ITALY: Cardofe, Ms Lucy SA
AUSTRALIA: Carlton, Miss Donna Leanne (Education Department of W.A.) WA AUSTRALIA:
Carpenter, Professor Thomas P. (University of Wisconsin) USA: Carr, Mr Murray R. (St John's
College) NEW ZEALAND: Carraher, Dr Terezinha N. (Mestrado em Psicologia) BRAZIL:
Carrington, Ms Anne (S.A.C.A.E.) SA AUSTRALIA: Carroll, Miss Lesley (All Saints' College)
NSW AUSTRALIA: Carss, Mrs Marjorie C. (University of Queensland) QLD AUSTRALIA:
Carter, Miss Anne Elizabeth (Annesley College) SA AUSTRALIA: Carter, Mr Denver Milton
(Edgewater College) NEW ZEALAND: Carter, Dr Michael Rowlinson (Massey University) NEW
ZEALAND: Carty, Mr John (Merici College) ACT AUSTRALIA: Case, Mr Brendan James
(Morialta High School) SA AUSTRALIA: Casey, Mr John Ernest (R.A.N. College) NSW
AUSTRALIA: Cassidy, John (Queensland Association of Maths Teachers) QLD AUSTRALIA:
Castelnuovo, Professor Emma ITALY: Causey, Ms Patsy (Liberty High School) USA: Cauty, Mr
Andre (Institut de Mathematiques) FRANCE: Chan, Mr Yiu-Hung (University of Hong Kong)
HONG KONG: Chang, Dr Ping-Tung (Laredo State University) REPUBLIC OF CHINA: Chang,
Mrs Swee Tong (Ministry of Education) SINGAPORE: Chapman, Ms Heather Marie (Narrabundah
College) ACT AUSTRALIA: Chapple, Mr Neil William (Peninsula School) VIC AUSTRALIA:
Chastrette, Mr Francois (French Embassy) ACT AUSTRALIA: Chen, Mr Fen (University of
Maryland) TAIWAN: Cheung, Mr Y.L. (University of Hong Kong) HONG KONG: Chiba, Ms S.
(Japan Tours International) JAPAN: Chittleborough, Mr Jeffery D (St Peters College) SA
AUSTRALIA: Chiu, Prof. Dr Sou-Yung (National Taiwan College of Education) TAIWAN:
Chong, Professor Frederick (Macquarie University) NSW AUSTRALIA: Christ, Professor Lily E.
(John Jay College, CUNY) USA: Christiansen, Professor Bent (Royal Danish School of
Education) DENMARK: Christie, Ms Carol R. (Forrestfield Senior High School) WA
AUSTRALIA: Christofferson, Dr Stig (Department of Mathematics Uppsala) SWEDEN:
Christopher, Miss Kay (Mary McKillop College) SA AUSTRALIA: Cibich, Mr Geoffrey Ronald
(Risdon Park High School) SA AUSTRALIA: Ciupryk, Mrs Frances Anne (W.A.C.A.E.) WA
AUSTRALIA: Claffey, Mr Jim (Methodist Ladies College) WA AUSTRALIA: Clark, Mr Donald
Gregory (Enfield High School) SA AUSTRALIA: Clark, Mr John L. (Toronto Board of
Education) CANADA: Clark, Mrs Julie Ann (Salesian College) SA AUSTRALIA: Clark, Mr
Stanley Edward (John Wycliffe Christian School) NSW AUSTRALIA: Clarke, Mrs Barbara Anne
(Monash University) VIC AUSTRALIA: Clarke, Mr David John (Monash University) VIC
AUSTRALIA: Clarke, Mr Douglas McLean (Monash University) VIC AUSTRALIA: Clarke, Mr
Francis John (Hawker College) ACT AUSTRALIA: Clarkson, Mr Philip (Inst. of Catholic
Education) VIC AUSTRALIA: Clav. Mr Bruce Richard (Lithgow High School) NSW
AUSTRALIA: Clayton, Ms Deborah Joy (C.I.A.E.) QLD AUSTRALIA: Cleaves, Mr Paul Joseph
(Burwood Heights H.S.) VIC AUSTRALIA: Clegg, Mr Arthur (Dept Education and Science)
ENGLAND: Clunas, Miss Helen (Department of Education) NSW AUSTRALIA: Cnop-
Grandsard, Dr Francine (Vrije Universiteit Brussel) BELGIUM: Coaker, Mr Peter Brian
(Greenslades) ENGLAND: Cockcroft, Dr Sir Wilfred Halliday (Secondary Examinations Council)
ENGLAND: Cocking, Mr Malcolm S. (Sunshine High School) VIC AUSTRALIA: Cocking, Mr
Robert (Victoria College) VIC AUSTRALIA: Coffey, Mr Vincent (Thomas More College) SA
AUSTRALIA: Cohen, Ms Louise Carol NSW AUSTRALIA: Cohen, Miss Sally Louise
(Melbourne Grammar School) VIC AUSTRALIA: Cohors-Fresenborg, Professor Elmar
(Universiteat Osnabrueck) W.GERMANY: Cole, Mr Michael (The Friends' School) TAS
AUSTRALIA: Coleman, Mr Edwin (S.A.C.A.E., Underdale) SA AUSTRALIA: Collette, Mr
Thomas Warren USA: Collis, Professor Kevin Francis (University of Tasmania) TAS
AUSTRALIA: Colomb, Dr Jacky (J.N.R.P.) FRANCE: Coman, Miss Helen Anne (Maths Centre)
QLD AUSTRALIA: Comer, Mr Peter (Marlborough Boys College) NEW ZEALAND: Comiti,
Mrs Claude (Universite De Grenoblei) FRANCE: Connew, Mrs Karen J. (St Ignatius' College)

NSW AUSTRALIA: Conroy, Dr John Short (Macquarie University) NSW AUSTRALIA: Conti, Miss Elizabeth (Ascham School) NSW AUSTRALIA: Conway, Mr Edwin (Melbourne C.A.E.) VIC AUSTRALIA: Cook, Miss Frances M. (Tenison College) SA AUSTRALIA: Cook, Mr Francis William (Central Institute of Technology) NEW ZEALAND: Cook, Dr Ian T. (University of Essex) ENGLAND: Cook, Ms Susan Flora (Silverstream Kindergarten) NEW ZEALAND: Cooley, Dr Robert (Wabash College) USA: Coolsaet, Mr Arnold (I.T.P.K.) BELGIUM: Cooney, Professor Thomas J. (University of Georgia) USA: Cooper, Ms Jemima Jane TAS AUSTRALIA: Cooper, Mr Leonard John (Department of Education) NEW ZEALAND: Cooper, Professor Martin (University of NSW) NSW AUSTRALIA: Cooper, Dr Thomas James (Brisbane C.A.E.) QLD AUSTRALIA: Cordey, Professor Pierre SWITZERLAND: Cornu, Dr Bernard (Universite Grenoble 1) FRANCE: Cornwell, Mr Raoul Morrow (Rodney College) NEW ZEALAND: Corrieu, Mr Louis (Ministere de l'Education Nationale) FRANCE: Costello, Ms Leanne (Mt Barker Primary School) SA AUSTRALIA: Costello, Mr Maurice John (Salisbury East High School) SA AUSTRALIA: Costello, Mr Patrick William (Swinburne Inst. of Technology) VIC AUSTRALIA: Cotter, Miss Margaret J. (Dwight Morrow High School) USA: Couch, Miss Tui (All Soul's School) QLD AUSTRALIA: Couchman, Mr Kenneth Edgar (Sydney Institute of Education) NSW AUSTRALIA: Coupland, Dr Jonathan (College of St Mark & St John) ENGLAND: Coupland, Miss Mary (Ascham School) NSW AUSTRALIA: Courtney, Mrs Jo (Distance Education Centre) WA AUSTRALIA: Cowling, Mr Gordon (STAGGS) VIC AUSTRALIA: Cozens, Ms Roslyn (Ringwood High School) VIC AUSTRALIA: Craine, Mrs Anne Marie (All Saints' College) NSW AUSTRALIA: Crameri, Mr Leo (D.D.I.A.E.) QLD AUSTRALIA: Crane, Ms Elizabeth Ann (Moreland School Dist./Baker School) USA: Craven, Dr Sherralyn (Central MO State University) USA: Crawford, Dr Douglas Houston (Queen's University) CANADA: Crawford, Ms Kathryn Patricia (Canberra C.A.E.) ACT AUSTRALIA: Crawford, Mr Raymond John (University of Wollongong) NSW AUSTRALIA: Crepin, Professor Roger FRANCE: Crepps, Mrs Sandra (Anderson Elementary) USA: Cribb, Mr Peter (Melton High School) VIC AUSTRALIA: Crimmins, Mr James C. USA: Cromie, Dr Robert G. (St. Lawrence University) USA: Cross, Dr James Manning (W.A. C.A.E.) WA AUSTRALIA: Crossley, Professor John N. (Monash University) VIC AUSTRALIA: Crosswhite, Mr Joe (The Ohio State University) USA: Cuculiere, Professor Roger FRANCE: Cuevas, Dr Gilbert J. (University of Miami) USA: Culhane, Mr Barry William (Gifted Children Task Force) VIC AUSTRALIA: Cumberbatch, Professor Ellis (Claremont Graduate School) USA: Cummins, Mr Jerry (Proviso West High School) USA: Curran, Professor Ann M. (Eastern Connecticut State University) USA: Cushing, Dr Robert G. (Sociology Arts A.N.U.) ACT AUSTRALIA: Cuthbert, Mrs Mae SA AUSTRALIA: Cuttle, Mr Richard James (Burra Community School) SA AUSTRALIA: Czapalay, Mrs Joan CANADA: Czapalay, Mr Stephen (Barrington High School) CANADA: Dadds, Mr Kenneth Jury (Kingscote Area School) SA AUSTRALIA: Dahan, Ms Amy (Universite de Picardie) FRANCE: Dahle, Dr Mary USA: Dalla Torre, Mr Paolo ITALY: Dalman, Mr Peter (Regional Computer Education Centre, Ballarat East) VIC AUSTRALIA: Dalton, Mr LeRoy C. (Wauwatosa West High School) USA: D'Ambrosio, Miss Beatriz S. (Indiana University) BRAZIL: D'Ambrosio, Professor Ubiratan BRAZIL: Damerow, Dr Peter (Max-Planck-Institut fur Bildungsforschung) W.GERMANY: Daniels, Mr Stanley (Grace Lutheran College) QLD AUSTRALIA: Danielsen, Mr Frederick Graham (St Peters College) SA AUSTRALIA: Dansie, Mr Brenton Ronald (S.A.I.T.) SA AUSTRALIA: D'Antuono, Mr Vincent Nino (Corpus Christi College) WA AUSTRALIA: Darche, Professor Michel (E.P.P.V.) FRANCE: Darche-Giorgi, Mrs Marie Laure (APMEP-Orleans-Tours) FRANCE: Das, Dr Amal K. (Joint Director of School Education) INDIA: Davey, Mr Geoffrey Muir (Brisbane C.A.E.) QLD AUSTRALIA: Davey, Mr Roger Glen (Adelaide High School) SA AUSTRALIA: Davidson, Mrs Avelyn Mima (Te Papapa School) NEW ZEALAND: Davis, Mr Brian Peter (Newman S.H.S.) WA AUSTRALIA: Davis, Mrs Lorraine Joy (Tintern CEGGS) VIC AUSTRALIA: Davis, Ms Patricia Joy (Open University) WALES: Davis, Professor Philip J. (Brown University) USA: Davis, Professor Robert B. USA: Dawe, Dr Lloyd Charles (Sydney Institute of Education) NSW AUSTRALIA: Day, Ms Lorraine Frances (Christchurch Grammar School) WA AUSTRALIA: De Corte, Professor Erik (Department of Education Vesaliusstraat 2) BELGIUM: De Klerk, Mr Martin (Springs Boys High School) SOUTH AFRICA: De Lange, Dr Jan HOLLAND: Deakin, Dr Michael A.B. (Monash University) VIC AUSTRALIA: Dear, Mr Russell Alfred (Southland Boys High School) NEW ZEALAND: Deeble, Miss Karen Lynette (Campbell Primary School) ACT AUSTRALIA: DeGuire, Dr Linda J. (University of North Carolina) USA: Del Grande, Mr John J. (Ontario Institute for Studies in Education) CANADA: del Regato, Prof. John Carl (Butler University) USA: Del Riego, Dr Lilia (Universidad Autonoma Metropolitana-Iztapalapa) MEXICO: Deleforge, Mr Daniel (IREM de Lille) FRANCE: Dempster, Mr Russell Paul (Marrara Christian School) NT AUSTRALIA: Dengate, Mr Robert William (Professional Services Centre) NSW AUSTRALIA: Denham, Miss Jennifer (Cannington S.H.S.) WA AUSTRALIA: Dennison, Mrs Christina Marie (Glenunga High School) SA AUSTRALIA: DeRidder, Dr Charleen M. (Knox County Schools) USA: Deschamps, Professor Claude (Lycee Louis Le Grand) FRANCE: Dessart, Dr Donald J. (University of Tennessee) USA: Detwiler, Professor Bettie C. (Western Kentucky University) USA: Dhombres, Professor Jean Guy (U.E.R. Math.) FRANCE: Diaz, Mr Marc Antione (Kambalda Senior High School) WA AUSTRALIA: Diaz R., Professor Jaime

(Facultad de Matematicas) CHILE: Dieschbourg, Professor Robert (Institut Superior d'etudes et de Recherches Pedagogiques) LUXEMBOURG: Dionysopoulou, Ms Triantaphyllia GREECE: Dixon, Ms Margaret Ann (Sydney Technical College) NSW AUSTRALIA: Djordjevic, Mrs Ruth NSW AUSTRALIA: Docwra, Mr David (St Peters College) SA AUSTRALIA: Doherty, Mr Ian Eric (St Josephs College) QLD AUSTRALIA: Doig, Mr Brian A. (Phillip Institute of Technology) VIC AUSTRALIA: Dolmans, Dr Frans H. (Interstudie MO) THE NETHERLANDS: Donovan, Dr Brian Francis (Phillip Institute of Technology) VIC AUSTRALIA: Doodie, Mr Geoffrey Ross (St Francis Xaviers College) VIC AUSTRALIA: Doolan, Mr Lawrence (Melbourne Grammar School) VIC AUSTRALIA: Dorfler, Professor Willibald (Universitat fur Bildungswissenschaften) AUSTRIA: Downes, Mr Barry Francis (B.C.A.E.) QLD AUSTRALIA: Downes, Mrs Judith Susan VIC AUSTRALIA: Dowsey, Mr John (Melbourne C.A.E.) VIC AUSTRALIA: Doyle, Mr Barry Joseph (Wellington Teachers College) NEW ZEALAND: Doyle, Ms Joanna (Los Angeles Unified School District) USA: Doyle, Mr K.D. SOUTH AFRICA: Doyle, Mrs Virginia Eleanor (Jahoe - Truckee High School) USA: Draisma, Mr Jan (I.N.D.E.) MOZAMBIQUE: Drangert, Dr Jan-Olof (Linkoping University) SWEDEN: Drobot, Professor Vladimir (University of Santa Clara) USA: Du Toit, Mr Dawid Johannes (Natal Education Department) SOUTH AFRICA: Dubiel, Sister Josephine (Mary Mackillop College) SA AUSTRALIA: Dubisch, Professor Roy (University of Washington) USA: Dubriel, Dr John B. (F.V.S.C.) USA: Duffin, Ms Janet M. ENGLAND: Duffy, Ms Pat WA AUSTRALIA: Dufour-Janvier, Professor Bernadette CANADA: Dugdale, Ms Sharon (University of Illinois) USA: Dullard, Mr Michael (Wesley College) VIC AUSTRALIA: Dumont, Dr Bernard (Universite Paris 7) FRANCE: Duncan, Mrs Aileen Paterson (Jordanhill College of Education) SCOTLAND: Duncan, Professor Ernest R. USA: Dunigan, Dr Nancy C. (University of Southern Mississippi) USA: Dunkels, Dr Andrejs (University of Lulea) SWEDEN: Dunkley, Dr Mervyn Eric (Macquarie University) NSW AUSTRALIA: Dunkley, Mr Ronald G. (University of Waterloo) CANADA: Dunn, Ms Joan M (Hawker College) ACT AUSTRALIA: Dunne, Mrs Trudy Anne (Gawler High School) SA AUSTRALIA: Dupuis, Mrs Claire (Department of Maths) FRANCE: Duxbury, Mr Desmond F. (Gardens Commercial High School) SOUTH AFRICA: Dzwonnik, Mrs Wanda Zarajczyk (Instytut Ksztalcenia) POLAND: Easson, Mr Tony B. (Reidy Park Primary School) SA AUSTRALIA: Ebeid, Professor William (Ain Shams University) EGYPT: Eckermann, Mr Brian Malcolm (Immanuel College) SA AUSTRALIA: Edmunds, Mrs Florence Joy (Maths Workshop) TAS AUSTRALIA: Edwards, Dr Allen PAPUA NEW GUINEA: Edwards, Mr Anthony James (W.A. C.A.E.) WA AUSTRALIA: Edwards, Ms Josephine (Canberra C.A.E.) ACT AUSTRALIA: Edwards, Mrs Margaret VIC AUSTRALIA: Edwards, Mr Russell (Cleeland High School) VIC AUSTRALIA: Egarr, Mr Alan John SA AUSTRALIA: Egsgard, Mr John (Twin Lakes Secondary School) CANADA: Ehr, Dr Carolyn (Fort Hays State University) USA: Eisenberg, Professor Theodora (Ben Gurion University) ISRAEL: El Tom, Professor Mohamed E.A. (University of Khartoum) SUDAN: Eleazar, Mrs Maria (Manila Science High School) PHILIPPINES: Elix, Mrs Joan Ann (Belconnen High School) ACT AUSTRALIA: Eljoseph, Mr Nathan (Tel-Aviv University) ISRAEL: Elkind, Mrs Elaine (Columbia High School, Maplewood) USA: Ellerton, Dr Nerida F. (Deakin University) VIC AUSTRALIA: Ellery, Mr Philip (Tuart College) WA AUSTRALIA: Elliott, Ms Alison (School of Education) NSW AUSTRALIA: Elliott, Professor David (University of Tasmania) TAS AUSTRALIA: Ellis, Mr John H (Pulteney Grammar School) SA AUSTRALIA: Ellis, Mr Robert James (Paralowie R-12 School) SA AUSTRALIA: Elvey, Ms Anne Frances (Brigidine Convent) VIC AUSTRALIA: Emery, Mrs Francoise (Universite Paris-Sud) FRANCE: Emmer, Professor Michele (Universita La Sapienza) ITALY: Endo, Mr Ari JAPAN: Engel, Professor Arthur (University of Frankfurt) W.GERMANY: Engeler, Professor Erwin SWITZERLAND: English, Miss Lyndall Denis (Brisbane College of Advanced Education) QLD AUSTRALIA: Ernest, Mr Paul (University of West Indies) JAMAICA: Ervynck, Professor Gontran (Ku Leuven Campus Kortryk) BELGIUM: Eshiwani, Professor George S. (Bureau of Educational Research Nairobi) KENYA: Evans, Mr Kenneth McRobert (Scotch College) VIC AUSTRALIA: Evans, Dr Michael (St Michaels Grammar School) VIC AUSTRALIA: Falayajo, Dr Wole (Institute of Education) NIGERIA: Falcetti, Ms Carol (Hillcrest High School) USA: Fandry, Mr Norbert WA AUSTRALIA: Farmanesh, Dr Mounesali IRAN: Farrall, Mr Michael William (Cardijn College) SA AUSTRALIA: Farrell, Mr Brian (W.A. College) WA AUSTRALIA: Farrell, Mr Gregory Kavanagh NSW AUSTRALIA: Farrell, Mr Kevin Matthew (Mt Lawley Technical College) WA AUSTRALIA: Farrell, Dr Margaret (State University of New York Albany) USA: Farzan, Dr Masoud (University for Teacher Education) IRAN: Fasanelli, Dr Florence (Sidwell Friends School) USA: Feghali, Dr Issa (Corning Community College) USA: Fenby, Mr Barrie VIC AUSTRALIA: Fensham, Professor Peter J (Monash University) VIC AUSTRALIA: Fenton, Mrs Alwine (Eden Gardens School) USA: Fergusson, Mr Andrew Louis (Prince Alfred College) SA AUSTRALIA: Fergusson, Ms Barbara SA AUSTRALIA: Ferres, Dr Graham (Victoria College) VIC AUSTRALIA: Fey, Professor James (University of Maryland) USA: Filsell, Mrs Jennifer (Port Adelaide High School) SA AUSTRALIA: Findlay, Ms Janice Gwen (Rochedale S.H.S.) QLD AUSTRALIA: Finlay, Mr James (Trinity Grammar School) VIC AUSTRALIA: Firth, Mr Donald Edward (La Trobe University) VIC AUSTRALIA: Fischer, Professor Walther (Universitat Erlangen-Nurnberg) W.GERMANY: Fisher, Mr Lyle James (Redwood High School) USA: Fitzgerald, Dr Des (Tasmanian CAE) TAS AUSTRALIA: Fitzgerald, Professor William (Michigan

State University) USA: Fitzpatrick, Mr J. Bernard VIC AUSTRALIA: Flake, Professor Janice L. (Florida State University) USA: Fleischer, Mrs Gerhilde (Seminar Fur Schulprak) GERMANY: Fletcher, Miss Sally Annette (Ashburton College) NEW ZEALAND: Fletcher, Dr Trevor James ENGLAND: Flores Penafiel, Prof. Alfinio (Centro de Investigacion en Matematicas) MEXICO: Fogarty, Mrs Kath (Clonard College) VIC AUSTRALIA: Fong, Mr Ho Kheong (Institution of Education) SINGAPORE: Foong, Mr Thai-Hong (Inspectorate of Schools, Perak) MALAYSIA: Fopp, Mrs Wendy Margaret SA AUSTRALIA: Forbes, Mr William M (Cranbrook School) NSW AUSTRALIA: Ford, Mr Kevin (Professional Services Centre) NSW AUSTRALIA: Forte, Mr Stephen Gerard (W.A.I.T.) WA AUSTRALIA: Fowler, Mr Geoffrey (Hodge Hill Comprehensive) ENGLAND: Foxman, Mr Derek David (National Foundation for Educational Research) ENGLAND: Francis, Ms Robyn Denise (Council of Adult Education) VIC AUSTRALIA: Frantz, Mr Dana Thomas (Mt Clear Technical High School) VIC AUSTRALIA: Fraser, Ms Rosemary Ellen (University of Nottingham) ENGLAND: Fraser, Ms Sherry (University of California) USA: Frederick, Mr Harold Richard (Our Lady of Mercy College) NSW AUSTRALIA: Freeman, Mr Chris SA AUSTRALIA: Freislich, Mrs Mary Ruth (University of NSW) NSW AUSTRALIA: Frey, Mr Peter (Christ Church Grammar School) VIC AUSTRALIA: Friis, Mr John Stewart (County Hall, Glenfield) ENGLAND: Frossinakis, Mr Tom A. (Mannum High School) SA AUSTRALIA: Froustet, Miss Mary Elizabeth (Caldwell College) USA: Fuenlabrada, Master Irma (Depto. Investigacoines Educativas) MEXICO: Fuentes, Prof. Wilda G. (Instituto Profesional de Vldivia) CHILE: Fujieda, Miss Michiko (The Association of Mathematical Instruction) JAPAN: Fujimori, Mr Sadaharu (Keio Gijyuku Senior High School) JAPAN: Fujioka, Mr Okimichi (Izumi Senior High School) JAPAN: Fujita, Professor Hiroshi (University of Tokyo) JAPAN: Fujita, Professor Teruo (Minatogawa-Women's Junior College) JAPAN: Fukuda, Mr Shoichiro JAPAN: Fuller, Mr Christopher John (Yewlands School) ENGLAND: Fuller, Mr Milton L (C.I.A.E.) QLD AUSTRALIA: Furusawa, Mr Hiroshi (Narashino High School) JAPAN: Fyles, Mr Andrew (Dripstone High School) NT AUSTRALIA: Gaffney, Mr Robert John (Curriculum Directorate Education Dept) SA AUSTRALIA: Gagen, Professor Terence M. (University of Sydney) NSW AUSTRALIA: Gagnon, Mrs France (G.R.M.S.) CANADA: Galbraith, Dr Peter Lawrence (University of Qld) QLD AUSTRALIA: Gale, Dr Barbara (Prince Georges College) USA: Gale, Mr Gus Neville John (Hornby High School) NEW ZEALAND: Gale, Miss Kathryn SA AUSTRALIA: Galvin, Mr Peter (Bendigo C.A.E.) VIC AUSTRALIA: Galvin, Mrs Selma SA AUSTRALIA: Gan, Mrs Maurine Anne NSW AUSTRALIA: Ganderton, Mr Paul Richard (Cannington Senior High) WA AUSTRALIA: Garden, Mr Robert Alastair (Department of Education, New Zealand) NEW ZEALAND: Gardiner, Dr Anthony (University of Birmingham) ENGLAND: Garner, Mrs Susan Margaret (Ave Maria College) VIC AUSTRALIA: Garnett, Dr Emma W. (Ball State University) USA: Garrett, Mr Colin James (Brisbane C.A.E.) QLD AUSTRALIA: Gauld, Professor David B. (University of Auckland) NEW ZEALAND: Gaulin, Professor Claude (Laval University) CANADA: Gaulke, Mr John Neville (Monash University) VIC AUSTRALIA: Gazzard, Mr Jeffery (Wantirna High School) VIC AUSTRALIA: Gee, Mr James Matthew (Scotch College) WA AUSTRALIA: Geoghegan, Mr Noel (Correspondence School) NSW AUSTRALIA: Gerdes, Dr Paulus (Eduardo Mondlane University) MOZAMBIQUE: Gerleman, Dr Sherry (University of Wyoming) USA: Gibson, Mr Gary QLD AUSTRALIA: Gibson, Dr Robert Dennis (Queensland Institute of Technology) QLD AUSTRALIA: Gilbert, Mr Hugh VIC AUSTRALIA: Giles, Mr Geoff (University of Stirling) SCOTLAND: Gilks, Mr Arthur Joseph (Deakin University) VIC AUSTRALIA: Gilmer, Dr Gloria (University of Maryland, Baltimore County) USA: Ginbayashi, Professor Ko (The Association of Mathematical Instruction) JAPAN: Ginbayashi, Mrs Mieko (The Association of Mathematical Instruction) JAPAN: Gleeson, Mr Ronald Leo (W.A. College) WA AUSTRALIA: Glencross, Mr Michael John (University of Witwatersrand) SOUTH AFRICA: Glover, Mr Owen Hugh (University of Port Elizabeth) SOUTH AFRICA: Goffree, Dr Fred (S.L.O.) THE NETHERLANDS: Gojak, Miss Linda (Hawken School) USA: Goldflam, Ms Barbara (Hollywood S.H.S.) WA AUSTRALIA: Gonzales Davila, Mr Manuel (S.A.P.M. Thales) SPAIN: Goodson, Mrs Mary Jo (McKinley School, Redwood City) USA: Goodwin, Mr Brian Thomas (Narrabundah College) ACT AUSTRALIA: Goodwin, Mr Jeffrey Alan (Hertfordshire College of Higher Education) ENGLAND: Gordon, Mrs Barabara E.R. (Roseville College) NSW AUSTRALIA: Gordon, Mr John (Erindale College) ACT AUSTRALIA: Gori-Giorgi, Professor Claudio (Dept of Information & Systems) ITALY: Gori-Giorgi, Professor Daniela ITALY: Goto, Mr Kenichi (Koyodai High School) JAPAN: Gower, Mr Gregory John VIC AUSTRALIA: Grace, Mr Neville (Curriculum Services Branch) QLD AUSTRALIA: Graf, Professor Klaus-D (Freie Universitat Berlin) GERMANY: Graham, Mrs Beth Evelyn (Professional Services Branch) NT AUSTRALIA: Grant, Mr James Forsyth (Professional Services Centre, Parramatta) NSW AUSTRALIA: Grasser, Professor H.-Siegfried P. (University of South Africa) SOUTH AFRICA: Gray, Dr Alistair (La Trobe University) VIC AUSTRALIA: Gray, Dr Jack David (University of N.S.W.) NSW AUSTRALIA: Gray, Mr Michael John (St Pius X College) NSW AUSTRALIA: Greaves, Mrs Beatrice Louise (Seaford Carrum High School) VIC AUSTRALIA: Greaves, Mr Brian VIC AUSTRALIA: Green, Dr Kevin Norman (Sydney Institute of Education) NSW AUSTRALIA: Green, Dr Walter (W.A. College of Advanced Education) WA AUSTRALIA: Greenbank, Mrs Christine Betty (Wellington Teachers College) NEW ZEALAND: Greenbury, Mr Garnet Jack QLD

AUSTRALIA: Greenleaf, Mr George (Suffield High School) USA: Greger, Professor Karl (University of Gothenburg) SWEDEN: Greig, Mr David Francis (Catholic Education Office, Brisbane) QLD AUSTRALIA: Greitzer, Emeritus Professor Samuel L (Rutgers University) USA: Griffin, Dr Patrick Edward (Education Department of Victoria) VIC AUSTRALIA: Griffiths, Professor Hubert Brian (The University Southampton) ENGLAND: Grimison, Mr Lindsay Athol (Catholic College of Education) NSW AUSTRALIA: Grouws, Professor Douglas A. (University of Missouri) USA: Groves, Dr Susie (Victoria College - Burwood) VIC AUSTRALIA: Grugnetti, Dr Lucia (Istituto di Matematica per Ingegneri) ITALY: Guckin, Ms Alice Mae (University of Minnesota, Duluth) USA: Guy, Mrs Dorothy L. (Huntington Beach High School) USA: Hagger, Mr Brian Gordon SA AUSTRALIA: Haggerty, Mrs Maire-Anne Jane (Education Dept of W.A.) WA AUSTRALIA: Haimes, Mr David Harold (Duncraig Senior High School) WA AUSTRALIA: Haimo, Professor Deborah Tepper (University of Missouri) USA: Hall, Mrs Gwendolyn (New York City Board of Education) USA: Hall, Mr Neil (Riverina C.A.E.) NSW AUSTRALIA: Hall, Mr Robert Edwin (S.A.I.T.) SA AUSTRALIA: Halpin, Mr Terence A. (Brisbane C.A.E.) QLD AUSTRALIA: Hamann, Mr Keith M. (Morialta High School) SA AUSTRALIA: Hamid bin Ahmad Kabir, Mr Shahul (Ministry of Education) MALAYSIA: Hamlett, Mrs Brenda (W.A.I.T.) SA AUSTRALIA: Hanbury, Mr Laurie (B.C.A.E.) QLD AUSTRALIA: Hanna, Dr Gila (Ontario Institute for Studies in Education) CANADA: Hansen, Ms Janette Margaret (Kotara High School) NSW AUSTRALIA: Hargreaves, Mr Graham Morris (Mawson High School) SA AUSTRALIA: Harker, Ms Lorraine (S.A.C.A.E.) SA AUSTRALIA: Harman, Mr Bruce Raymond (Hurstbridge High School) VIC AUSTRALIA: Harman, Dr Christopher (Mitchell C.A.E.) NSW AUSTRALIA: Harmer, Mr Jon (ACTAFE) SA AUSTRALIA: Harris, Mr Donald Allan (R.M.I.T.) VIC AUSTRALIA: Harris, Mrs Elizabeth Anne SA AUSTRALIA: Harris, Ms Pam (Warlpiri Bilingual Schools) NT AUSTRALIA: Harrison, Mr Alan Leslie (Daylesford Technical High School) VIC AUSTRALIA: Hart, Dr Kathleen (Chelsea College) ENGLAND: Hasegawa, Professor Takashi (Oita University) JAPAN: Hasemann, Dr Klaus (Universitat Hannover) W.GERMANY: Hashimoto, Mr Hideo JAPAN: Hashimoto, Mr Hirakazu (Takii Elementary School) JAPAN: Hashimoto, Professor Yoshihiko (Yokohama National University) JAPAN: Hashimoto, Mr Yuichiro (Soegami High School) JAPAN: Hassan, Dr Bakr A. (University of Petroleum & Minerals) SAUDI ARABIA: Hassani, Dr Akbar (Iran University of Science & Technology) IRAN: Hatfield, Professor Evelyn F. (Mankato State University) USA: Hatfield, Professor Francis (Mankato State University) USA: Hatfield, Professor Larry L. (The University of Georgia) USA: Hatherly, Ms Sue (TAFE Staff Development Division) NSW AUSTRALIA: Hatori, Ms Asako (University of Tokyo) JAPAN: Hatton, Mr Leslie Gordon (The Geelong College) VIC AUSTRALIA: Hayashi, Mr Etsuo (Eiko High School) JAPAN: Hayes, Mr Robert (Hawthorn Institute of Education) VIC AUSTRALIA: Hayter, Mr John (School of Education, Bristol) ENGLAND: Healey, Mr Clive (Moonta Area School) SA AUSTRALIA: Heard, Mr Terence James (City of London School) ENGLAND: Heaton, Ms Susan Marie (Tintern CEGGS) VIC AUSTRALIA: Hebenstreit, Ms Rachel (IREM Universite) FRANCE: Heinke, Professor Clarence H. (Capital University) USA: Hemsley, Mr Darryl John SA AUSTRALIA: Henderson, Mrs Jennifer (University of N.S.W.) NSW AUSTRALIA: Henderson, Dr Le Roy Freame (University of Sydney) NSW AUSTRALIA: Henn, Dr Hans-Wolfgang (Lessing-Gymnasium) W.GERMANY: Henry, Mr James Bruce (Victoria College) VIC AUSTRALIA: Henry, Millicent Mary (Wesley College) VIC AUSTRALIA: Henry, Mrs Renee (Florida Dept of Education) USA: Herbert, Mr Peter SA AUSTRALIA: Herman, Dr Maureen (Hunter College, C.U.N.Y.) USA: Herrington, Mr Anthony John (Tasmanian College Advanced Education) TAS AUSTRALIA: Herron, Mr Ian (Birchip Community Education Complex) VIC AUSTRALIA: Herscovics, Professor Nicolas (Concordia University) CANADA: Hersee, Professor John William (Westfield College) ENGLAND: Hershkowitz, Mrs Riva (Weizmann Institute) ISRAEL: Hesamaddinny, Mr Mohssen (Ministry of Education) IRAN: Hess, Ms Patricia M. (Albuquerque Public Schools) USA: Hickling, Mr Ron (Queensland Dept of Education) QLD AUSTRALIA: Hidaka, Professor Keizo (Teikoku Women's University) JAPAN: Hiddleston, Dr Pat (St Margarets School) SCOTLAND: Hiebert, Professor James (University of Delaware) USA: Higginson, Dr William (Queen's University) CANADA: Hight, Mrs Betty J. (USO 250) USA: Hight, Professor Donald W. (Pittsburg State University) USA: Higuchi, Professor Teiichi (Yokohama National University) JAPAN: Hill, Mr Anthony Edmond (Canberra Grammar School) ACT AUSTRALIA: Hill, Dr Donald M (Florida A & M University) USA: Hill, Mrs Kerri Dawn (Merici College) NSW AUSTRALIA: Hill, Mrs Norrene Joyce (Presbyterian Ladies' College) VIC AUSTRALIA: Hill, Professor Shirley A. (University of Missouri - Kansas City) USA: Hine, Mr David John SA AUSTRALIA: Hirabayashi, Mr Hiroaki (Tennoji Senior High School) JAPAN: Hirabayashi, Professor Ichiei (Hiroshima University) JAPAN: Hirdjan, Dr (National Secondary Maths Teacher Upgrading Centre) INDONESIA: Hirose, Mr Kazuaki (Keika High School) JAPAN: Hirota, Mr Hitoshi (Taisei High School) JAPAN: Hirvonen, Miss Soili Sirkku Inkeri (Hyrylan Lukio) FINLAND: Hitotumatu, Professor Sin (Kyoto University) JAPAN: Hlatshwayo, A.F. SWAZILAND: Hocquenghem, Professor Marie-Louise (IREM de Paris VII) FRANCE: Hocquenghem, Mr Serge (C.N.A.M.) FRANCE: Hodges, Mr Colin Charles (Community College of Central Australia) NT AUSTRALIA: Hodgson, Mr Brian (Yarra Valley Anglican School) VIC AUSTRALIA: Hoffman, Dr Nathan (Education Dept of W.A.) WA AUSTRALIA: Hoffman,

Professor Ruth Irene (University of Denver) USA: Hogan, Mr John Stuart (Mandurah Senior High) WA AUSTRALIA: Hoggan, Mr John (Lurnea High School) NSW AUSTRALIA: Holder, Mrs Anne (Catholic Education Office) NSW AUSTRALIA: Holliday, Mr John (Brisbane C.A.E.) QLD AUSTRALIA: Hollis, Ms Mary USA: Holmes, Mr Peter (University of Sheffield) ENGLAND: Holten, Mr Leslie Roy (Northam Senior High School) WA AUSTRALIA: Homma, Mr Toshio (Nihon University Fujisawa Senior High School) JAPAN: Hong, Professor Im Sik (Nippon University) KOREA: Hood, Mr Donald James (Brighton High School) SA AUSTRALIA: Hood, Ms Gail (Chisholm Institute of Technology) VIC AUSTRALIA: Hooper, Miss Sue (Watsonia Technical School) VIC AUSTRALIA: Horain, Mrs Yvette Denis Solange (Lycee Watteau) FRANCE: Horibe, Miss Hiroko (The Association of Mathematical Instruction) JAPAN: Horne, Mrs Marjorie (Victoria College) VIC AUSTRALIA: Horsley, Mr Anthony Rowley (The Armidale School) NSW AUSTRALIA: Horton, Mr Bruce Walter (Mitchell College of Advanced Education) NSW AUSTRALIA: Horvath, Mr Sandor (Elizabeth West High School) SA AUSTRALIA: Hoskins, Mr James Dene (Auckland Grammar School) NEW ZEALAND: Hough, Mr Peter SA AUSTRALIA: Houghton, Mrs Jill F. (Loreto Senior School) NSW AUSTRALIA: House, Professor Peggy A. (University of Minnesota) USA: Howarth, Mrs Anne (Stand VIth Form College) ENGLAND: Howchin, Ms Jennifer Anne (Northam Senior High School) WA AUSTRALIA: Howell, Mrs Alison (Beverly Hills High School) USA: Howell, Mr Laurie (Curriculum Development Centre) ACT AUSTRALIA: Howson, Dr Geoffrey (The University Southampton) ENGLAND: Hsieh, Professor Simon C. (Soochow University) REP. OF CHINA: Hubbard, Ms Ruth Frances (Queensland Institute of Technology) QLD AUSTRALIA: Hughes, Miss Jill (Education Department of W.A.) WA AUSTRALIA: Hulse, Ms Anita M (Eltham High School) VIC AUSTRALIA: Human, Professor Petrus Gerhardus (University of Stellenbosch) SOUTH AFRICA: Hume, Ms Betty Jane (Carnamah D.H.S.) WA AUSTRALIA: Humphries, Mr Alan Desmond VIC AUSTRALIA: Huneke, Dr Harold (University of Oklahoma) USA: Hunt, Mrs Janet Margaret (Churchlands Senior High School) WA AUSTRALIA: Hunter, Mrs Carolyn (Northgate High School) USA: Hunter, Dr Charles (Concord High School) USA: Hunter, Professor John (University of Glasgow) SCOTLAND: Hunter, Mr Maxwell Norman (La Trobe University) VIC AUSTRALIA: Hunting, Dr Robert (W.A.I.T.) WA AUSTRALIA: Hurburgh, Mrs Lesley Margaret TAS AUSTRALIA: Hurn, Mr Richard Walter (Prince Alfred College) SA AUSTRALIA: Hussain, Mr Mansour (Ministry of Education) KUWAIT: Hutchens, Mrs Helen (Melbourne College of Advanced Education) VIC AUSTRALIA: Hutter, Dr Rudolf USA: Hutter, Mrs Ruth (The Dalton School) USA: Hyde, Mr Hartley SA AUSTRALIA: Iacomella, Mr John (Kewdale S.H.S.) WA AUSTRALIA: Iacono, Mr John Robert (Swinburne Institute) VIC AUSTRALIA: Iacono, Mrs Rae Ellen (Nunawading State School) VIC AUSTRALIA: Ichihashi, Mr Kimio (The Association of Mathematical Instruction) JAPAN: Ichikawa, Mr Ryo (The Association of Mathematical Instruction) JAPAN: Iida, Professor Toshi (Kisarazu Technical College) JAPAN: Iizuka, Mr Yutaka (The Association of Mathematical Instruction) JAPAN: Ikari, Mrs Kuniko (The Association of Mathematical Instruction) JAPAN: Ikeda, Mr Takeshi (Kanagawa Prefectural Samukawa High School) JAPAN: Ikin, Mrs Joan Elsa (Mitcham High School) VIC AUSTRALIA: Inglis, Mr Norman John (St Peters College) SA AUSTRALIA: Inui, Mr Haruo JAPAN: Ioannakis, Miss Irene (Wanneroo Senior High School) WA AUSTRALIA: Ireland, Mr Dennis Victor (Mandurah Senior High School) WA AUSTRALIA: Irons, Dr Calvin James (Brisbane C.A.E.) QLD AUSTRALIA: Irving, Mr Stephen Earl (James Cook High School) NEW ZEALAND: Isaacs, Mr Ian (School of Education U.W.I.) JAMAICA: Ishida, Ms Mitsuko (Tokyo Metropolitan Mukojima Kommercial High School) JAPAN: Ishida, Professor Tadao (Hiroshima University) JAPAN: Ishii, Mr Yutaka JAPAN: Ishikura, Mr Toshio (Fuchu Technical High School) JAPAN: Iwata, Mr Kazuo JAPAN: Iwata, Mr Kisamuro (Teikoku Owada High School) JAPAN: Izumi, Mr Kimizo JAPAN: Izumori, Mr Hitoshi (The Association of Mathematical Instruction) JAPAN: Izushi, Professor Takashi (Fukuoka University of Education) JAPAN: Jacob, Mrs Lorraine Florence (Yilgarn Regional Educ.) WA AUSTRALIA: Jacobsen, Mr Edward (UNESCO) FRANCE: Jaeger, Professor Arno (Ruhr University) W.GERMANY: Jalili, Mr Mirza (Ministry of Education) IRAN: Jansen van Vuuren, Mr Andries Barend (Potchefstroom University for C.H.E.) SOUTH AFRICA: Jariyavidyanont, Dr Patrakoon THAILAND: Jaworski, Mrs Barbara (Open University) ENGLAND: Jaworski, Mr John (BBC/Open University) ENGLAND: Jean, Professor Roger V. (University of Quebec in Rimouski) CANADA: Jennings, Mr Trevor Charles (Centenary Heights High) QLD AUSTRALIA: Jensen, Mr Jorgen Cort (Royal School of Education Copenhagen) DENMARK: Jensen, Ms Ulla Kurstein (Vallensbaek Statsskole) DENMARK: Jimbo, Vice-principal Katsuro (Seisen Jogakuin Senior High School) JAPAN: Johansson, Dr Bengt (University of Goteborg) SWEDEN: John, Mr Gregory Frederick (Eudunda Area School) SA AUSTRALIA: Johnson, Mr Allyn (Sydney Institute of Education) NSW AUSTRALIA: Johnson, Mrs Avril (Crown St Public School) NSW AUSTRALIA: Johnson, Dr David C. (Eastern Michigan University) USA: Johnson, Prof. Dr Howard (Syracuse University) USA: Johnston, Mrs Annette (Kingscote Area School) SA AUSTRALIA: Johnston, Ms Jayne Elizabeth (Kelmscott S.H.S.) WA AUSTRALIA: Johnston, Dr Lindsay Collinge (University of Wellington) NEW ZEALAND: Jones, Dr Graham Alfred (Brisbane C.A.E.) QLD AUSTRALIA: Jones, Dr John James VIC AUSTRALIA: Jones, Dr Peter Lloyd (Swinburne Inst. of Technology) VIC AUSTRALIA: Joubert, Dr Stephan Victor

(Technikon Pretoria) SOUTH AFRICA: Joyce, Professor Donald (University of Papua New Guinea) PAPUA NEW GUINEA: Juister, Mrs Barbara (Elgin Community College) USA: Jungwirth, Mr Fred (Eltham College) VIC AUSTRALIA: Jurdak, Professor Murad (American University of Beirut) LEBANON: Kahane, Professor Jean-Pierre (Universite de Paris-Sud) FRANCE: Kaiser, Ms Gabrielle (Gesamthochschule Kassel) W.GERMANY: Kajikawa, Dr Yuji (Osaka Prefectural Ibaraki High School) JAPAN: Kajiwara, Mr Mamoru JAPAN: Kakiya, Miss Hiroko JAPAN: Kalfus, Mr Alfred (Babylon Jr-Sr High School) USA: Kalin, Professor Robert (Florida State University) USA: Kaliszewicz, Dr Olga Szkilnik POLAND: Kalla, Ms Hellevi (OKL, The University of Helsinki) FINLAND: Kam, Ms Rosalind (Ministry of Education) MALAYSIA: Kametani, Mrs Michiko (The Association of Mathematical Instruction) JAPAN: Kametani, Mr Yoshitomi (The Association of Mathematical Instruction) JAPAN: Kaminaga, Miss Ikuko JAPAN: Kaminski, Mr Eugene (McAuley College) QLD AUSTRALIA: Kanemoto, Professor Yoshimichi (Kagoshima Junior College) JAPAN: Kantowski, Professor E.L. (University of Florida) USA: Kapadia, Dr Ramesh (Polytechnic of the South Bank) ENGLAND: Kapur, Professor Jagat N. (Indian Institute of Technology) INDIA: Kari, Mr Francis Masu (UPNG-Goroka Teachers' College) PAPUA NEW GUINEA: Karmelita, Mr Will (St Hilda's Anglican School for Girls) WA AUSTRALIA: Kasetani, Mr Kozo JAPAN: Katagiri, Professor Shigeo (Yokohama National University) JAPAN: Kawaguchi, Ms K. (Japan Tours International) JAPAN: Kawaguchi, Honorary President Tadasu (Japan Society of Mathematical Education) JAPAN: Kayizzi, Mr Abudu (Ministry of Education) UGANDA: Kazim, Professor Maassouma (Qatar University) EGYPT: Keane, Mr Leonard William SA AUSTRALIA: Keegan, Mr Stephen PAPUA NEW GUINEA: Keeves, Dr John Philip (Radford House) VIC AUSTRALIA: Kelly, Mr Edward Stanley (Scottish Education Department) SCOTLAND: Kelly, Mr James Patrick (Mathematics Science Resource Centre) TAS AUSTRALIA: Kelly, Mr Michael Thomas (Queensland Institute of Technology) QLD AUSTRALIA: Kelly, Mr Paul Douglas (Ballarat C.A.E.) VIC AUSTRALIA: Kelly, Miss Sandra Elizabeth (Ingle Farm East Junior Primary) SA AUSTRALIA: Kemp, Mrs Marian Elizabeth F. (Penrhos College) WA AUSTRALIA: Kennedy, Mr Denis C (Broadford High School) VIC AUSTRALIA: Kennedy, Mrs Janice A. (Buckley Park High School) VIC AUSTRALIA: Kennelly, Mr Paul Anthony (University of Melbourne) VIC AUSTRALIA: Kenney, Dr Margaret (Boston College) USA: Kennington, Mr Raymond William (The University of Adelaide) SA AUSTRALIA: Keogh, Mrs Erica Clare (University of Zimbabwe) ZIMBABWE: Keranto, Professor Tapio (University of Tampere) FINLAND: Kerslake, Dr Daphne (Bristol Polytechnic) ENGLAND: Kieran, Ms Carolyn (Univ. du Quebec a Montreal) CANADA: Kieren, Dr Thomas E. (University of Alberta) CANADA: Kikuchi, Mr Isamu (Showa-diichi Senior High School) JAPAN: Kilborn, Mr Wiggo (W & B Educ. Prod.) SWEDEN: Kilpatrick, Professor Jeremy (University of Georgia) USA: Kimura, Miss Toshiko (The Association of Mathematical Instruction) JAPAN: Kimura, Mr Yoshio (The Association of Mathematical Instruction) JAPAN: Kimura, Professor Yuzo (Tokyo College of Music) JAPAN: Kindt, Mr Martin THE NETHERLANDS: King, Mr Declan (Copland College) ACT AUSTRALIA: Kinnersley, Mr Ian William David (Daylesford Tech/High School) VIC AUSTRALIA: Kirkpatrick, Mrs Connie (McAuley High School) NEW ZEALAND: Kiroff, Mrs Irene Elizabeth (St Marys College) SA AUSTRALIA: Kirsche, Dr Peter (Universitat Augsburg) GERMANY: Kissane, Mr Barry (The University of Western Australia) WA AUSTRALIA: Klamkin, Professor Murray S. (University of Alberta) CANADA: Kleiner, Professor Israel (York University) CANADA: Klika, Dr Manfred (Hodischule Hildesheim) W.GERMANY: Knight, Dr Gordon Henry (Massey University) NEW ZEALAND: Knipp, Professor Emeritus John (University of Pittsburgh) USA: Knox, Mrs Nora Constance ENGLAND: Kobayashi, Professor Ugoro (Niigata University) JAPAN: Koeswachjoeni, Ms (Senior Secondary School) INDONESIA: Kok, Mr Douwe (Free University) THE NETHERLANDS: Komano, Mr Makoto (Tokyo Metropolitan Koiwa Senior High School) JAPAN: Kondo, Mr Toshiji (The Association of Mathematical Instruction) JAPAN: Konig, Mr Gerhard (Zentralblatt f. Didaktik der Mathematik) W.GERMANY: Koop, Dr Anthony J. (Macquarie University) NSW AUSTRALIA: Korbosky, Mr Richard (Western Australia College) WA AUSTRALIA: Koss, Ms Roberta (Redwood High School) USA: Koster, Mr John (New York City Board of Education) USA: Kota, Professor Osamu (Rikkyo University) JAPAN: Kotagiri, Mr Tadato (The Association of Mathematical Instruction) JAPAN: Kotler, Dr Martin (Pace University, Pleasantville) USA: Koumi, Mr Jack (B.B.C. Open University) ENGLAND: Koura, Mr Toshihide (Kasei Junior High School) JAPAN: Koyama, Mr Masataka (Hiroshima University) JAPAN: Krawcewicz, Mrs Zinaida Lowicka (Instytut Ksztalcenia) POLAND: Kressen, Mr David P. (Polytechnic School) USA: Kriegsman, Dr Helen (Pittsburg State University) USA: Kriel, Mr Dawid Jacobus (University of Port Elizabeth) SOUTH AFRICA: Krulik, Professor Stephen (Temple University) USA: Kulshrestha, Dr Devendra Kumar (Flinders University) SA AUSTRALIA: Kumar, Ms Mia Amanda (Burwood Girls High) NSW AUSTRALIA: Kunimoto, Assistant Professor Keiyu (Kochi, University) JAPAN: Kupari, Mr Pekka Antero (University of Jyvaskyla) FINLAND: Kusumoto, Supervisor Zennosuke (Tokyo Metropolitan Inst. for Educational Research) JAPAN: Laborde, Dr Colette (IMAG, University of Grenoble) FRANCE: Labrie, Professor Jean-Marie (Les Entreprises culturelles enr.) CANADA: Labrousse, Professor Jean-Philippe (Dept de Mathematiques Parc Valrose) FRANCE: Lacampagne, Professor Carole (University of Michigan-Flint) USA: Lai, Professor Hang-Chin (National Tsing-Hua University)

REP. OF CHINA: Laine, Professor Ilpo Ensio (University of Joensuu) FINLAND: Lamshed, Mr Murray SA AUSTRALIA: Landwehr, Dr James (A.T. & T Bell Laboratories) USA: Langdon, Mr Nigel (Middle Row School) ENGLAND: Langevin, Mr Michel (Ecole Normale Superieure) FRANCE: Langham, Mr Russell Arthur (Education Department, Durham) ENGLAND: Langston, Mr Garth Lawrence (Onslow College) NEW ZEALAND: Lanier, Prof. Perry E. (Michigan State University) USA: Lantry, Mr Anthony (St Joseph's College) NSW AUSTRALIA: Laridon, Mr Paul Edward (Witwatersrand University) SOUTH AFRICA: Larkin, Mr Alan Thomas (S.A.C.A.E) SA AUSTRALIA: Larson, Mr Larry (Green River Community College, Auburn) USA: Lavallade, Mr Didier FRANCE: Lavcov, Prof. Dr Igor (State Commete of Science & Technic) USSR: Lawrence, Mrs Christina (Kewdale S.H.S.) WA AUSTRALIA: Lawrence, Mr Denis (Rossmoyne S.H.S.) WA AUSTRALIA: Lawrence, Mrs Jennifer Anne (Ave Maria College) VIC AUSTRALIA: Lawrence, Ms Joyce Margaret (Victorian Education Department) VIC AUSTRALIA: Lawson, Mrs Barbara (Mary MacKillop College) SA AUSTRALIA: Lawson, Mr Julian Bruce (St Pius X College) NSW AUSTRALIA: Lawton, Mr Alexander Vivian (Brisbane Grammar School) QLD AUSTRALIA: Layton, Ms Katherine Puckett (Beverly Hills High School) USA: Le Berre, Mr Guy (L.E.P. Batiment) FRANCE: Lean, Mr Glendon Angove (University of Technology) PAPUA NEW GUINEA: Leder, Dr Gilah (Monash University) VIC AUSTRALIA: Lee, Professor David (S.A. Institute of Technology) SA AUSTRALIA: Lee, Dr James VIC AUSTRALIA: Lee, Ms Norma (Seymour College) SA AUSTRALIA: Lees, Mr Kevin (W.I.A.E.) VIC AUSTRALIA: Leeson, Mr Neville James Frederick (N.R.C.A.E.) NSW AUSTRALIA: LeGrand, Mr Pierre-Olivier (I.R.E.M.) TAHITI: Lehto, Professor Olli (University of Helsinki) FINLAND: Leigh-Lancaster, Mr David (Yarra Valley Anglican School) VIC AUSTRALIA: Leino, Professor Jarkko (University of Tampere) FINLAND: Leipnik, Professor Roy Bergh (University of California, Santa Barbara) USA: Leitch, Ms Julie Adeline ACT AUSTRALIA: Lenchner, Dr George (Math Olympiads for EL Schools) USA: Lerman, Mr Stephen (Chelsea College Ctre for Science & Mathematics Education) ENGLAND: Lesh, Professor Richard (Northwestern University) USA: Lester, Mr David (Korringal High School) NSW AUSTRALIA: Lester, Professor Frank K. (Indiana University) USA: Levick, Mr Bruce Robert (Rathkeale College) NEW ZEALAND: Levick, Mrs Gwyneth Frances (St Matthew's Collegiate School) NEW ZEALAND: Lewis, Miss Clair Barbara (Lameroo Area School) SA AUSTRALIA: Lewis, Mr Edward (Correspondence School) NSW AUSTRALIA: Lewis, Dr Gillian M. (Brisbane C.A.E.) QLD AUSTRALIA: Lewit, Mrs Anita Pnina VIC AUSTRALIA: Lichtenberg, Dr Betty K. (University of South Florida) USA: Lichtenberg, Dr Donovan R. (University of South Florida) USA: Lilienthal, Ms Robyn Ann VIC AUSTRALIA: Lillie, Mrs Margaret (George Street Normal School) NEW ZEALAND: Lim, Mr Chee-Lin SINGAPORE: Lim, Dr Chong-Keang (University of Malaya) MALAYSIA: Lim, Dr Tek An (Senior Secondary School) INDONESIA: Lincoln, Mr Gregory Brian (University of Tasmania) TAS AUSTRALIA: Lind, Dr Ingemar (Linkoping University) SWEDEN: Lindquist, Dr Mary M. (Columbus College) USA: Lindsay, Mr John Joseph (Eltham College) VIC AUSTRALIA: Linfoot, Mrs Joyce Jones (Lucy Cavendish College) ENGLAND: Ling, Mr John Francis (Westfield College) ENGLAND: Linton, Dr Matthew David (University of Hong Kong) HONG KONG: Little, Dr John Jeffrey (Armidale C.A.E.) NSW AUSTRALIA: Livingstone, Mr Ian David (New Zealand Council for Educational Research) NEW ZEALAND: Llewellyn, Mr Bruce (Directorate of Studies) NSW AUSTRALIA: Lluis, Professor Emilio (Institutode Matematicas) MEXICO: Lobb, Mr Lawson (Mitchell College) NSW AUSTRALIA: Long, Professor Calvin T. (Washington State University) USA: Long, Mrs Eleanor Margaret (University of Adelaide) SA AUSTRALIA: Lopez-Yanez, Dr Alejandro (Civdad Universitaria) MEXICO: Lorenz, Dr Dan H. (Israel Institute of Technology) ISRAEL: Lorimer, Professor Peter James (University of Auckland) NEW ZEALAND: Losada, Prof. Mary (Universidad Nacional) COLOMBIA: Losada, Prof. Ricardo (Universidad Antonio Narino) COLOMBIA: Lott, Professor Johnny W. (University of Montana) USA: Lounesto, Professor Pertti (Helsinki University of Technology) FINLAND: Lovitt, Mr Charles Joseph (Mooroolbark High School) VIC AUSTRALIA: Low, Dr Brian Charles (Macquarie University) NSW AUSTRALIA: Low, Dr Lewis (University of Adelaide) SA AUSTRALIA: Lowe, Mr Ian Roy VIC AUSTRALIA: Lowenthal, Professor Francis D. (University of Mons) BELGIUM: Lowman, Dr Pauline (Western Kentucky University) USA: Luckie, Mrs Margaret (Correspondence School) NSW AUSTRALIA: Lunn, Dr Adrian Daniel (The Open University) ENGLAND: Luscombe, J.P. (Education Department, Adelaide) SA AUSTRALIA: Lynch, Mrs Barbara Joy (Lauriston Girls' School) VIC AUSTRALIA: Maarschalk, Professor Jan (Rand Afrikaans University) SOUTH AFRICA: Machida, Professor Shoichiro (Saitama University) JAPAN: Mack, Dr John (Sydney University) SA AUSTRALIA: MacKenzie, Mr Douglas Edward (University of NSW) NSW AUSTRALIA: MacKenzie, Mr Robert SA AUSTRALIA: MacKinlay, Mr Malcolm John (Institute of Catholic Education) VIC AUSTRALIA: Macrides, Dr George (W.A. College) WA AUSTRALIA: Mahanti, Dr P.K. (University of Ilorin) NIGERIA: Mahon, Brother James Michael (Catholic College of Education) NSW AUSTRALIA: Mailler, Mrs Carol Ann ACT AUSTRALIA: Maioni, Mr Robert (Curriculum Development Centre, Boroko) PAPUA NEW GUINEA: Makela, Mr Heikki FINLAND: Makino, Mr Masahiro (Keio Gijuku Junior High School) JAPAN: Malakunas, Mr Henri (Melbourne C.A.E.) VIC AUSTRALIA: Mallinson, Mr Philip (Bush School) USA: Malone, Dr John Anthony (Western Australia Institute of Technology) WA AUSTRALIA:

Maloney, Ms Madeleine (Sacre Coeur) VIC AUSTRALIA: Manabe, Assistant Professor Tatsuki (Kochi University) JAPAN: Mangan, Miss Jane (Bendigo C.A.E.) VIC AUSTRALIA: Mann, Mr Wilfrid J.A. (Dept of Education & Science) ENGLAND: Mansfield, Dr Helen Margaret (W.A. C.A.E.) WA AUSTRALIA: Marar, Dr Raman P. (Swinburne Institute of Technology) VIC AUST. Marosz, Professor Alma (San Diego State University) USA: Marshall, Mr Stephen John (Sacred Heart College Senior) SA AUSTRALIA: Martin, Mr David Charles (Launceston Church Grammar School) TAS AUSTRALIA: Martin, Professor Jean Claude (Academie de Bordeaux) FRANCE: Martin, Mrs Kathy (Hawker College) ACT AUSTRALIA: Martinez, Professor Arturo G. (Facultad de Matematicas) CHILE: Maskell, Mr Frederick G.B. (Algonquin College) CANADA: Mason, Dr John H. (Open University) ENGLAND: Matovu, Mr F.K. (National Curriculum Development Ctre) UGANDA: Matsoha, S.T. LESOTHO: Matsumaru, Mr Mitsuo (Ichikawa Gakuen High School) JAPAN: Matsumoto, Mrs Hiroko (Hiroshima Municipal Gion Junior High School) JAPAN: Matsumura, Mr Toro (Japan Tours International) JAPAN: Matsuo, Professor Yoshitomo (Science University of Tokyo) JAPAN: Matsuoka, Professor Motohisa (Bunkyo University) JAPAN: Matthes, Mr Harold Thomas (Curtis High School) USA: Matthes, Mrs Laney Dahl (Curtis High School & Sigma) USA: Matthews, Mr Denis Bede (St Marys College) NSW AUSTRALIA: Matthews, Miss Eleanor (Chicago Board of Education) USA: Mauk, Dr Cherry C. (St Edward's University) USA: Maurer, Ms Ann Irene (Shelford C.E.G.G.S.) VIC AUSTRALIA: Maurin, Mr Roger (16 bis Rue du General) FRANCE: Maxwell, Mr Kelvin Matthew (Catholic Education Office) NSW AUSTRALIA: May, Mr Robert Leonard (R.M.I.T.) VIC AUSTRALIA: Maynard, Mr Leslie Howard (Kadina Memorial High School) SA AUSTRALIA: Mayo, Mrs Helen Elaine (Hillmorton High School) NEW ZEALAND: McAvaney, Dr Kevin Lawrence (Deakin University) VIC AUSTRALIA: McBay, Ms Shirley M. (Massachussetts Institute of Technology) USA: McCabe, Mrs Christine Mary SA AUSTRALIA: McCall, Mrs Gwynnyth Ada (Sacred Heart Regional Girls' College) VIC AUSTRALIA: McCallum, Mr Brian NSW AUSTRALIA: McCloskey, Mrs Elinor NSW AUSTRALIA: McComb, Ms Sandra VIC AUSTRALIA: McCombe, Mr John (S.A. Institute of Teachers) SA AUSTRALIA: McCreddin, Mr Robert (Education Department, East Perth) WA AUSTRALIA: McDonald, Professor Anita D. (University of Missouri) USA: McDonald, Mr Kenneth S. (Catholic High School) NSW AUSTRALIA: McDonald, Miss Margaret (Bomaderry High School) NSW AUSTRALIA: McDonell, Ms Win (V.I.S.E.) VIC AUSTRALIA: McDonough, Ms Denise (St Marys High School) NSW AUSTRALIA: McDougall, Dr David (W.A. C.A.E.) WA AUSTRALIA: McGee, Professor Ian J. (University of Waterloo) CANADA: McGuckin, Miss Lisa Mary (All Hallows' School) QLD AUSTRALIA: McInnes, Mr Donald Ross (Waikerie High School) SA AUSTRALIA: McIntosh, Mr Alistair James (W.A.C.A.E.) WA AUSTRALIA: McKenzie, Mr Dennis Francis (Huntly College) NEW ZEALAND: McKenzie, Mr Rodney John (C.S.R.E.O.) SA AUSTRALIA: McKnight, Dr Curtis C. (University of Oklahoma) USA: McLean, Mr Roderick John (P.I.T.) VIC AUSTRALIA: McLellan, Mrs Janette (School of Mines & Industries) VIC AUSTRALIA: McLennan, Ms Anne (Banyule High School) VIC AUSTRALIA: McLeod, Professor Douglas B. (San Diego State University) USA: McLone, Dr Ronald Redman (University of Southampton) ENGLAND: McManus, Professor Maurice (University of NSW) NSW AUSTRALIA: McNabb, Dr William (Skyline Center) USA: McNally, Ms Corinne (Bendigo C.A.E.) VIC AUSTRALIA: McQualter, Mr John William (University of Newcastle) NSW AUSTRALIA: McWhirter, Ms Wendy (St Thomas Aquinas School) WA AUSTRALIA: Mead, Mr Ray Francis NSW AUSTRALIA: Medigovich, Mr William M. (Redwood High School) USA: Medlin, Mrs Diana de E. (Pembroke College) SA AUSTRALIA: Mehaffey, Ms Judith (College of External Studies) NSW AUSTRALIA: Meiklejohn, Mr Graham Keith (Curriculum Services, Brisbane) QLD AUSTRALIA: Meissner, Professor Hartwig (Universitat Munster) GERMANY: Melick, Miss Robin (Dover Heights High School) NSW AUSTRALIA: Mellor, Mr David (Chisholm Institute of Technology) VIC AUSTRALIA: Mendicuti, Dr Teresa Navarro De MEXICO: Menon, Mr Ramakrishnan (M.P.S.I.) MALAYSIA: Menzel, Miss Ute GERMANY: Merrell, Miss Margaret Anne (Abbotsford Primary School 1886) VIC AUSTRALIA: Meserve, Dr Bruce E. (University of Vermont) USA: Meserve, Mrs Dorothy T. USA: Metenier, Professor Jacqueline (Sodexport/Unipresse) FRANCE: Meyer, Mr Joerg W.GERMANY: Michaels, Mr Fahmy (Copland College) ACT AUSTRALIA: Michalak, Mr Pierre FRANCE: Michel-Pajus, Ms Annie (Lycee Joffre, Allee de la Citadelle) FRANCE: Michener, Mr Norman John (Tech/High School) VIC AUSTRALIA: Michiwaki, Professor Yoshimasa (Gunma University) JAPAN: Mientka, Professor Dr Walter E. (University of Nebraska) USA: Mikhail, Dr Khairy A. Sedrak EGYPT: Miller, Mrs Diane (University of Missouri) USA: Miller, Mr Keith Raymond (Prince Alfred College) SA AUSTRALIA: Miller, Mr Leonard Mark (Sahuarita Junior High School) USA: Mills, Dr Graham (S.A.I.T.) SA AUSTRALIA: Milton, Mr Kenneth George (University of Tasmania) TAS AUSTRALIA: Mina, Dr Fayez M. (University College of Bahrain) BAHRAIN: Minato, Professor Saburo (Akita University) JAPAN: Mitchell, Dr Merle (University of New Mexico) USA: Mitchell, Mr Peter Raymond (Millicent High School) SA AUSTRALIA: Mitchelmore, Dr Michael C. (University of The West Indies) JAMAICA: Mitsutsuka, Professor Naomi (Faculty of Education Toyama University) JAPAN: Miwa, Professor Tatsuro (The University of Tsukuba) JAPAN: Miyai, Ms N. (Japan Tours International) JAPAN: Miyamoto, Professor Kohei (Mukogawa Women's University) JAPAN: Miyasato, Mr Yasuhisa JAPAN: Miyeke, Dr R. JAPAN: Modesitt, Mrs

Maxine D. (Rosemary School) USA: Mogi, Professor Isamu (Bunkyo University) JAPAN: Mohyla, Dr Jury (S.A.C.A.E.) SA AUSTRALIA: Mok, Miss Mo Ching Magdalena (University of Hong Kong) HONG KONG: Mollehed, Mr Ebbe SWEDEN: Monakhov, Prof. Dr Vadim (Moscow Makazenko 5/16) USSR: Moncur, Mr David R.S. (Buttershaw Upper School) ENGLAND: Monthubert, Mr Bernard FRANCE: Moore, Dr Charles G. (Northern Arizona University) USA: Moortgat, Dr Luke R. (De La Salle University) PHILIPPINES: Moran, Professor W. (University of Adelaide) SA AUSTRALIA: Morden, Mrs Margaret G. (Los Angeles Unified School District) USA: More, Mr Roger Vivian (Geelong Grammar School) VIC AUSTRALIA: Morello, Ms Lorenza Maria SA AUSTRALIA: Moresh, Dr Stephen (Francis Lewis H.S.) USA: Morgan, Mrs Elsie (Peninsula School) VIC AUSTRALIA: Morgan, Mr Francis William (Sydney Institute of Education) NSW AUSTRALIA: Morin, Professor Christiane (Lycee Technique) FRANCE: Morioka, Mr Shigeyoshi (Hyogo University of Education) JAPAN: Moriya, Mrs Mieko (Seika Girls' Senior High School) JAPAN: Morley, Mrs Daphne (University of Melbourne) VIC AUSTRALIA: Morley, Mr Derek T. (Glengowrie High School) SA AUSTRALIA: Moroney, Mr Ronald John (John Therry High School) NSW AUSTRALIA: Morony, Mr Laurie William (Glengowrie High School) SA AUSTRALIA: Morris, Mr Peter George (Ingle Farm High School) SA AUSTRALIA: Morris, Dr Sidney A (LaTrobe University) VIC AUSTRALIA: Morrow, Miss Lorna Jane (Brookbanks Centre) CANADA: Mortlock, Professor Roland S. (University of W.A.) WA AUSTRALIA: Moule, Mrs Carolyn SA AUSTRALIA: Moushovitz-Hadar, Dr Nitsa ISRAEL: Mowchanuk, Mr Timothy (Brisbane C.A.E.) QLD AUSTRALIA: Mullen, Dr Gail (University of Guam) GUAM: Mulligan, Mrs Joanne T. (Catholic College of Education) NSW AUSTRALIA: Mundy, Dr Joan (University of New Hampshire) USA: Munro, Mr Ian (St Peters College) SA AUSTRALIA: Mura, Dr Roberta (Laval University) CANADA: Murakami, Professor Haruo (Kobe University) JAPAN: Murakami, Mrs Nancy A. (LAU School District) USA: Murakami, Mr Toshio (The Association of Mathematical Instruction) JAPAN: Murphy, Miss Julianne SA AUSTRALIA: Murray, Mrs Ann (St Peter's Lutheran College) QLD AUSTRALIA: Murray, Mrs Jenny (Mitchell College of Advanced Education) NSW AUSTRALIA: Murray, Mr Robert Scott (N.C.A.E.) NSW AUSTRALIA: Murray, Mr Stephen QLD AUSTRALIA: Myers, Ms Jane Clare (St Francis Xavier College) VIC AUSTRALIA: Nagai, Miss Keiko (Junior High School to Ochanomizu University) JAPAN: Nagasaki, Mr Eizo (National Institute for Educational Research) JAPAN: Nagata, Ms Chigusa (Doshisha High School) JAPAN: Nakajima, Mr Shigenori (Yamagataminami High School) JAPAN: Nakamura, Mr Tadashi (Ottemon Otemae High School) JAPAN: Nakata, Mr Motokuni JAPAN: Nancarrow, Mr Rodney George (Central Western Reo) SA AUSTRALIA: Napper, Mrs Margaret Ruth (S.A.C.A.E.) SA AUSTRALIA: Nason, Ms R.A. (The University of Sydney) NSW AUSTRALIA: Natera, Mr Julius (Dept of Education, Curriculum Unit) PAPUA NEW GUINEA: Neal, Mr David Marshall (Birmingham Polytechnic) ENGLAND: Neale, Mrs Jillian (Education Dept W.A.) WA AUSTRALIA: Nebres, Professor Bienvenido PHILIPPINES: Needham, Mr Peter (C.E.R.E.O.) SA AUSTRALIA: Neill, Mr Henry (National Aboriginal Education Commission) AUSTRALIA: Neilson, Mr Jack (Knox Grammar School) NSW AUSTRALIA: Nelson, Mr John R. (The Geelong College) VIC AUSTRALIA: Nener, Mr Kevin (Education Department of W.A.) WA AUSTRALIA: Nestle, Professor Fritz (Padagogische Hochschule) Nestle W.GERMANY: Nettle, Mrs Brenda (Annesley College) SA AUSTRALIA: Nettur, Mr Mithran Krishnan (Mount St Joseph College) VIC AUSTRALIA: Neufeld, Professor Evelyn M. (San Jose State University) USA: Neumann, Emeritus Professor Bernhard Hermann (Australian National University) ACT AUSTRALIA: Newberry, Mr Ronald Arthur (Newton Moore S.H.S.) WA AUSTRALIA: Newman, Dr Michael F. ACT AUSTRALIA: Newman, Mrs Mira (Copland College) ACT AUSTRALIA: Newton, Mr William David (Curriculum Programs Branch) VIC AUSTRALIA: Nicholls, Mr Paul Michael VIC AUSTRALIA: Nicholls, Ms Penny Ann (Ascham School) NSW AUSTRALIA: Nicholson, Mr Michael John (Hilton College) SOUTH AFRICA: Nickson, Dr Marilyn (Chelmer-Essex Inst of Higher Education) ENGLAND: Niknam, Dr A (Mashad University) IRAN: Nikol'si, Academician USSR: Nisbet, Mr Steven (Brisbane C.A.E.) QLD AUSTRALIA: Nishimiya, Professor Han (Saitama University) JAPAN: Nishitani, Mr Izumi (Osaka Prefectural Izumiohtsu High School) JAPAN: Niss, Professor Mogens (Roskilde University Centre) DENMARK: Nitobe, Mr Kuniyuki (Yamagataminami High School) JAPAN: Nkabinde, Mr Washington Beacon F. (Dept of Education & Training) SOUTH AFRICA: Noda, Mr Yorinobu JAPAN: Noguchi, Professor Hiroshi JAPAN: Noguchi, Mr Yasufumi (Yuge Mercantile Marine College) JAPAN: Nohda, Assistant Professor Nobuhiko (The University Of Tsukuba) JAPAN: Nojima, Ms Junko (Nihon University Busan Women's High School) JAPAN: Nolan, Mr Peter William (Scotch College) SA AUSTRALIA: Nomachi, Mr Tadashi (The Association of Mathematical Instruction) JAPAN: Noyes, Dr Beatrice (N.Y.C. Board of Education) USA: Nozawa, Mr Shigeru (The Association of Mathematical Instruction) JAPAN: Nuesch, Professor Peter E. SWITZERLAND: Nulty, Mr Bruce Raymond (Sadadeen High School) NT AUSTRALIA: Oberholzer, Mr Guillaume Johannes SOUTH AFRICA: O'Brien, Mr Dennis John (Brisbane C.A.E.) QLD AUSTRALIA: O'Brien, Mr Mark Raymond (Cannington Senior High) WA AUSTRALIA: O'Brien, Ms Patricia (Glenaven Primary School) NEW ZEALAND: O'Connor, Mr Gregory James (St John's College) NSW AUSTRALIA: Ogura, Mr Shohei (Doshisha Junior High School) JAPAN: O'Halloran, Mr Peter Joseph (Canberra C.A.E. School of Information Science) ACT AUSTRALIA: Ohshio, Mr Yoshiaki (Asahi Denko Kogy. Co Ltd)

JAPAN: Okada, Mr Yoshio (Hiroshima University) JAPAN: Okamori, Professor Hirokazu (Osaka University of Education) JAPAN: O'Keefe, Mr Edward Donald (Macquarie University) NSW AUSTRALIA: Okiyama, Mr K. (The Association of Mathematical Instruction) JAPAN: Okuda, Dr Shogo JAPAN: Okuno, Mr Hiroshi (The Association of Mathematical Instruction) JAPAN: Oldham, Ms Elizabeth Evelyn (Trinity College) IRELAND: Olivecrona, Mr Carl W.S. (Department of Education, Wanganui) NEW ZEALAND: Oliver, Mr Jack (Darwin Community College) NT AUSTRALIA: Olivier, Mr Alwyn Ivo (University of Stellenbosch) SOUTH AFRICA: Ollongren, Professor Alexander (Institute of Applied Math and Comp. Science) THE NETHERLANDS: Olsen, Mr Christopher Marius (Wanneroo Senior High School) WA AUSTRALIA: Olssen, Mr Kevin Harry (Glengowrie High School) SA AUSTRALIA: Olsson, Mr Stig SWEDEN: Olstorpe, Mrs Kristina SWEDEN: Omatsu, Mr Eiichi (Koishikawa Upper Secondary High School) JAPAN: O'Neill, Mr Desmond John (Brisbane Grammar School) QLD AUSTRALIA: O'Neill, Rev Patrick John (St Johns College) NEW ZEALAND: Oosthuizen, Dr Wynand Louw (University of Pretoria) SOUTH AFRICA: Ormiston-Smith, Mr Robert (St Michaels Grammar School) VIC AUSTRALIA: Osborne, Professor Alan (Ohio State University) USA: Oscarsson, Mr Edor SWEDEN: Otake, Mr Noboru (The Upper Secondary School, Tsukuba University) JAPAN: Outhred, Ms Lynne (Macquarie University) NSW AUSTRALIA: Owens, Dr Douglas T. (University of British Columbia) CANADA: Owens, Mrs Marilyn B. (Silveyville Primary School) USA: Ozawa, Mr Kenichi (The Association of Mathematical Instruction) JAPAN: Ozawa, Mr Yoshiaki JAPAN: Packer, Mr Christopher (Blackfriars Priory School) SA AUSTRALIA: Padberg, Professor Friedhelm (Universitat Bielefeld) W.GERMANY: Page, Mr Warren (New York City Technical College) USA: Palanisamy, Mr K.V. (Ministry of Education, Pesiaran Duta) MALAYSIA: Palmer, Mr Michael (Maths/Science Resource Centre, Hobart) TAS AUSTRALIA: Palmer, Mr William Douglas (Sydney Institute of Education) NSW AUSTRALIA: Pantua, Miss Remedios (Patrician Bros. College) NSW AUSTRALIA: Parker, Mr Raymond George (Morphett Vale High School) SA AUSTRALIA: Parris, Mr Alan (Linwood High School) NEW ZEALAND: Parrott, Dr David Leslie (University of Adelaide) SA AUSTRALIA: Partridge, Mr Barry Douglas (Brisbane C.A.E.) QLD AUSTRALIA: Parzysz, Mr Bernard (IREM) FRANCE: Pascoe, Miss Linley (Cannington Senior High) WA AUSTRALIA: Pateman, Mr Neil Arthur (University of Georgia) USA: Patterson, Professor Edward McWilliam SCOTLAND: Payne, Professor Joseph N. (University of Michigan) USA: Paynter, Mrs Jennifer Dawn (Walford School) SA AUSTRALIA: Peard, Mr Robert (Brisbane C.A.E.) QLD AUSTRALIA: Pearson, Mrs Yvonne Loraine (Sacred Heart College Senior) SA AUSTRALIA: Pedler, Ms Angela (Magill Junior Primary School) SA AUSTRALIA: Pedler, Mr Pender James (W.A. College) WA AUSTRALIA: Peerboom, Miss Eugenie Louise (C.C.E.G.G.S) ACT AUSTRALIA: Pegg, Mr John E. (University of New England) NSW AUSTRALIA: Peiris, Mr Anthony (La Chataigneraie School) SWITZERLAND: Pellerey, Professor Michele (Universita Salesiana) ITALY: Pence, Professor Barbara J. (San Jose State University) USA: Pengelly, Ms Helen Jean SA AUSTRALIA: Penman, Mr John Macleay (Dept of Education, Tamworth) NSW AUSTRALIA: Perham, Professor Bernadette (Ball State University) USA: Perham, Mrs Joyce (Nundah Remedial Centre) QLD AUSTRALIA: Perry, Dr Bob (Institute of Early Childhood Studies) NSW AUSTRALIA: Persson, Dr Carl-Gustav (Linkoping University) SWEDEN: Petkovic, Mr Peter ACT AUSTRALIA: Peyrot, Mrs Jacqueline (Lycee la Coliniere) FRANCE: Phillips, Mr Brian Richard (Swinburne Inst. of Technology) VIC AUSTRALIA: Phillips, Professor E. Ray (University of South Florida) USA: Phillips, Mr John W. (Brisbane C.A.E) QLD AUSTRALIA: Phillips, Dr Richard Julian (Shell Centre for Mathematical Education) ENGLAND: Pillsbury, Mrs Virginia (Milwaukee Area Technical College) USA: Pimm, Mr David (The Open University, England) ENGLAND: Pirie, Dr Susan Elizabeth B. (University of Warwick) ENGLAND: Pitman, Dr Jane (University of Adelaide) SA AUSTRALIA: Pittick, Mr Ian Anthony (Safety Bay High School) WA AUSTRALIA: Plaister, Ms Robyn (Directorate of Studies) NSW AUSTRALIA: Pollak, Dr Henry O. (Bell Laboratories) USA: Pollingher, Mr Adolf (Technion) ISRAEL: Pollock, Mr Gerard Joseph (Scottish Council for Research in Education) SCOTLAND: Porteous, Dr Ian Robertson (University of Liverpool) ENGLAND: Porter, Professor Gerald J. (University of Pennsylvania) USA: Poteralski, Mr Peter (Adelaide College of T.A.F.E.) SA AUSTRALIA: Potts, Professor Renfrey Burnard (University of Adelaide) SA AUSTRALIA: Poulton, Mr Christopher (Wesley College) VIC AUSTRALIA: Power, Mr James Patrick (St Columba's High School) SCOTLAND: Praeger, Professor C. (University of Western Australia) WA AUSTRALIA: Praeger, Prof. Cheryl Elizabeth (University of Western Australia) WA AUSTRALIA: Prawiroatmodjo, Mr Sugiman (Senior Secondary School) INDONESIA: Price, Dr Jack (Palos Verdes Peninsula U.S.D.) USA: Priest, Mr Donald Keith (Eudunda Area School) SA AUSTRALIA: Pruzan, Ms Lynn USA: Purcell, Mrs Leah (Adelaide College of Technical & Further Education) SA AUSTRALIA: Purser, Miss Prudence Meredith (Burnside High School) NEW ZEALAND: Putt, Dr Ian John (James Cook University) QLD AUSTRALIA: Puygrenier, Mr Michel (Ministere de l'Education Nationale) FRANCE: Quail, Mr Michael Merrett (Bordeaux Primary) SOUTH AFRICA: Quinlan, Brother Cyril (Catholic College of Education Sydney) NSW AUSTRALIA: Quinn, Mr Michael John (I.C.E. Oakleagh Campus) VIC AUSTRALIA: Quirk, Mrs Jackie (Catholic College of Education) NSW AUSTRALIA: Raass, Mr Asa (St John's High School) TONGA: Rachlin, Dr Sidney (University of Hawaii) USA: Rade, Dr Lennart (Chalmers Univ of Technology) SWEDEN:

Raftery, Mr John (S.A.C.A.E.) SA AUSTRALIA: Rahilly, Dr Alan John VIC AUSTRALIA:
Rajan, Mrs Radha (United Nations International School) USA: Ralston, Professor Anthony
(SUNY at Buffalo) USA: Ramaswami, Mr S. (Mathematical Association of Botswana)
BOTSWANA: Randall, Mr Robert William (Education Department of W.A.) WA AUSTRALIA:
Randall, Mrs Susan Jane (Seacombe High School) SA AUSTRALIA: Rankin, Mrs Leanne
(Elizabeth Field J.P. School) SA AUSTRALIA: Ransley, Dr Wayne (University of Tasmania) TAS
AUSTRALIA: Rasmussen, Mr Duncan (Phillip Institute) VIC AUSTRALIA: Rasoulian,
Professor Amid (Tehran University) IRAN: Rasschaert, Mrs Yvette (R.M.S. II) BELGIUM:
Ratsch, Mr Colin (Seaton High School) SA AUSTRALIA: Rautio, Ms Arninne USA: Rayment,
Dr Philip Robert (Gippsland I.A.E.) VIC AUSTRALIA: Reardon, Miss Marina Ann (W.A.C.A.E.)
WA AUSTRALIA: Redden, Mr Michael George (Riverina C.A.E.) NSW AUSTRALIA:
Redonnet, Mr Jean-Claude (French Embassy) ACT AUSTRALIA: Reeler, Miss Susan Fiona
(Gardens Commercial High) SOUTH AFRICA: Reeves, Ms Noelene (W.A. College of Advanced
Education) WA AUSTRALIA: Reilly, Mrs Barbara Joy (University of Auckland) NEW
ZEALAND: Reilly, Professor Ivan Leon (University of Auckland) NEW ZEALAND: Reiss, Dr
Matthias B. (Universitat, IDM) W.GERMANY: Remilton, Mrs Elizabeth (Annesley College) SA
AUSTRALIA: Resek, Professor Diane (San Francisco State University) USA: Retallack, Mr
Kenneth (W.A.C.A.E.) WA AUSTRALIA: Reuille-Irons, Mrs Rosemary (Brisbane C.A.E.) QLD
AUSTRALIA: Reyes, Dr Laurie Hart (University of Georgia) USA: Reynolds, Mr Peter (Suffolk
Education Department) ENGLAND: Rhodes, Professor Mary (Bishops University) CANADA:
Rice, Ms Sally (S.A. Institute of Technology) SA AUSTRALIA: Rice, Mr Stephen Donald
(Kambalda Senior High School) WA AUSTRALIA: Richards, Ms Anthea Ellen (Dripstone High
School) NT AUSTRALIA: Richardson, Dr Jim (University of Sydney) NSW AUSTRALIA:
Richardson, Mrs Kathy (Maple School) USA: Ridgway, Dr James Edward (University of
Lancaster) ENGLAND: Riel, Mrs Maryjean Allan (Truckee Meadows Community College) USA:
Rigby, Dr John Frankland (University College, Cardiff) WALES: Robb, Mrs Kay Valerie
(Ashburton College) NEW ZEALAND: Robbie, Dr Desmond Alexander (Melbourne C.A.E.) VIC
AUSTRALIA: Roberts, Mr Robert Edwin QLD AUSTRALIA: Robertson, Mrs Denise (Walford
School) SA AUSTRALIA: Robinson, Miss Belinda Jane (Billabong H.S.) NSW AUSTRALIA:
Robinson, Dr David Francis (University of Canterbury) NEW ZEALAND: Robitaille, Dr David
Ford (University of British Columbia) CANADA: Rodriguez, Dr Argelia (U.S.A. Department of
Education) USA: Rogers, Miss Sandra Lee (District 14 - Board of Education N.Y.C.) USA:
Rogerson, Dr Alan Thomas (Scotch College) VIC AUSTRALIA: Rolfe, Ms Edith Mary (Methodist
Ladies College) WA AUSTRALIA: Romanoski, Mrs Prudence Ann (Murdoch University) WA
AUSTRALIA: Romberg, Professor Thomas (University of Wisconsin) USA: Ronshausen, Dr
Nina L. (Texas Tech University) USA: Roos, Mr Richard Clair (Dept of Education & Training)
SOUTH AFRICA: Rose'Meyer, Mr Peter (St Leo's College) VIC AUSTRALIA: Rosenthal,
Professor Julius I. (Kingsborough Community College) USA: Roseveare, Mr David ENGLAND:
Ross, Mr Kenneth Joseph (Board of Secondary Education) WA AUSTRALIA: Roussel, Mr Jean
(Universite Paris Sud) FRANCE: Roussel, Mr Yves (A.D.C.S.) FRANCE: Ruberu, Mr Jathiratne
(La Trobe University) SRI LANKA: Rubinstein, Professor Zalman (University of Colorado)
ISRAEL: Rudder, Mr Kimberley Moulton (Cranbrook School) NSW AUSTRALIA: Rudnick,
Professor Jesse A. (Temple University) USA: Ruffini, Mr Sergio (Melton High School) VIC
AUSTRALIA: Russotti, Mr Vincent S. (North Shore Schools) USA: Ruthmell, Dr Edward C.
(University of Northern Iowa) USA: Ryan, Sister Angela (Brigidine Convent) VIC AUSTRALIA:
Ryan, Mrs Suzanne (Sunraysia College of Technical and Further Education) VIC AUSTRALIA:
Ryan, Mr William John (Malak School) NT AUSTRALIA: Rydberg, Dr Lars Gosta (Linkoping
University) SWEDEN: Sadr, Mr Mohammad T. (Tehran University) IRAN: Saengcharoenrat, Dr
Porama (Prince of Songkla University) THAILAND: Sag, Mrs Lidia SA AUSTRALIA: Sagara,
Mr Kazimasa (Miyazaki-kita High School) JAPAN: Sagnet, Professor Jean-Pierre (College de
Chicoutimi) CANADA: Sakagami, Mr Eisho (Jyohoku High School) JAPAN: Sakaki, Mr Tadao
(The Association of Mathematical Instruction) JAPAN: Sakane, Ms Hizuru (Higashi Jojo Junior
High School) JAPAN: Sakimoto, Mr Kenji (Kure Board of Education) JAPAN: Sakitani,
Professor Shinya (Hyogo University of Education) JAPAN: Salagaras, Dr Stan (Risdon Park High
School) SA AUSTRALIA: Salmon, Mr Barry James (Curriculum Services Branch, Brisbane) QLD
AUSTRALIA: Sampson, Mr Cecil Kenneth SOUTH AFRICA: Samuels, Mr Peter James (Wesley
College) WA AUSTRALIA: Saruwatari, Assistant Professor Seiju (Ariake Technical College)
JAPAN: Sasaki, Professor Gentaro (Hyogo University of Teacher Education) JAPAN: Sasaki, Mr
Shuei (Tateoka Senior High School) JAPAN: Sasaki, Mr Tetsuro (Hiroshima University) JAPAN:
Sato, Mr Katsuhiko (The Association of Mathematical Instruction) JAPAN: Sato, Professor
Shuntaro (Fukushima University) JAPAN: Sattler, Mr John Anthony (St Ives High School) NSW
AUSTRALIA: Sauer, Mr Graeme Ronald (Christchurch Teachers College) NEW ZEALAND:
Sauer, Mrs Jennifer Joy NEW ZEALAND: Sauviney, Dr Randall (University of California) USA:
Sawada, Mr Toshio (National Institute for Educational Research) JAPAN: Sawyer, Mr Edward
(Catholic College of Education Sydney) NSW AUSTRALIA: Saxe, Professor Geoffrey (Graduate
School of Education) USA: Saxena, Professor Subhash C. (University of South Carolina) USA:
Scales, Mrs Dora VIC AUSTRALIA: Scanlon, Mrs Helen (Monte Sant' Angelo College) NSW
AUSTRALIA: Scheding, Dr John (St George Institute of Education) NSW AUSTRALIA:

Schildkamp-Kuendiger, Dr Erika (University of Windsor, Canada) W.GERMANY: Schleiger, Mr Howard Edward (Victoria College) VIC AUSTRALIA: Schleiger, Dr Noel William (Phillip Institute of Technology) VIC AUSTRALIA: Schliemann, Dr Analucia Dias (Mestrado em Psicologia) BRAZIL: Schmitt, Professor Frederick G. (College of Marin) USA: Schoenfeld, Dr Alan H (University of Rochester) USA: Schofield, Ms Deborah Ellen (Murraylands Region Education Office) SA AUSTRALIA: Schramm, Dr Ruben (Tel-Aviv University) ISRAEL: Schroder, Dr F.T. Mark (University of Waikato) NEW ZEALAND: Schroeder, Dr Thomas L. (University of Calgary) CANADA: Schuring, Mr Henk N. (C.I.T.O.) THE NETHERLANDS: Schwartzman, Mrs Pauline (University of the Witwatersrand) SOUTH AFRICA; Schwarzenberger, Professor Rolph (University of Warwick) ENGLAND: Schweiger, Professor Fritz (University Salzburg) AUSTRIA: Scoltock, Mr David (Canberra Grammar School) ACT AUSTRALIA: Scotland, Miss Joan (District 14 - Board of Education N.Y.C.) USA: Scott, Mrs Janice (C.C.E.S. Mount Saint Mary Campus) NSW AUSTRALIA: Scott, Mrs Joy (Mathematics Learning Centre) WA AUSTRALIA: Scott, Dr Paul Raymond (University of Adelaide) SA AUSTRALIA: Scott, Mr Roger (Eltham High School) VIC AUSTRALIA: Secker, Mr David Martin (Balga Senior High School) WA AUSTRALIA: Sedgman, Mr David Burgess NSW AUSTRALIA: Seeger, Dr Falk (Institut fur Didaktik der Mathematik) W.GERMANY: Seifried, Ms Ethne (Seymour College) SA AUSTRALIA: Sein Min, U (University of Rangoon) BURMA: Selden, Mr Rodney Charles (PNG University of Technology) PAPUA NEW GUINEA: Selkirk, Mr Keith Edward (School of Education University Park Nottingham) ENGLAND: Semadeni, Professor Zbigniew (Polish Academy of Sciences) POLAND: Sendov, Mr Blagovest C. (Bulgarian Academy of Science) BULGARIA: Senk, Professor Sharon L. (Syracuse University) USA: Seydi, Professor Hamet (Universite de Dakar) SENEGAL: Shahvarani-Semnani, Dr Ahmad (Teacher Training University Tehran) IRAN: Shanks, Ms Anne Isabella (Para Vista High School) SA AUSTRALIA: Shannon, Dr Anthony G. (N.S.W.I.T.) NSW AUSTRALIA: Shaw, Mrs Pamela Frances (Macquarie University) NSW AUSTRALIA: Shaw, Mr Robin A.E. (Guildford Grammar School) WA AUSTRALIA: Shefi, Mrs Yael (Hebrew University) ISRAEL: Shelley, Ms Nancy ACT AUSTRALIA: Shepherd, Ms Leesa Jayne (Houghton Primary School) SA AUSTRALIA: Shepherd, Mrs Rosslyn SA AUSTRALIA: Sherer, Mrs Ada (Technion - Israel Inst of Technology) ISRAEL: Sherman, Dr Brian Francis (University of Adelaide) SA AUSTRALIA: Shibata, Mr Hidekazu (Shijonawate-kita Senior High School) JAPAN: Shida, Mr Masao (Nagayama Junior High School) JAPAN: Shidfar, Dr Abdullah (Iran University of Science & Technology) IRAN: Shield, Mr Paul Gregory (Brisbane C.A.E.) QLD AUSTRALIA: Shigematsu, Professor Keiichi (Nara University of Education) JAPAN: Shimizu, Mr Ken-Ichi JAPAN: Shinatrakool, Dr Somchai (Bansomdej Teachers' College) THAILAND: Shipton, Ms Sheila Frances (Southend College of Technology) ENGLAND: Shkil, Prof. Dr Nikolai (State Pedagogical Institute) USSR: Shoda, Mr Ryo (Musashi High School) JAPAN: Short, Mr Greg NSW AUSTRALIA: Shravah, Mr Ved WESTERN SAMOA: Shuard, Miss Hilary B (Homerton College) ENGLAND: Shulte, Dr Albert P. (Oakland School) USA: Shumway, Professor Richard (Ohio State University) USA: Siebuhr, Mr Edward George (D.D.I.A.E.) QLD AUSTRALIA: Siemon, Mrs Dianne E. (Monash University) ENGLAND: Silver, Professor Edward A. (San Diego State University) USA: Sim, Dr Wong Kooi (Institute of Education) SINGAPORE: Simon, Mr Jacques (Lycee Charlemagne) FRANCE: Sin, Mr Kwai Meng (Curriculum Development Institute of Singapore) SINGAPORE: Sinha, Mr D.K. (Jadavpur Institute of Calcutta) INDIA: Sinha, Prof. D.K. (Jadavpur University) INDIA: Sinha, Professor Sri Rama (Mehta Research Institute & University of Allahabad) INDIA: Siu, Dr Man-Keung (University of Hong Kong) HONG KONG: Sizer, Ms Terry Louise (Pt Pirie R.E.O.) SA AUSTRALIA: Skoogh, Dr Lennart SWEDEN: Skovsmose, Professor Dr Ole (Aalborg University Centre) DENMARK: Skypek, Dr Dora Helen (Emory University) USA: Slinn, Mr Keith (Sydney Institute of Education) NSW AUSTRALIA: Smale, Mr Terry William VIC AUSTRALIA: Smallman, Mr Donald E.F. (W.A.C. Advanced Education) WA AUSTRALIA: Smit, Dr Cornelis Petrus (University of Botswana) BOTSWANA: Smith, Mr Brian John (Melrose High School) ACT AUSTRALIA: Smith, Mrs Christine (Bendigo C.A.E.) VIC AUSTRALIA: Smith, Mr Donald Alfred SA AUSTRALIA: Smith, Miss Doreen Helen (Naremburn Professional Services) NSW AUSTRALIA: Smith, Professor Eugene P. (Wayne State University) USA: Smith, Mr Greg (St Leo's College) NSW AUSTRALIA: Smith, Mr J. Clifford (University of Durban Westville) SOUTH AFRICA: Smith, Mr Michael T. (Alice Springs High School) NT AUSTRALIA: Smith, Mr Ronald William (Livingstone Primary School) VIC AUSTRALIA: Smith, Ms Wendy Anne (Seaford-Carrum High) VIC AUSTRALIA: Soderstrom, Dr Kerstin (Norregardskolan) SWEDEN: Soderstrom, Dr Ulf (Hogskolan i Vaxjo) SWEDEN: Soedjadi, Dr INDONESIA: Soma, Mr Sumihiko JAPAN: Sourdillat, Ing. Frederic (C.N.A.M.) FRANCE: Southwell, Dr Beth (Nepean C.A.E.) NSW AUSTRALIA: Sowey, Dr Eric Richard (University of N.S.W.) NSW AUSTRALIA: Spargo, Ms June Elaine (Pennsville Public Schools) USA: Spark, Mr Philip George (Cleeland High School) VIC AUSTRALIA: Sparke, Mr Donald Adrian (Dept of Education, Windsor) NSW AUSTRALIA: Spiro, Ms Lea (Levinsky Teachers' College) ISRAEL: Srinivasan, Mr Panchalam K. (University of Madras) INDIA: Srivastava, Dr Dinesh Mohan (Correspondence School) VIC AUSTRALIA: Stacey, Dr Kaye (Victoria College - Burwood) VIC AUSTRALIA: Stacey, Dr Peter (La Trobe University) VIC AUSTRALIA: Stack, Mr Bruce Gavin (Rosny College) TAS AUSTRALIA: Stamp, Mr Michael Arthur (Wallington High School for Boys)

ENGLAND: Starritt, Mr Alan J. (Scottish Curriculum Development Service) SCOTLAND: Stebbing, Mr Leslie Paul (Yeshivah College) VIC AUSTRALIA: Steel, Mrs Roslyn (Council Adult Education) VIC AUSTRALIA: Stees, Mrs Janet (Beverley Hills Calif, School Dist) USA: Stees, Dr Larry (Beverley Hills Calif. School Dist.) USA: Steffe, Professor Leslie (The University of Georgia) USA: Stein, Prof. Dr Gunter (Technische Hochschule) W.GERMANY: Steinbring, Dr Heinz (IDM/Uni.) W.GERMANY: Steiner, Professor Hans-Georg (Institute for the Didactics of Mathematics) W.GERMANY: Stephens, Dr Walter Maxwell (Education Department - Curriculum Branch Carlton) VIC AUSTRALIA: Stewart, Mr Ian (Mitchell College) NSW AUSTRALIA: Stewart, Mr John Prescot (D.D.I.A.E) QLD AUSTRALIA: Stodart, Miss Barbara (Adelaide High School) SA AUSTRALIA: Stoddart, Mr Henry James (Otago Boys' High School) NEW ZEALAND: Stoessiger, Ms Margaret (H.M.C.) TAS AUSTRALIA: Stokes, Ms Victoria Burnard (NT Department of Education) NT AUSTRALIA: Stone, Mr David William (Colo High School) NSW AUSTRALIA: Storck, Mr David L. (The Westbourne Grammar School) VIC AUSTRALIA: Stoutemyer, Professor David R. (University of Hawaii) USA: Stowasser, Professor Roland (TU Berlin) W.GERMANY: Straesser, Dr Rudolf (University of Bielefeld) W.GERMANY: Strathie, Mr Malcolm Graeme (Education Office, Clare) SA AUSTRALIA: Streefland, Dr Leen THE NETHERLANDS: Street, Dr Anne Penfold (University of Queensland) QLD AUSTRALIA: Streeter, Mrs Sandra Ann VIC AUSTRALIA: Strzelec, Mr Stephen Zbieniew (Glengowrie High School) SA AUSTRALIA: Stubbs, Mrs Pamela (Kelmscott S.H.S.) WA AUSTRALIA: Suffolk, Mr John Arnold (University of Papua New Guinea) PAPUA NEW GUINEA: Sujono, Mr INDONESIA: Sullivan, Mr Peter ACT AUSTRALIA: Sullivan, Mr Peter (I.C.E. (Oakleigh)) VIC AUSTRALIA: Sumner, Dr Raymond (Nat Foundation for Educ. Research) ENGLAND: Sunouchi, Professor Haruo JAPAN: Surch, Mrs Eileen Frances VIC AUSTRALIA: Sutcliffe, Professor John Philip (University of Sydney) NSW AUSTRALIA: Sutherland, Miss Christine (Bulahdelah Central School) NSW AUSTRALIA: Sutton, Miss Joan Elizabeth (Department of Education, Seven Hills) NSW AUSTRALIA: Suzuki, Mr Minoru (The Association of Mathematical Instruction) JAPAN: Swan, Mr Kenneth Campbell (Paisley High School) VIC AUSTRALIA: Swan, Mr Malcolm (Shell Centre for Math Education) ENGLAND: Swann, Mrs Elizabeth Anne (Ringwood High School) VIC AUSTRALIA: Swe, Professor Chit (University of Rangoon) BURMA: Sweeney, Mr Ross William (Kambolda Senior High School) WA AUSTRALIA: Sweller, Mr John (University of NSW) NSW AUSTRALIA: Swift, Mr Jim (Nanaimo Senior Sec. School) CANADA: Swinson, Mr Kevan V. (Brisbane C.A.E.) QLD AUSTRALIA: Szetela, Dr Walter (The University of British Columbia Vancouver) CANADA: Szibrowski, Mr Rudolf (Greystanes High) NSW AUSTRALIA: Tacey, Mr Brad (Nudgee College) QLD AUSTRALIA: Tadman, Mrs Christine (Eltham College) ENGLAND: Taffe, Mr John Raymond (Australian National University) ACT AUSTRALIA: Takada, Mrs Kayoko (The Association of Mathematical Instruction) JAPAN: Takahashi, Mr Shigeji JAPAN: Takasai, Mr Toshio (High School attached to Nippon University) JAPAN: Takenouchi, Mr Osamu (Osaka University) JAPAN: Takimoto, Miss Reiko (Fujimi Junior High School) JAPAN: Tame, Ms Linda (Aranui High School) NEW ZEALAND: Tamlin, Mr Eric Arthur (S.A.I.T.) SA AUSTRALIA: Tamura, Miss Paula M. (Prospect High School) USA: Tanabe, Mr Syohe (The Association of Mathematical Instruction) JAPAN: Taole, Dr James K. (National University of Lesotho) LESOTHO: Tapia, Professor Carlos Alberto (Independencia 2625) ARGENTINA: Taylor, Mr David (Epsom College) ENGLAND: Taylor, Dr Francis B. (Manhattan College) USA: Taylor, Dr Harold (National Council of Teachers of Mathematics) USA: Taylor, Mr Jim (S.A. C.A.E.) SA AUSTRALIA: Taylor, Dr John Robert (Statistics Section, Dept of Education) VIC AUSTRALIA: Taylor, Dr Peter J. (C.C.A.E.) ACT AUSTRALIA: Taylor, Mr Robert Richard (Wattle Park Teachers Centre) SA AUSTRALIA: Telfer, Mrs Raelene June (Adelaide High School) SA AUSTRALIA: Teljakovskii, Dr S.A. USSR: Tenido, Mr Cristito (Marist Bros.) NSW AUSTRALIA: Terada, Professor Fumiyuki JAPAN: Edward Bruce Chicheley (W.A.C.A.E.) WA AUSTRALIA: Thornton, Mr Stephen John (Reynella East High School) SA AUSTRALIA: Thorpe, Mr Dennis (O'Connell Education Centre) ACT AUSTRALIA: Tilly, Mr Graham (Sydney University) NSW AUSTRALIA: Tobin, Dr Kenneth (W.A.I.T.) WA AUSTRALIA: Tomkins, Mr Robert Geoffrey (Bell Park High School) VIC AUSTRALIA: Tong, Prof. Dr Monbill (National Chengchi University) TAIWAN: Toohey, Miss Margaret Ann (S.E.C.W.) QLD AUSTRALIA: Toomer, Mrs Ruth Grace NEW ZEALAND: Toovey, Mrs Lyn (University of Canterbury) NEW ZEALAND: Toure, Mr Saliou (University of Abidjan) IVORY COAST: Towns, Mr Peter George (Papatoetoe High School) NEW ZEALAND: Trafton, Dr Paul (National College of Education, Evanston) USA: Travers, Professor Kenneth J. USA: Travis, Dr Betty S.P (University of Texas) USA: Treilibs, Mr Vern (Wattle Park Teachers Centre) SA AUSTRALIA: Trembath, Mr David (Brighton High School) SA AUSTRALIA: Trend, Mrs Judy WA AUSTRALIA: Trenerry, Dr Dennis W. (University of NSW) NSW AUSTRALIA: Trestrail, Mr Colin (Kambalda Senior High School) WA AUSTRALIA: Trewenack, Mrs Toni (M.C.A.E. I.E.C.D.) VIC AUSTRALIA: Trudinger, Dr Peter Lawrence (University of Sydney) NSW AUSTRALIA: Truran, Mr John Maxwell SA AUSTRALIA: Truran, Mrs Kathleen SA AUSTRALIA: Tsagas, Professor Grigorios (University of Thessaloniki) GREECE: Tsubota, Ms Etsuko (Sapporo Higashi High School) JAPAN: Tsuji, Professor Yoshio (Shiga University) JAPAN: Turner, Mr Alan Rodney (Hornsby College of TAFE) NSW AUSTRALIA: Turner, Dr John C. (University of Waikato) NEW ZEALAND:

Tyler, Mrs Shirley May WA AUSTRALIA: Tyrnell, Mrs Davina (National Aboriginal Education Commission) AUSTRALIA: Uche, Dr Peter Ikechukwu (University of Nigeria) NIGERIA: Uegaki, Mr Wataru (The Association of Mathematical Instruction) JAPAN: Uetake, Professor Tsuneo (Asia University) JAPAN: Uibo, Mrs Dadamain (National Aboriginal Education Commission) AUSTRALIA: Ujiie, Professor Katsumi (Tokai University) JAPAN: Umakoshi, Mr Yoichi (Tamagawa Senior High School) JAPAN: Usiskin, Ms Karen (Scott, Foresman and Co.) USA: Usiskin, Dr Zalman (University of Chicago) USA: Vaidya, Professor Arun M. (South Gujarat University) INDIA: Valida, Mr Abelardo C. PHILIPPINES: van Barneveld, Mr Gert (S.L.O.) THE NETHERLANDS: van de Craats, Professor Jan (K.M.A.) THE NETHERLANDS: Van den Brink, Mr Frederik Jan (University of Utrecht) THE NETHERLANDS: Van Dormolen, Dr Joop (University of Utrecht) THE NETHERLANDS: Van Hamelsveld, Mr Ronald NEW ZEALAND: Van Lint, Prof. Dr Jack H. (Eindhoven University of Technology) THE NETHERLANDS: Van Pinxteren, Mr Peter (Ballarat C.A.E.) VIC AUSTRALIA: Van Rensburg, Mr Ben SOUTH AFRICA: Van Rooy, Mr Martinus Paul (Technikon Pretoria) SOUTH AFRICA: Van 't Riet, Mr Nol THE NETHERLANDS: Vazquez de Tapia, Professor Nelly Esther (Independencia 2625) ARGENTINA: Vellard, Mr Dominique FRANCE: Verbyla, Mr John (Kildare College) SA AUSTRALIA: Vere-Jones, Professor David (Victoria University) NEW ZEALAND: Verghese, Mr Ninan (Catholic Regional College) VIC AUSTRALIA: Vergnaud, Dr Gerard (Centre d'Etude du Processus Cognitifs) FRANCE: Vermandel, Professor Alfred (University of Antwerp) BELGIUM: Vermeulen, Professor J (University of Western Cape) SOUTH AFRICA: Verstappen, Dr Piet F.L. (S.L.O.) THE NETHERLANDS: Vervoort, Prof. Dr Gerardus (Lakehead University) CANADA: Villani, Professor Vinicio (Dipartimento di Matematica, Pisa) ITALY: Vincent, Mrs Julie Kaye (W.A.C.A.E.) WA AUSTRALIA: Vinner, Dr Shlomo (Hebrew University) ISRAEL: Voges, Mr Kenneth Charles WA AUSTRALIA: Vollugi, Mr Peter Robert VIC AUSTRALIA: Von Harten, Dr Gerd F. (IDM/University) W.GERMANY: Vonk, Mr Gustav A. (University of Utrecht) THE NETHERLANDS: Vugts, Mrs Mary Philomena (St Marys College) SA AUSTRALIA: Wachsmuth, Dr Ipke (Universitat Osnabruck) W.GERMANY: Wah, Mr Richard FIJI: Wain, Mr Geoffrey Thomas (University of Leeds) ENGLAND: Wakefield, Mr Jeffrey H (Phillip Institute of Technology) VIC AUSTRALIA: Wall, Professor Gordon Elliott (University of Sydney) NSW AUSTRALIA: Wallace, Miss Robyn Ann (Longanlea State High School) QLD AUSTRALIA: Walmsley, Mr Mike (Colac Education Centre) VIC AUSTRALIA: Walter, Mr Roger Graham (Paisley High School) VIC AUSTRALIA: Wanby, Professor Goran (University of Lund) SWEDEN: Wangerin, Mr Orville F. (Sunshine North Technical School) VIC AUSTRALIA: Warrick, Mr Malcolm Douglas (Marryatville High School) SA AUSTRALIA: Watanabe, Mr Noriaki (Showa-daiichi Senior High School) JAPAN: Watanabe, Mr Tanimasa (Fukuyama Municipal Senior High) JAPAN: Watkins, Dr Ann (Los Angeles Pierce College) USA: Watkins, Dr William (California State University) USA: Watson, Dr Helen Ruth (Institute of Education) NIGERIA: Watson, Dr Jane Marie (University of Tasmania) TAS AUSTRALIA: Watson, Mr Rodney Reginald (Moorleigh High School) VIC AUSTRALIA: Watt, Mr Stanley (Mt Scopus College) VIC AUSTRALIA: Watterson, Mrs Marion Lucy (Brentwood High School) VIC AUSTRALIA: Wearne-Hiebert, Dr Diana (University of Delaware) USA: Webb, Professor John Harold (University of Cape Town) SOUTH AFRICA: Webber, Mr Cameron Paul (Charnwood High School) ACT AUSTRALIA: Weber, Dr Jeannine (Ecole Normale de Colmar) FRANCE: Weinzweig, Professor Dr Aurum Israel (University of Illinois) USA: Wellington, Ms Kay VIC AUSTRALIA: Welsh, Mr Ronald James (Melbourne C.A.E.) VIC AUSTRALIA: Wenger, Professor Ronald H. (University of Delaware) USA: Weremy, Mr John David (Trinity Grammar School) NSW AUSTRALIA: Werry, Mr Bevan William (Department of Education, Wellington) NEW ZEALAND: West, Mr Errol (National Aboriginal Education Commission) AUSTRALIA: Westbye, Superintendent Oivind (Radhuset) NORWAY: Westermann, Mr Leo R.J. (Econometrics Institute) THE NETHERLANDS: Weston, Mr Melvyn WA AUSTRALIA: Wheal, Mr Michael (Wattle Park Teachers Centre) SA AUSTRALIA: Wheatley, Professor Grayson (Purdue University) USA: Wheaton, Mrs Anita Rose SA AUSTRALIA: Wheaton, Mr Neville Kingsley (Campbelltown High School) SA AUSTRALIA: Wheeler, Professor David H. (Concordia University, Montreal) CANADA: Whippy, Mrs Helen J.D. (Papua New Guinea University of Technology) PAPUA NEW GUINEA: White, Dr Arthur L. (Ohio State University) USA: White, Miss Barbara Maida (All Hallows' School) QLD AUSTRALIA: White, Mr Gerald K (Richmond Primary School) SA AUSTRALIA: White, Mr John David (G.I.A.E.) VIC AUSTRALIA: White, Mr Miles (Kapunda High School) SA AUSTRALIA: Whiteley, Ms Molly (Reed Elementary School) USA: Whitfield, Miss Julie Anne (Bendigo C.A.E.) VIC AUSTRALIA: Whitford, Dr Anthony Kenneth (S.A.C.A.E.) SA AUSTRALIA: Whitford, Dr Heather Joy SA AUSTRALIA: Whitman, Professor Nancy C. (University of Hawaii) USA: Whitney, Professor Hassley (Institute for Advanced Study, Princeton) USA: Wickins, Mr Eugene Paul (Cabra College) SA AUSTRALIA: Wicks, Ms Suzanne Jane (Waverley College) NSW AUSTRALIA: Widmer, Dr Constance C. (Northern Kentucky University) USA: Wieman, Mr Theo Henry (Rossmoyne S.H.S.) WA AUSTRALIA: Wiese, Mr Gregory John (Grant High School) SA AUSTRALIA: Wilcox, Ms Joan Frances (Sydney Institute of Education) NSW AUSTRALIA: Wilcox, Mrs Marie S. USA: Wilkins, Mrs Naomi (University of Technology) PAPUA NEW GUINEA: Williams, Mr Douglas Desmond (Bimbadeen Heights P.S.)

VIC AUSTRALIA: Williams, Mr James L (Sydney University) NSW AUSTRALIA: Williams, Mr Leonard George (Fulham Gardens Primary School) SA AUSTRALIA: Williams, Mr Raymond John (Wanneroo Senior High School) WA AUSTRALIA: Williams, Dr Rosemary (Math Learning Center) USA: Williams, Mr William Richard (S.S.A.B.S.A.) SA AUSTRALIA: Williamson, Mr Jeffrey Robert NEW ZEALAND: Willis, Dr Sue (Murdoch University) WA AUSTRALIA: Willis, Dr Susan Gay (Murdoch University) WA AUSTRALIA: Willis, Mrs Vivienne VIC AUSTRALIA: Willoughby, Professor Stephen S. (New York University) USA: Wilmot, Mr Eric (National Aboriginal Education Commission) AUSTRALIA: Wilson, Professor James W. (University of Georgia) USA: Wilson, Mr Jeffrey Raymond (Coomandook Area School) SA AUSTRALIA: Wilson, Mr John Desmond (Sheffield City Polytechnic) ENGLAND: Wilson, Mrs Patricia Jean (Coniston Public School) NSW AUSTRALIA: Wiltshire, Mrs Zaiga (La Trobe University) VIC AUSTRALIA: Wily, Mrs Helen Mary NEW ZEALAND: Winston, Mr Mark Andrew (Woomera Area School) SA AUSTRALIA: Winter, Mrs Ana Maria (Ivanhoe Girls' Grammar School) VIC AUSTRALIA: Winteridge, Mr David John (Westhill College) ENGLAND: Wirszup, Professor Izaak (University of Chicago) USA: Wither, Mr David Peter (Pulteney Grammar School) SA AUSTRALIA: Wittmann, Professor Erich Christian (Universitat Dortmund) GERMANY: Wood, Mr Garry Leonard (Gilles Plains High School) SA AUSTRALIA: Wootten, Miss Amy (Cartmel College) ENGLAND: Wright, Mr Robert John (Northern Rivers C.A.E.) NSW AUSTRALIA: Wrigley, Mr Jack (Brisbane C.A.E.) QLD AUSTRALIA: Wyatt, Mr Ken (National Aboriginal Education Commission) AUSTRALIA: Wynne-Wilson, Dr William (University of Birmingham) ENGLAND: Yadav, Mr S.P. (Adarsh Hindi High School) INDIA: Yamagishi, Professor Yohsuke (Mukogawa Women's University) JAPAN: Yamaguchi, Mrs Chisato (Shitennoji High School) JAPAN: Yamaguti, Professor Kiyosi (Hiroshima University) JAPAN: Yamano, Mr Masayoshi (Kaizuka Senior High School) JAPAN: Yamashina, Mr Shiko (Keisen Jogaku-En Junior High School) JAPAN: Yamashita, Professor Akira (Fukuoka University Of Education) JAPAN: Yamashita, Mr Koji (Sumiyoshi Gakuen High School) JAPAN: Yates, Brother Bede (St Joseph's College) NSW AUSTRALIA: Yates, Mr Erin Dominic NSW AUSTRALIA: Yates, Mr Sean Vincent (St Clare's College) NSW AUSTRALIA: Yokochi, Professor Kiyoshi (Yamanashi University) JAPAN: Yoshida, Mr Hajime (The Association of Mathematical Instruction) JAPAN: Yoshida, Dr Hajime (Miyazaki University) JAPAN: Yoshida, Mrs Yuko JAPAN: Yoshikawa, Mr Masaki (Tokyo Metropolitan Kitano High School) JAPAN: Young, Dr Jerry (Boise State University) USA: Young, Dr Sharon L. (Louisiana State University) USA: Young, Mrs Valis Betty TAS AUSTRALIA: Yu, Professor Hi-Se (Korea University) KOREA: Zaare-Nahandi, Dr Rahim (University of Tehran) IRAN: Zasada, Mrs Bozenna Rajewska (Instytut Ksztalcenia) POLSKA: Zeleznik, Mrs Myra (Westhill High School) USA: Zimmermann, Dr Bernd F.R. GERMANY: Zofchak, Miss Veronica Ann (Los Angeles Unified School District) USA: Zulauf, Professor Achim (University of Waikato) NEW ZEALAND: Zweng, Prof. Dr Marilyn J. (University of Iowa) USA:

KOOKABURRA